Optimization and Computational Fluid Dynamics

Dominique Thévenin · Gábor Janiga
Editors

Optimization and Computational Fluid Dynamics

Springer

Editors:

Prof. Dr.-Ing. Dominique Thévenin
Dr.-Ing. Gábor Janiga

Otto-von-Guericke-Universität Magdeburg
Institut für Strömungstechnik und Thermodynamik (ISUT)
Universitätsplatz 2
39106 Magdeburg
Germany

thevenin@ovgu.de
janiga@ovgu.de

ISBN 978-3-540-72152-9 e-ISBN 978-3-540-72153-6

DOI 10.1007/978-3-540-72153-6

Library of Congress Control Number: 2007941795

Typesetting: supplied by the authors
Production: LE-TEX Jelonek, Schmidt & Vöckler GbR, Leipzig, Germany
Cover design: eStudioCalamar S.L., F. Steinen-Broo, Girona, Spain

Printed on acid-free paper

9 8 7 6 5 4 3 2 1

springer.com

Preface

The idea of this book was born during the "Conference on Modelling Fluid Flow" held in Budapest at the beginning of September 2006. During this occasion, we had decided to propose and thus hold a workshop entitled "Coupling CFD with Optimisation", based on our rapidly increasing experience with this highly interesting topic. We were nevertheless surprised to see the resonating enthusiasm displayed throughout the workshop by the conference participants.

From the discussions with all the speakers present at this workshop as well as the survey of the scope of the available books and review articles on this subject, it became easier to understand this great interest. While there is a wealth of new research projects that deal with the coupling of *Computational Fluid Dynamics* (CFD) and modern *Optimization* techniques, it is however difficult to find reference publications on this topic. There are indeed a few, excellent books available (see also the Introduction), but they are mostly restricted to aerodynamics, since this has been the first field of CFD for which optimization has become a tool of major importance. Moreover, the connection between CFD and Evolutionary Algorithms, often required when considering more complex systems of equations and physical models, has not been documented extensively.

Therefore we decided, together with the support of almost all workshop participants and a few internationally renowned newcomers, to gather and recount our experience concerning Optimization based on evaluations obtained through Computational Fluid Dynamics (a procedure abbreviated in this book as *CFD-O*), in order to prepare a book covering most of the relevant aspects and issues. Thanks to the hard work and constant support of all contributors, it has been finally possible to release this publication almost exactly one year after the workshop in Budapest. We hope that the interested readers will find here appropriate answers to the main questions: "What is indeed CFD-O? What simulation is today possible using CFD-O? How can I rely on CFD-O for my own applications and which approach should I choose?"

Our first research project on CFD-O was connected with the Ph.D. supervision of Mr. R. Baron at the École Centrale in Paris. He is the creator of our Optimization library (Opal) and must be thanked here for the quality of his work and for his unsurpassed motivation. The authors would furthermore like to thank Ms. Imelda Pasley for her thorough corrections of the manuscript. The quality of many graphical illustrations has been greatly enhanced by Mr. Imre Ferencsin.

Magdeburg,
August 2007

Dominique Thévenin
Gábor Janiga

Contents

Part I Generalities and methods

List of Contributors

Hans Georg BOCK
Universität Heidelberg, Interdisziplinäres Zentrum für Wissenschaftliches Rechnen (IWR), Im Neuenheimer Feld 368, D - 69120 Heidelberg, Germany, e-mail: scicom@iwr.uni-heidelberg.de

René A. Van den BRAEMBUSSCHE
von Kármán Institute for Fluid Dynamics, Turbomachinery and Propulsion Department, Waterloose steenweg, 72, B - 1640 Sint-Genesius-Rode, Belgium, e-mail: vdb@vki.ac.be

Laurent DUMAS
Paris 6 University, Laboratory Jacques-Louis Lions, Boîte 187, 4, place Jussieu, F - 75252 Paris Cedex 05 - France, e-mail: dumas@ccr.jussieu.fr

Nicolas R. GAUGER
German Aerospace Center (DLR), Institute of Aerodynamics and Flow Technology, Lilienthalplatz 7, D - 38108 Braunschweig, Germany, e-mail: Nicolas.Gauger@dlr.de

Kyriakos C. GIANNAKOGLOU
National Technical University of Athens, Lab. of Thermal Turbomachines, P.O. Box 64069, GR - Athens 15710, Greece, e-mail: kgianna@central.ntua.gr

Jari HÄMÄLÄINEN
Department of Physics, University of Kuopio, P.O. Box 1627, FI-70211 Kuopio, Finland, e-mail: jari.hamalainen@uku.fi

Taija HÄMÄLÄINEN
Department of Physics, University of Kuopio, P.O. Box 1627, FI-70211 Kuopio, Finland, e-mail: taija.hamalainen@uku.fi

Gábor JANIGA
University of Magdeburg "Otto von Guericke", Lab. of Fluid Dynamics and Technical Flows, Universitätsplatz 2, D - 39106 Magdeburg, Germany,
e-mail: janiga@ovgu.de

Elina MADETOJA
Department of Physics, University of Kuopio, P.O. Box 1627, FI-70211 Kuopio, Finland,
e-mail: elina.madetoja@uku.fi

Marco MANZAN
Università di Trieste, Dipartimento di Ingegneria Navale, del Mare e per l'Ambiente, Via A. Valerio 10, I - 34127 Trieste, Italy,
e-mail: manzan@units.it

Enrico NOBILE
Università degli Studi di Trieste, Dipartimento di Ingegneria Navale, del Mare e per l'Ambiente, Via A. Valerio 10, I - 34127 Trieste, Italy,
e-mail: nobile@units.it

Dimitrios PAPADIMITRIOU
National Technical University of Athens, Lab. of Thermal Turbomachines, P.O. Box 64069, GR - Athens 15710, Greece,
e-mail: dpapadim@mail.ntua.gr

Stefano PIERI
Danieli & C. Officine Meccaniche Spa, Buttrio (UD), Italy,
e-mail: stpieri@danieli.it

Francesco PINTO
Università di Trieste, Dipartimento di Ingegneria Navale, del Mare e per l'Ambiente, Via A. Valerio 10, I - 34127 Trieste, Italy,
e-mail: fpinto@units.it

Henri RUOTSALAINEN
Department of Physics, University of Kuopio, P.O. Box 1627, FI-70211 Kuopio, Finland,
e-mail: henri.ruotsalainen@uku.fi

Volker SCHULZ
University of Trier, Department of Mathematics, Building E, D - 54286 Trier, Germany, e-mail: Volker.Schulz@uni-trier.de

Dominique THÉVENIN
University of Magdeburg "Otto von Guericke", Lab. of Fluid Dynamics and Technical Flows, Universitätsplatz 2, D - 39106 Magdeburg, Germany,
e-mail: thevenin@ovgu.de

Acronyms

AGA	Approximated Genetic Algorithms
AHM	Adaptive Hybrid Method
ANN	Artifical Neural Network
BFGS	Broyden, Fletcher, Goldfarb and Shanno
CC	Cross-Corrugated
CFD	Computational Fluid Dynamics
DNS	Direct Numerical Simulation
DOE	Design of Experiment
DOF	Degrees Of Freedom
EA	Evolutionary Algorithm
ES	Evolution Strategies
FEA	Finite Element Analysis
FOPD	Fiber Orientation Probability Distribution
GA	Genetic Algorithm
HOCS	Headbox Optimization Control Simulator
LES	Large-Eddy Simulation
LSD	Low Solidity Diffuser
MD	Machine Direction
MDO	Multi-disciplinary Design Optimization
MG	Multigrid
MOEA	Multi-objective Evolutionary Algorithm
MOGA	Multi-objective Genetic Algorithm
NSGA	Nondominated Sorting Genetic Algorithm
NURBS	Non-Uniform Rational Basic Splines
OF	Objective Function
PDE	Partial Differential Equation
POF	Pareto Optimal Frontier
RANS	Reynolds-Averaged Navier Stokes
RBF	Radial Basis Function
RSM	Reynolds-Stress Model
SA	Simulated Annealing
VEGA	Vector Evaluation Genetic Algorithm

Part I
Generalities and methods

Chapter 1
Introduction

Dominique Thévenin

A book dedicated to optimization applied to practical engineering configurations must probably start with a warning: "optimization" means much more than "improvement"! It is indeed a pity that so many researchers and engineers still employ the terminology "optimization" in the title or abstract of their publications when they simply mean in practice that starting from a non-satisfactory configuration, they have tried two or three other ones and chosen at the end the best case. This is undoubtedly related to optimization, but in a very minimalistic sense! In the present book optimization means

> the design and operation of a system or process to make it *as good as possible* in some defined sense

which is the definition proposed by the WiKi dictionary, that can be consulted under http://en.wiktionary.org/wiki/optimization. As a consequence, the *best possible solution* constrained by appropriate conditions should be found, and not simply a "better" one.

Mathematical optimization methods allowing to identify such a constrained, best possible solution have been known for a long time, but have not permeated all engineering disciplines yet. Concerning fluid dynamics more specifically, the first applications of optimization are found for aeronautical problems (see, e.g., as a starting point the relevant chapters of the recent publication by Capasso and Périaux [2]), in particular to improve wing profile and flight properties (typically, reduce drag). This is a problem with a high added-value and involves "only" the basic equations of fluid dynamics (Euler or Navier-Stokes equations, depending on the investigated properties). This

Dominique Thévenin
Lab. of Fluid Dynamics and Technical Flows,
University of Magdeburg "Otto von Guericke", Germany
(e-mail: thevenin@ovgu.de)

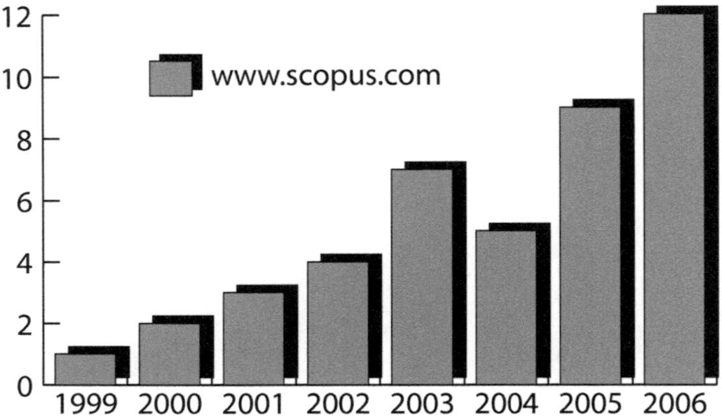

Fig. 1.1 Number of publications regarding CFD-O based on the search tool Scopus. The year 1999 is chosen as reference year and is thus associated to a value of 1

explains why most available books and articles dealing with Optimization relying on evaluations obtained by Computational Fluid Dynamics (what we abbreviate in this book as CFD-O, or *CFD-based Optimization*) concern such situations. Even then, the number of such books and review articles remains quite limited. We have so far only found two books published since 1999 that deal exclusively with CFD-O (obviously, a large number of books consider this issue, but only for a single chapter or a few pages). The work of Mohammadi and Pironneau [5] sets the emphasis on shape optimization for aeronautical applications using adjoint methods, a topic that will be mostly considered in Chapters 3 to 5 of the present book. The book by Gunzburguer [4] covers both optimization and control of flows relying on CFD, but concentrates mostly on control issues in practice and is mainly written as a tutorial for students. No review article could be found on this topic at all.

In order to get a more global albeit inexact scope of the literature, an Internet-based literature search has been carried out using the professional scientific literature search tool Scopus (see http://www.scopus.com). The year 1999 is retained as reference year and is thus associated to a value of 1. By comparison, the number of references found by Scopus for all the years until 2006 is represented in Fig. 1.1. The search criterion considers only title and abstracts of all corresponding publications and looks for ((CFD OR Computational Fluid Dynamics) AND (Optimization OR Optimisation)). This is obviously a very nonspecific search and hence will leave out a few relevant publications as well as identify several publications for which "Optimization" simply means "Improvement", as explained previously. It is nevertheless interesting to identify the trend associated with CFD-O. As seen in Fig. 1.1, the number of corresponding publications has multiplied by a factor of 12 within 7 years demonstrating the rapid development of this field.

It seems therefore appropriate to prepare an extensive work on CFD-O. For this purpose, both theoretical foundations and practical engineering applications of optimization relying on CFD evaluations should be presented in a unified framework. This is the main challenge of the present publication. In what follows the first three chapters illustrate the considered issues and introduce most mathematical tools needed to tackle this problem, in particular when considering adjoint methods which are mathematically much more demanding. The subsequent five chapters cover a variety of specific engineering applications (aerospace, turbomachines, automotive, heat transfer, papermaking) illustrating different possibilities to carry out successfully CFD-O depending on varying requirements concerning accuracy, computing times and/or complexity. This should allow most readers to find both the needed theoretical background and examples of practical realizations very similar to her/his own problems, so that the first few steps on the long way toward a successful CFD-O should be greatly facilitated.

In order to make it even easier, let us try now to illustrate the challenges of CFD-O using only basic concepts. In Fig. 1.2 the simplest possible optimization problem is considered: find the input parameter (sometimes called degree of freedom) that minimizes the objective function (OF, sometimes also called cost function) for an analytically known, smooth function. This is a case with a single parameter (or degree of freedom, the x-coordinate in Fig. 1.2) and a single objective (the y-coordinate in Fig. 1.2) with an OF involving only a single minimum. It is therefore an extremely simplified configuration almost never found in practice. It could nevertheless represent a problem found when thermodynamically optimizing a process [1], e.g., minimizing exergy loss when varying one process parameter, knowing analytically all thermodynamical properties. In such a case, the optimization could in principle be carried out by hand, computing the full functional behavior, since this will be quite trivial. A standard gradient-based (also called "steepest descent") algorithm [3] would easily find the solution of this basic problem. Otherwise, *any* optimization technique presented in this book would also be able to identify the optimum within a few seconds of computing time.

The picture presented in Fig. 1.2 is somewhat misleading, since it suggests that the full function $OF(x)$ is known. If this would be the case, the engineer in charge could obviously identify the optimal solution at first glance without resorting to any optimization algorithm! In practice, only a discrete set of points $OF(x_i)$ with $i \ldots N_\mathrm{p}$ will be known at the end of the process: these are the points for which a CFD-based evaluation has been requested. This issue is illustrated in Fig. 1.3 where the successive points obtained by a simple steepest-descent algorithm are shown schematically using $*$-symbols on top of the (in fact unknown) objective function represented by the solid line in the background.

To further increase the relevance of the considered problem, Figure 1.4 should now be considered. It involves again a single parameter (x-axis) and a single objective (y-axis) for an analytically known, smooth function, but

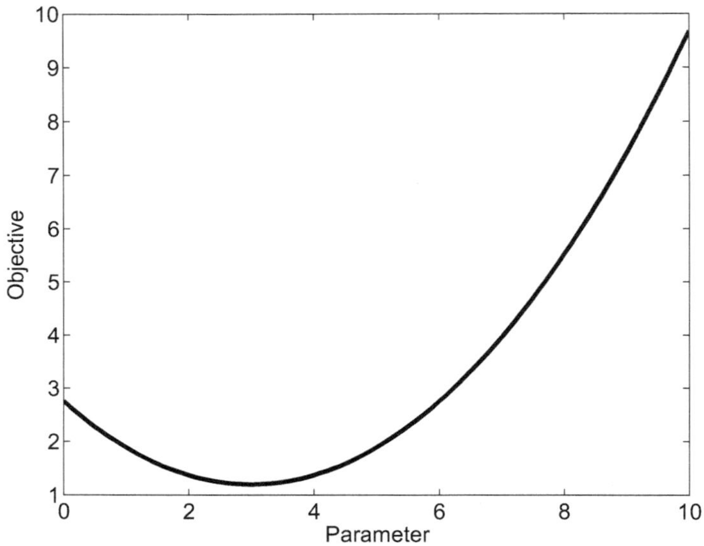

Fig. 1.2 Schematic representation of the simplest possible optimization problem involving a single parameter, a single objective and an objective function with a single minimum

shows this time two minima. One of those is obviously only a local minimum, while the other one corresponds to the optimal parameter which should be found. This configuration illustrates one major difficulty of optimization, not specific to CFD-O: a robust optimization algorithm must be able to cope with local minima, avoiding to get "stuck" into them. For a classical, steepest-descent search, the success would, in the case illustrated in Fig. 1.4 usually depend on the starting point for the search algorithm: starting from Point 1 would lead to the optimal solution, while a search initiated at Point 2 would end at the local minimum. Obviously, for practical applications, it is not a good idea to rely too much on luck and hope that the starting point will always be the right one! Therefore, solutions have to be found to take the optimization algorithm "out" of possible local minima. Practical solutions have been proposed for most existing optimization methods, but will be usually problem-dependent. In practice, the issue of local minima hinders most classical gradient-based methods and simple ad-hoc procedures like Simplex optimization [6]. A combination of different optimization algorithms (gradient method and Evolutionary Algorithms) could for example be employed to solve this issue.

When trying to get even closer to practical optimization problems involving CFD, it is necessary now to abandon the idea of an *exact* evaluation of the Objective Function. From now on each evaluation is itself the result of an (approximate) numerical simulation, obtained typically by solving the

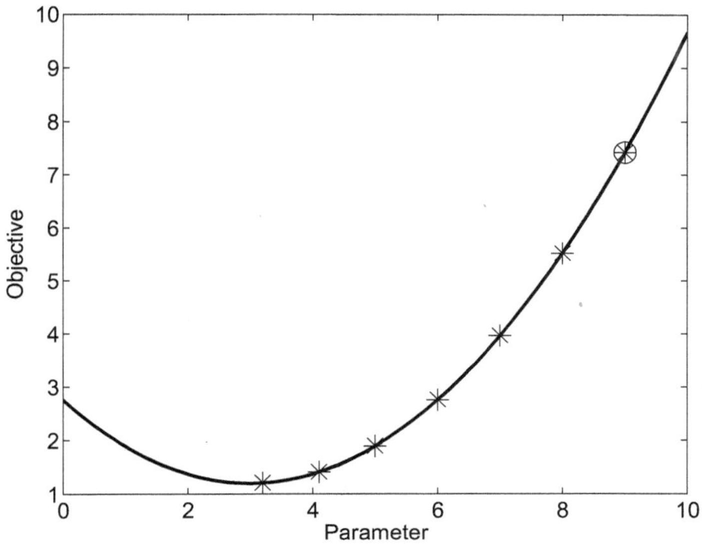

Fig. 1.3 Schematic representation of the simplest possible optimization problem involving a single parameter, a single objective, and an objective function with a single minimum. The points really known through a CFD-based evaluation are shown using $*$-symbols (the starting point for the algorithm is identified by a $\circ$) on top of the (in fact unknown) objective function, represented by the solid line in the background

Navier-Stokes equations together with appropriate physical models used for example to describe turbulent, multiphase or reacting flows. This leads directly to two new issues:

1. First, many scientific questions are still unsolved concerning such complex flows. This means that the CFD solution might be indeed quite far from the real, physical solution. This is of course an essential issue usually called model validation, but there is no solution for this problem: further studies are then required until very accurate models become available to describe complex flows. The optimal solution delivered by the methods considered in this book will implicitly but obviously rely on the hypothesis that the *CFD evaluations indeed describe correctly the physics of the considered flows.*
2. Second, and more important for this Introduction, the evaluation obtained by CFD will be subject to a certain amount of numerical uncertainty leading to a probably small but not necessarily negligible inaccuracy. Several reasons can be identified for this problem: a discretization grid that is too coarse in important regions of the flow; an insufficiently low threshold on the residuals leading to a premature interruption of the iterative solution of the flow equations; computer round-off errors; spatial or temporal

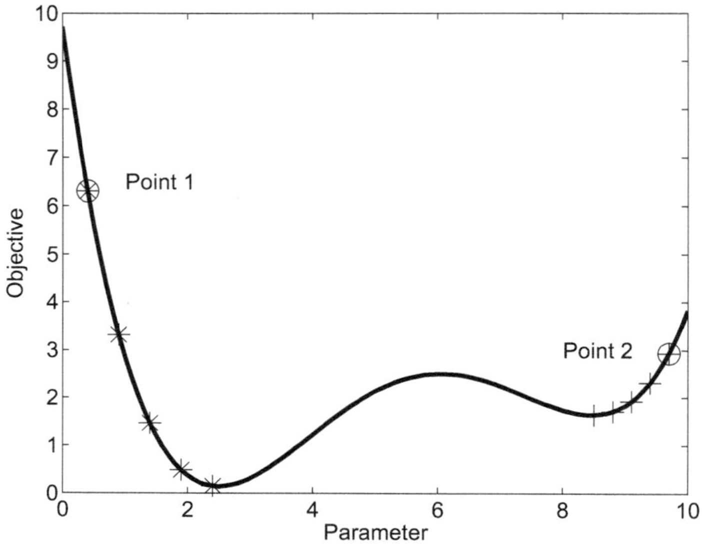

Fig. 1.4 Schematic representation of a simple optimization problem involving a single parameter and a single objective. Now, the objective function shows two minima. The points really known through a CFD-based evaluation are shown using symbols on top of the (in fact unknown) objective function represented by the solid line in the background. A gradient-based algorithm starting at Point 1 would probably find the right optimum (symbols $*$) while the same algorithm starting from Point 2 (symbols $+$) would most probably get stuck in the local minimum

discretization errors, etc. Such problems are often to be expected since the numerical cost of the evaluations is the main problem for CFD-O as explained at the end of the present Introduction. Therefore, in order to facilitate CFD-O, many users will try to speed-up the evaluation process as much as possible, thus using coarse grids or a very limited number of iterations. As a consequence, the evaluations will be typically associated with a certain amount of inaccuracy, which will depend on the configuration considered. Obviously, it should remain small enough to allow for a meaningful optimization. In many cases, it will lead to both a *systematic* and to a *stochastic* evaluation uncertainty as depicted in Fig. 1.5.

The systematic error will move the optimal solution obtained by CFD-O slightly away from the truly physical, optimal solution. There is no possibility to solve this problem in principle, apart from using an extremely accurate numerical procedure. The stochastic error will lead to a specific difficulty: the appearance of a potentially high number of non-physical, local minima. It becomes therefore even more important to be able to deal with this issue in practice, in order to come near enough to the true optimal solution.

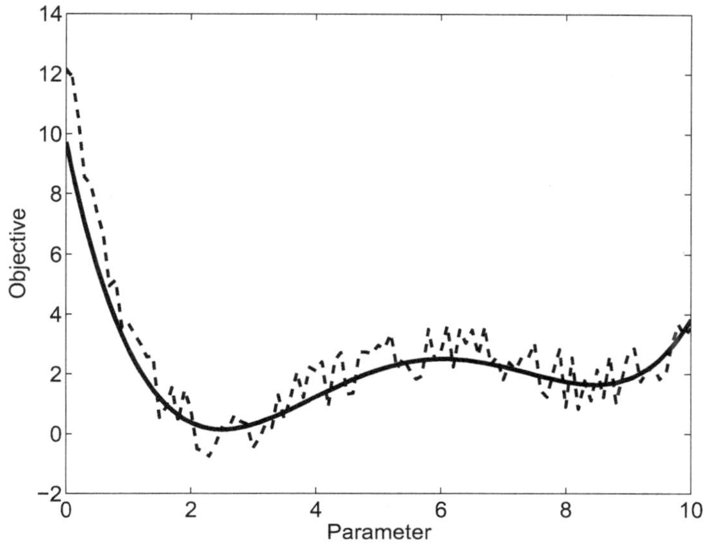

Fig. 1.5 Schematic representation of a simple optimization problem involving a single parameter and a single objective. The objective function shows two minima. The (unknown) exact objective function is represented by the solid line in the background while the (unknown) inaccurate CFD-based evaluation of the objective function is shown as a dashed line

Finally, real optimization problems found in engineering as well as in fundamental research will generally involve several (often too many!) input parameters (degrees of freedom) and several objectives. In general, these objectives will be furthermore concurrent, meaning that it is impossible to identify a single set of parameter that would be best for all objectives simultaneously. Instead, a so-called Pareto Optimal Frontier (POF) will appear, a concept that will be discussed several times in this book. The POF contains all the parameter sets that are *optimal in the sense of Pareto* that is nondominated by any other realization. In common language, this parameter set belongs to the elite (the group containing the best configurations), but this group contains many sets and not a single point any more. The concept of POF is illustrated schematically in Fig. 1.6.

Moreover, the input parameters are not always continuous, but can also be discrete numbers, or a combination of continuous and discrete entries. When considering several input parameters, it will often happen that some parameter sets do not correspond to any practical configuration as shown in Chapter 8 of this book. Usually, all the input parameters will be constrained to some acceptable subspace, possibly much smaller than the full parameter space.

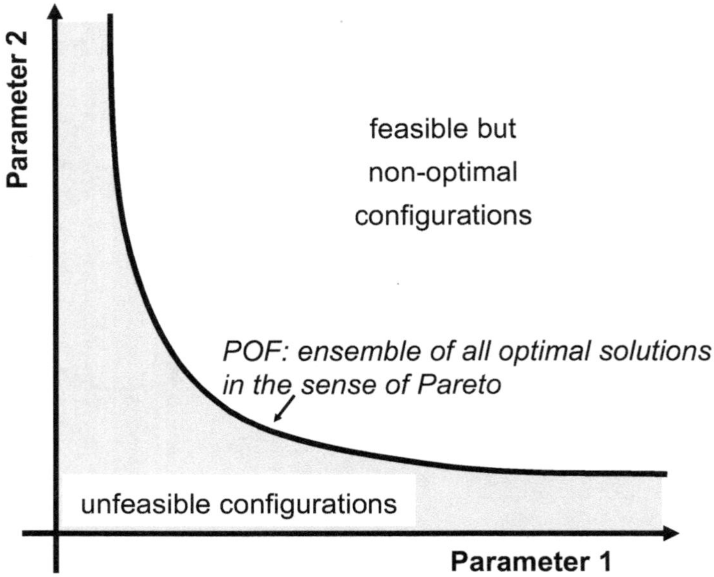

Fig. 1.6 Schematic representation of the concept of Pareto Optimal Frontier

After having introduced these notions, the practical limits of CFD-O in terms of computer requirements have to be discussed: what can be investigated today using CFD-O, and how can it be done? In order to investigate this point, the natural reference point for discussing the requirements in terms of computing time (CPU) and computer memory is of course the corresponding requirement for a *single evaluation*, i.e., for us one CFD simulation of the considered configuration. Admittedly, each CFD evaluation will usually lead to slightly different numerical costs (needed computing time and computer memory), but the variations are normally negligible. On the other hand, the typical cost of a single CFD evaluation depends *tremendously* on the considered problem. Looking for example at the examples presented in the next chapter of this book:

- one simulation of the burner considering multispecies transport and combustion takes typically about 20 hours of computing time and 1.8 Gigabytes (GB) of computer memory;
- one simplified simulation of the turbulent channel flows requires less than 5 seconds of computing time and 10 Megabytes (MB) of computer memory.

In other words, the computing times vary by 5 orders of magnitude and the computer memory by 3 orders of magnitude. This is typical of what will be found in practice: simple CFD problems can be solved within seconds on a standard PC; high-end CFD problems can require weeks of computing times and Terabytes (1 TB=1000 GB) of memory on supercomputers. In such cases, the limits of CFD-O are clear in principle: CFD-O is possible for simple CFD problems and almost impossible for high-end CFD configurations. This can be quantified somewhat more precisely by remembering that the optimization process will usually require a large number of evaluations (i.e., CFD computations). Considering again the examples of the next chapter:

- only 64 evaluations corresponding to realizable configurations have been possible for the burner involving multispecies transport, and this already at an extremely high computational cost;
- on the other hand, more than 5000 evaluations have been easily carried out for the simplified turbulent channel flow.

Since the dimension and size of the parameter space (or in other words, the number of degrees of freedom) directly conditions the requested number of evaluations to get an accurate estimation of the optimal solution(s), the feasibility of CFD-O is listed in Table 1.1 where the figures concerning number of parameters, computing time and memory should only be considered as typical values, and not as exact limits since this will be clearly problem-dependent:

Even if this is only a crude description, it corresponds quite well to what can be found in the literature. When looking for instance at all the examples presented in this book, no CFD-O has been carried out for cases where a single CFD-evaluation requires more than a day of computing time. Many examples require less than one hour of computing time for each evaluation (CFD-computation). As a rule of thumb for engineering practice, it seems therefore appropriate to state

CFD-O is possible when the duration of a single CFD computation does not exceed a few hours at most.

This is of course a disappointing information! There are indeed many interesting applications for which the CFD evaluation is fast enough, as shown in the present book. But, in many cases involving for example a large, complex,

Table 1.1 Limits of CFD-O

Evaluation cost	Optimization cost (# of parameters) ≤ 2 parameters	≥ 3 parameters
low cost (1 h CPU, 100 MB memory)	very easy	still possible
high cost (20 h CPU, 2 GB memory)	still possible	almost impossible

three-dimensional flow geometry discretized using millions of elements and involving complex physics, typical computing times will be days or weeks. Should we abandon the hope of any optimization for such interesting problems?

Not necessarily! Since the duration of the complete optimization process reads in principle "(number of evaluations)×(duration of a single CFD evaluation)", there are at least two clear possibilities to still use CFD-O for such problems:

1. the first one consists in speeding-up as much as possible the evaluation procedure;
2. the second one consists in reducing as much as possible the number of required evaluations.

To speed-up each CFD-based evaluation, different solutions might again be employed and intelligently combined. It is in principle possible to use, separately or simultaneously, a physical, a mathematical or an algorithmic point of view:

- From the physical point of view, speeding-up the evaluation means reducing the model complexity while keeping all (or most) of the needed coupling processes describing the important physics. This will for example be demonstrated in the last chapter of this book, where reduced equations will be partly used instead of the full Navier-Stokes equations, leading to a tremendous reduction in the needed computing time. It is also employed in the next chapter when considering optimization of turbulence model parameters: instead of solving the full multi-dimensional Reynolds-Averaged Navier-Stokes (RANS) equations to describe turbulent channel flows, a reduced model will be used instead. Model reduction is clearly a very efficient procedure and should be used every time this is possible. But it requires of course a clear understanding of all physical processes controlling the application considered!
- Applied mathematics should also be considered to reduce as much as possible the needed computing time and computer memory needed by CFD. This means that the most efficient solution procedures should be employed. This can lead for example to multigrid acceleration, perhaps to solve the pressure/velocity coupling equation; or to the implementation of a highly complex, three-dimensional adaptive unstructured grid or Adaptive Mesh Refinement, in order to drastically reduce the needed number of discretization elements; or to an efficient pre-conditioning of the system. None of these are specific to CFD-O. But clearly, an experienced CFD specialist has a better chance to succeed also in CFD-O!
- Finally, the practical coding of the CFD evaluation should also be improved as much as possible. From an algorithmic point of view, this will mean in particular the adaptation of the code to the properties of the employed computer: perhaps a vector, more often today a parallel system.

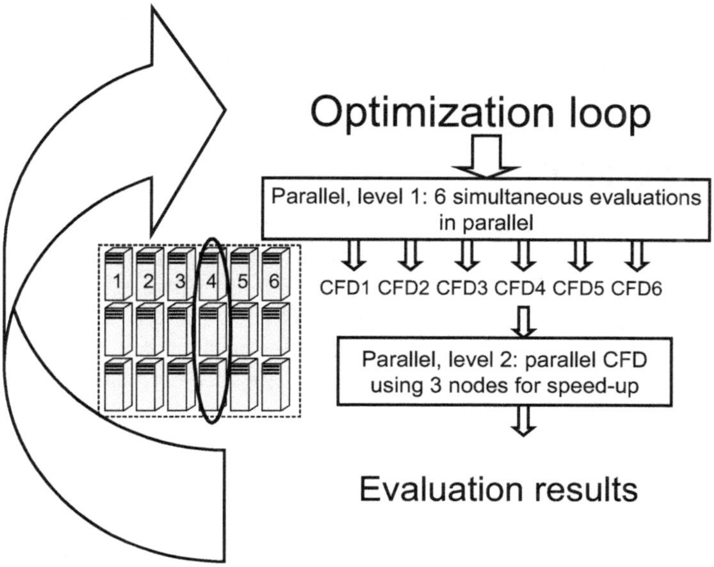

Fig. 1.7 Principle of a double-layer parallelization as might be used for example when carrying out CFD-O with Evolutionary Algorithms. A very small parallel cluster with 18 nodes is represented here to facilitate the graphical illustration, but this technique could be used without modification on a much larger parallel computer, using hundreds or thousands of nodes

Clearly, parallelizing the CFD evaluation using an adequate number of processors can drastically reduce the user waiting time, but not really the "computing" time, since the accounting of CPU hours usually relies on the cumulated computing time on all nodes. Nevertheless, CFD on parallel computers might easily be used to bring the duration of one single evaluation from 1 day down to 1 hour, just using perhaps 30 processors.

The advantage of parallelization for CFD-O can be even two-fold. It is indeed not only possible to parallelize each CFD evaluation, but very often, the optimization loop itself can be parallelized again. This is for example demonstrated in the next chapter considering Evolutionary Algorithms (EA). Using a farmer/worker paradigm on a multiprocessor machine, CFD-O relying on EA is in principle a so-called *embarassingly parallel* application for which a linear speed-up is expected when increasing the number of employed nodes. In this case, one can thus end up with the structure presented in Fig. 1.7 where a double layer of parallelization can be used on a parallel supercomputer with a large number of nodes that should lead to a huge reduction of the user waiting time. Such a structure, in principle easy to implement, could allow CFD-O based on EA for very complex configurations provided an appropriate parallel computer is available.

Let us come now to the second possibility mentioned previously: reduce as much as possible the needed number of evaluations. Here again, at least two different sub-methods can be identified for this purpose:

- the first one consists in replacing a CFD-based evaluation by an *appropriate approximation of it.* Depending on how this is done, this could also be implemented as a model reduction and is thus related to the first item listed in the previous list. But using Artificial Neural Networks, as done in the Chapter 6 of this book, or some other kind of intelligent "interpolation" in the broadest sense like the Response Surface Technique could be seen indeed as a reduction of the number of needed CFD-evaluations. As demonstrated in the present book, this method can be very efficient to speed-up the full optimization procedure. Nevertheless, it is obvious that great care must be taken, since the alternative evaluation method should not falsify in any way the evolution of the optimization algorithm. It is therefore not at all a trivial task to identify suitable approximation methods!
- the second one leads again back to applied mathematicians. They have already recognized a long time ago the issues associated with a large number of evaluations, and indeed proposed an alternative formulation, "optimal" from the point of view of numerical analysis: the *adjoint method.* Due to its importance for many practical problems, in particular in the aerospace industry, three chapters of this book (3 to 5) will be mostly dedicated to CFD-O based on adjoint methods. In principle, the adjoint approach requires not much more than a "single evaluation", which sounds almost too good to be true! One difficulty of the adjoint approach is its formal complexity for the common engineer, perhaps not familiar with all the needed mathematical concepts. It is our hope that the corresponding chapters will help such users understand how this method works. But two other major difficulties are more or less inherent to the adjoint approach itself:
  - a suitable consistent adjoint system must first be identified for the considered system of equations. While this is easily done (and well-documented in the literature), e.g., for the Euler equations, the task will become much more difficult when considering complex, multiphysics problems involving perhaps a turbulent multiphase flow with chemical reactions and concurrent objectives.
  - furthermore, the adjoint approach requires a full knowledge of the intermediate approximations on the way to the full solution of the system of equations solved by CFD. In simple words, this means that the adjoint approach, while reducing tremendously the requested number of evaluations (and thus to a large extent the needed computing time), will lead to a huge increase of the requested computer memory which will again become a major problem for complex, three-dimensional flows involving many unknowns at each discretization point.

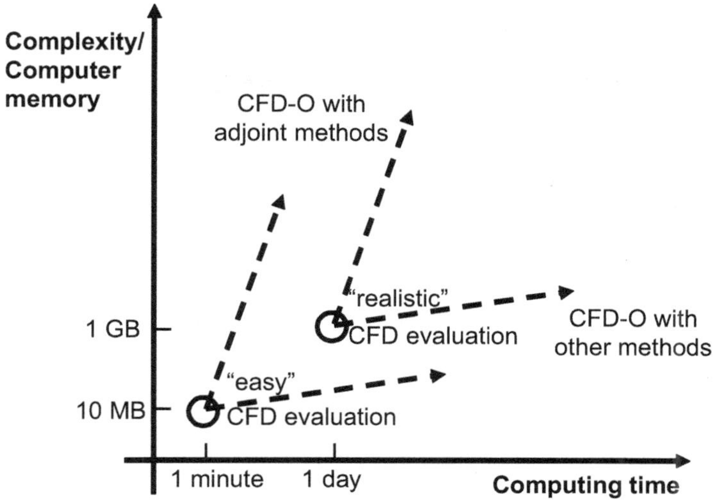

Fig. 1.8 Principle requirements of CFD-O in terms of computing time and computer memory. Again, the figures listed here are just orders of magnitude to help the reader, and should by no means be considered as exact limits

Fig. 1.9 What is the true connection between Optimization and Evolutionary Algorithms? (reproduction courtesy of Noémi Janiga)

To summarize, one ends up with the schematic picture shown in Fig. 1.8. This will hopefully help any interested user to decide *if* CFD-O is possible at all for a specific application and *how* to tackle the problem. If now you need to know more, well... just read the book!

References

1. Bejan, A.: Entropy Generation Minimization. CRC Press, Boca Raton, Florida (1996)
2. Capasso, V., Périaux, J.: Multidisciplinary Methods for Analysis Optimization and Control of Complex Systems. Mathematics in Industry. Springer (2005)
3. Fletcher, R.: Practical Methods of Optimization, 2 edn. John Wiley & Sons, New York (1987)
4. Gunzburguer, M.D.: Perspectives in Flow Control and Optimization. SIAM, Philadelphia (2003)
5. Mohammadi, B., Pironneau, O.: Applied Shape Optimization for Fluids. Numerical Mathematics and Scientific Computation. Oxford University Press (2001)
6. Nelder, J.A., Mead, R.: A simplex method for function minimization. Computer Journal **7**, 308–313 (1965)

Chapter 2
A Few Illustrative Examples of CFD-based Optimization
Heat Exchanger, Laminar Burner and Turbulence Modeling

Gábor Janiga

Abstract In this chapter, several multi-objective design optimizations are performed in order to illustrate major issues associated with CFD-based optimization. First, a heat exchanger configuration (Case A) is considered using the coupled solution of the flow/heat transfer processes. The aim of the procedure is to find the positions of the tubes most favorable to simultaneously maximize heat exchange while obtaining a minimum pressure loss.

Next, the optimization of the flame shape of a laminar burner is investigated when varying the fuel/air ratio in a primary and a secondary inlet (Case B). The objectives are to reduce the CO emission at a prescribed distance from the injection plane and to obtain the most homogeneous temperature profile at the same position. The flow involving chemical reactions is solved using the in-house Computational Fluid Dynamics (CFD) code UGC^+. These two cases are the continuation of our previous studies, introducing new results and new aspects.

The last case presented here is a new proposal to optimize the model parameters of an engineering turbulence model (Case C).

In all the presented cases, an Evolutionary Algorithm (EA) is applied to find the optimal configurations. An in-house computer package, called Opal, performs the optimization process in a fully automatic manner. The EA relies on a relatively large number of simulations which may result in a considerable computational effort, depending on the configuration. The procedure can thus be performed in parallel on a Linux PC-cluster to reduce user waiting time.

Gábor Janiga
Lab. of Fluid Dynamics and Technical Flows,
University of Magdeburg "Otto von Guericke", Germany
(e-mail: `janiga@ovgu.de`)

2.1 Introduction

Designing optimal shapes or optimal configurations for practical engineering applications has been the subject of numerous publications during the last decade. Generic and robust search methods inside the design space, such as Evolutionary Algorithms (EAs), offer several attractive features for such problems [38]. The basic idea associated with the EA approach is to search for optimal solutions using an analogy to the evolution theory. During the iteration (or "evolution" using EA terminology) procedure, the decision variables or genes are manipulated using various operators (selection, combination, crossover or mutation) to create new design populations, i.e., new sets of decision variables. For simpler optimization problems, more classical optimization methods, like the Simplex approach, are often better adapted to find the optimal solution within a small number of iterations [74].

The main goal of this work is to achieve cost-efficient design optimization of problems involving complex flows with heat transfer or chemical reactions using Computational Fluid Dynamics (CFD) codes for practical configurations, while keeping reasonable overall computing times.

Classical optimization techniques, like gradient-based methods are known for their lack of robustness and for their tendency to fall into local optima. Generic and robust search methods, such as EAs [16, 34], offer several attractive features and have been used widely for design shape optimization [1, 48, 55, 62, 64]. They can, in particular, be used for multi-objective multi-parameter problems. They have been successfully tested in many practical cases, for example for design shape optimization in the aerospace [1, 24, 54, 55, 64] and automotive industry [61]. The use of a fully automatic EA coupled with CFD for a multi-objective problem still remains limited by the computing time and is up to now far from being a practical tool for all engineering applications.

2.1.1 Purpose

The purpose of this chapter is to illustrate possible methodologies for the fully automatic optimization of various engineering problems involving CFD. We are not interested here in developing a new algorithm for optimization. We solely wish to demonstrate that it is possible to reach an optimal configuration for a case involving coupled fluid flow, heat transfer and chemical reactions, investigated using CFD within a reasonable computing time, for configurations very close to practical ones.

In Case A, we consider laminar flows because it corresponds to a realistic engineering problem, for example for low-power systems. A model configuration is chosen consisting of a cross-flow tube bank heat exchanger. The problem is to optimize the positions of the tubes so that the heat exchange is maximal while keeping a minimal pressure loss. A set of automatized numerical tools are used together to solve this problem involving mesh generation, CFD, an in-house C++ implementation of EAs, a shell-script and complementary C programs for automatization and parallelization (Sect. 2.3).

Case B introduces the optimization of the flame shape of a laminar burner. The fuel/air ratio in a primary and a secondary inlet vary and the objectives are to reduce the pollutant (CO) emission at a prescribed distance from the injection plane and to obtain the most homogeneous temperature at the same location (Sect. 2.4).

In Case C, the model parameters of the k–ω engineering turbulence model from Wilcox [79] are studied for Reynolds-Averaged Navier-Stokes (RANS) computation. The objective in this work is to fit the parameters of the model using optimization, in order to better predict the time-averaged turbulent velocity profiles in channel flows (Sect. 2.5).

The applied multi-objective Evolutionary Algorithm (MOEA), based on the concept of Pareto dominance, is described next. In the following sections, the model problems are introduced first, putting into evidence the requirements for the choice of an adequate optimization strategy. The practical computational methodology for mesh generation, CFD solution and parallelization are presented afterwards. Results are then shown and discussed followed by concluding remarks.

2.1.2 Heat Exchanger Optimization (Case A)

Improving the performance of an existing configuration often involves optimization. The optimal placement of the heat sources or sinks in a channel, a cavity or a heat exchanger may affect dramatically the performance of the device. In these circumstances, CFD have a high potential to easily explore a large number of different configurations. As a whole, optimization of configurations involving the coupled simulation of flow and heat transfer remains a fairly new field of research.

Heat exchange through smooth and corrugated walls has been for example investigated in [22]. Shape improvement of a cylinder with heat transfer was carried out in [15]. The optimal shapes of fins and pins inside heat exchangers have been examined by various authors [3, 12, 21, 23, 49]. Tiwari et al. [75] have studied different angles of attack for the delta winglets mounted on the fin-surface on top of oval-shaped tubes. Heat transfer of finned and non-finned circular and elliptic tubular arrangements are investigated numerically in [57] to maximize the total heat transfer rate. The flow through a heated pipe with an inserted twisted tape was examined in [46] for different slopes. This analysis is based on the entropy production minimization [9]. Multi-parameter optimization coupled with CFD was investigated in [76] to maximize the performance of a heat sink. Foli et al. [31] and Okabe et al. [66] have obtained

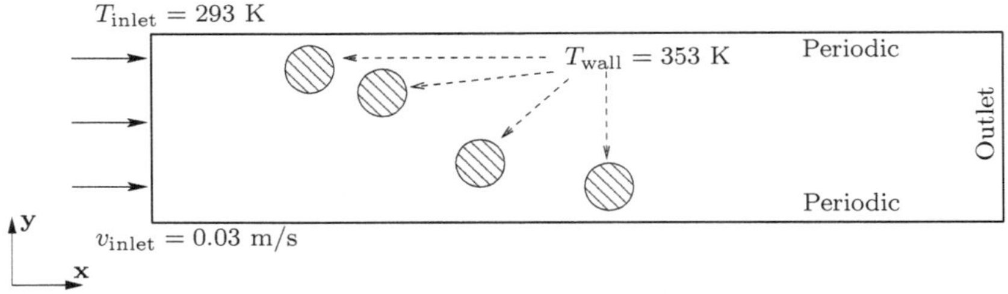

Fig. 2.1 Schematic description of the tube bank heat exchanger configuration considered in Case A

optimal results for a micro heat exchanger based on different multi-objective optimization methods.

The optimal spacing problem with three chips in an enclosure is reported in [51]. The optimal location of heat sources was investigated by da Silva et al. [70] for forced convection and in [71] for natural convection. The latter was examined by Dias and Milanez [17] using Genetic Algorithm (GA) to find the optimal location. Staggered finned circular and elliptic tubes in forced convection are studied by Matos et al. [56]. Bello-Ochende et al. [10] analyzed cylinders in cross-flow with up to three different sizes in row configurations for maximizing the heat transfer density.

Nevertheless, optimization based on an arbitrary positioning of the tubes in a cross-flow heat exchanger could not be found in the literature. This will be considered now. In Case A, parallel EA are coupled with a CFD code and a two-dimensional model of a cross-flow tube bank heat exchanger is considered. One possible simulated configuration is shown in Fig. 2.1. Air enters the domain at $T_{\text{inlet}} = 293$ K and is warmed up by passing between the tubes in which a warm fluid flows in the corresponding practical application. The tubes are supposed to have a constant outer wall temperature, $T_{\text{wall}} = 353$ K. The outlet is at atmospheric pressure.

The optimization problem consists of finding the best locations of the tubes to increase heat exchange while at the same time to limit the pressure loss. The two corresponding numerical parameters to optimize are the average temperature difference ΔT and pressure difference ΔP. These two objectives are obviously interrelated. If the exchange surface increases, the heat exchange will be favored and the temperature difference between inflow and outflow will be also enhanced. But, simultaneously for a given air flow rate, the pressure loss will increase and the heat exchanger loses efficiency.

2.1.3 Optimization Coupled with Chemical Reactions (Case B)

There is also a great interest at present to optimize complex flows involving chemical reactions. A detailed description of the chemical kinetics is generally needed to fully understand the combustion processes in such cases. MOEAs are often employed for determining and adjusting the reaction parameters or for reducing the number of reactions [18, 19, 20]. Furthermore, the reduction of pollutant emission (NOx, CO, soot) is of major practical interest. EAs have been applied in gas turbines for minimizing NOx emissions and/or for reducing combustion noise [13, 14, 67]. Mono-objective optimization of a laminar burner was investigated in [74] where the objective was to obtain a homogeneous temperature profile at a prescribed distance from the injection plane. As a whole, optimization of configurations involving the coupled simulation of flows and combustion processes remains a fairly new field of research.

Laminar flows involving chemical reactions play an important role in many practical applications, for example domestic burners. In this section, a two-dimensional simulation dedicated to solving the Navier-Stokes equations in the low-Mach number limit is presented using accurate models for chemistry, diffusion and thermodynamics.

A two-dimensional configuration is considered, involving a primary inlet in the center of the computational domain and a secondary inlet at the periphery. The optimization problem consists of finding the minimal mass-flow rate of the pollutant species CO while maintaining a homogeneous temperature distribution. The corresponding integral value and the variation are computed at a prescribed distance from the inlet. These two objectives are in this case the minimal concentration of CO along the corresponding horizontal cut through the solution and the temperature difference between the maximum and minimum value. The parameters modified by the optimization procedure are the fuel and oxidizer mass flows of the primary inlet while the total amount of fuel and oxidizer injected through both inlets is of course kept constant. There are, therefore, two parameters that may freely vary between a lower bound (0: no fuel or no oxidizer injected through this inlet) and an upper bound (all the available fuel or all the available oxidizer injected through this inlet).

The investigated configuration is depicted in Fig. 2.2. Due to the symmetry, only the right half of the domain is considered in the computation. Depending on the input parameters, methane/air mixtures with different compositions enter the domain through the primary and secondary inlets with a fresh gas temperature of 298 K. Atmospheric pressure is imposed at the outlet. The inlet wall temperature is constant and equal to 298 K.

It will be shown that it is possible using CFD to reach an optimal configuration for such a complex problem involving coupled fluid flow, heat transfer

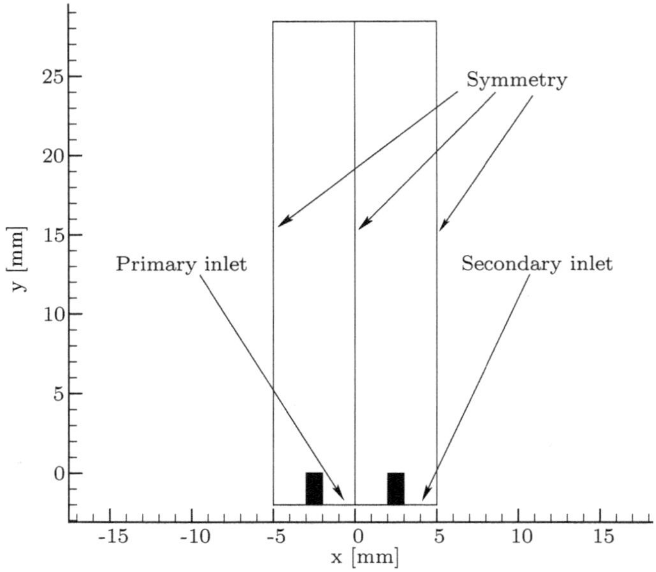

Fig. 2.2 General configuration of the laminar burner (Case B). A methane/air mixture enters through the primary and secondary inlets with a varying composition

and chemical reactions, and this within a reasonable computing time, even for configurations close to practical ones.

2.1.4 Determination of Turbulence Model Parameters Based on Optimization (Case C)

Turbulent flows are essential for a wealth of practical applications. Predictive numerical tools are of course deeply needed to improve existing devices and develop new configurations in the turbulent regime. The broad range of problems underlying turbulent flows cannot be solved with a unique method. It is therefore useful to identify different levels of modeling, describe their domain of application and some characteristic features. Direct Numerical Simulations (DNS) are only possible for simple configurations and low-Reynolds number flows. Large-Eddy Simulations (LES) resolve the large structures of the flow while the small scales are modeled. LES modeling becomes now feasible in many situations due to the increase of computing power but remains an expensive solution for industrial problems. In the RANS modeling, the balance equations are averaged in time with respect to the turbulent fluctuations of the flow variables. This averaging introduces higher-order moments which cannot be computed without approximate closure schemes. Nevertheless, numerical simulations based on RANS are still widely used today for engineering problems and complex geometries due to a higher computational efficiency. The objective in this case is to optimize the prediction of the time-averaged turbulent velocity distribution in channel flows.

A similar investigation was performed in [37], where a variable Schmidt number model for jet-in-crossflows was improved using GA optimization. Multi-objective EAs have been also employed for determining and adjusting reaction parameters in [18, 19, 20]. But, to our knowledge, no paper has been published up to now considering the optimization of the model constants of an engineering turbulence model.

2.2 Evolutionary Algorithms for Multi-objective Optimization

2.2.1 Multi-objective Optimization

Mathematically speaking, a multi-objective problem consists of optimizing (i.e., minimizing or maximizing) several objectives simultaneously, with a number of inequality or equality constraints. The problem can be formally written as follows:

$$\text{Find } \boldsymbol{x} = (x_i)\ \forall\, i = 1, 2, \ldots, N_{\text{param}} \text{ such as}$$
$$f_i(\boldsymbol{x}) \text{ is a minimum (resp. maximum)},\ \ \forall\, i = 1, 2, \ldots, N_{\text{obj}}$$

subject to:

$$g_j(\boldsymbol{x}) = 0, \qquad \forall\, j = 1, 2, \ldots, M \tag{2.1}$$
$$h_k(\boldsymbol{x}) \leq 0, \qquad \forall\, k = 1, 2, \ldots, K \tag{2.2}$$

where $\boldsymbol{x}$ is a vector containing the N_{param} design parameters, $(f_i)_{i=1\ldots N_{\text{obj}}}$ the objective functions and N_{obj} the number of objectives. In this study, only inequality constraints are considered and are prescribed as bounded domains. In other words, upper and lower limits are imposed on all parameters:

$$x_i \in [x_{i,\min}; x_{i,\max}] \quad i = 1 \ldots N_{\text{param}}\ . \tag{2.3}$$

The objective function $(f_i(\boldsymbol{x}))_{i=1\ldots N_{\text{obj}}}$ returns a vector containing the set of N_{obj} values associated with the elementary objectives to be optimized simultaneously. The number of input parameters and objective functions for the different cases considered in this chapter are summarized in Table 2.1.

A common practice to solve such a problem is to use a trade-off between the objectives by linearly combining them using some fixed weights prescribed by the user (see for example [16, 73]). The resulting single objective function

Table 2.1 Number of input parameters and objectives functions

Case	Section	# of parameters N_{param}	# of objectives N_{obj}
Case A (heat exchanger)	2.3	8	2
Case B (laminar burner)	2.4	2	2
Case C (turbulence model)	2.5	5	4

is optimized using for instance, a classical gradient-based or Simplex method [63]. The first limitation of this kind of approach is that the choice of the weights associated with each objective obviously influences the solution of the optimization problem. A bad choice can lead to completely sub-optimal results in comparison with the solution obtained by considering the interrelated objectives in an independent manner. Moreover, this method does not allow access to all the set of optimal solutions.

The EAs are semi-stochastic methods, based on an analogy with Darwin's laws of natural selection [34]. Each configuration $\boldsymbol{x}$ is considered as an individual. The parameters x_i represent its genes. The main principle is to consider a so-called population of N individuals, i.e., a set of individuals covering the search domain and to let it evolve along generations (or iterations) so that the best individuals survive and have offsprings, i.e., are taken into account and allow to find better and better configurations.

The characteristics of the EA used in the present study are based on the approach proposed by Fonseca and Fleming [32]. The genes (sometimes called characters) of the individuals (also called strings or chromosomes) are the N_{param} design parameters, encoded using a floating-point representation [59]. The initial population is a set of quasi-random configurations in the domain defined by the limits imposed on the parameters, Eq. (2.3). The creation of a new generation from the previous one is performed by applying genetic operators to the individuals of the present generation, as described below. One common way to select individuals to be parent in the next generation is the tournament selection [16]. In the present study, the fitness-based selection is applied. At each generation, the individuals are classified as a function of their corresponding objective values, leading to a rank within the population and finally to a fitness. The definition of the rank for our specific case is described later in Sect. 2.2.2. The probability for an individual to participate in the reproduction process is determined by a probability based on its fitness value, linearly calculated from its rank in the classification. For example, for individual number i in a group of N individuals:

$$\text{Fitness}(i) = \frac{N - \text{rank}(i) + 1}{\sum_j (N - \text{rank}(j) + 1)} \tag{2.4}$$

with index j varying from 1 to N.

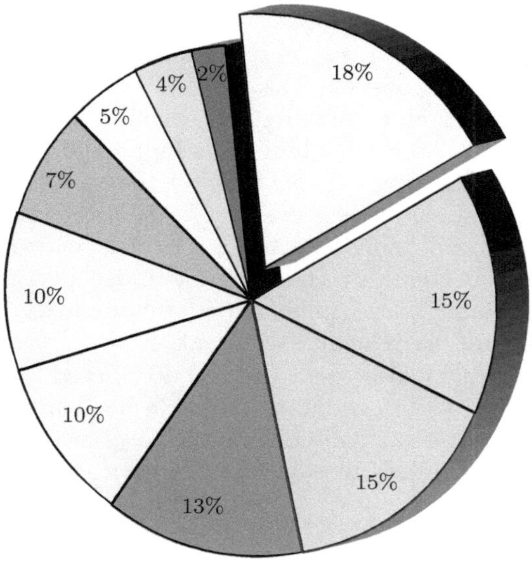

Fig. 2.3 Example for the rank of 10 individuals and the corresponding probability (Eq. (2.4)) to participate in the reproduction process, represented on a circular diagram

Figure 2.3 depicts a simple example showing for a group of 10 individuals, the rank values and the corresponding probability to participate in the reproduction process, directly based on its fitness. Using this technique, individuals with equal rank have an equal probability to reproduce. Individuals associated with a higher fitness value have a better chance to survive and to take part in the reproduction process, as explained in Sect. 2.2.3. In this way, better and better generations are generated step by step. EAs operate on the entire population. Thus, they offer a good potential to explore the whole search space and to avoid local optima. Their good robustness is mainly due to the fact that there is no evaluation of the derivatives of the objective function. Moreover, the process can iterate further even if some evaluations fail.

The main drawback associated to EAs in general remains their cost in terms of computing (CPU) time. On the other hand, due to the fact that the evaluations are performed independently, they are easily parallelizable as demonstrated later.

2.2.2 The Concept of Pareto Dominance

In a multi-objective problem, the set of parameters (the individuals in the EA terminology) can be compared according to Pareto's rule [32]: the individual

A dominates the individual B if, for at least one of the objectives, A is strictly better adapted than B and if, for all other objectives, A is not worse than B. An individual will be considered as optimal if it is non-dominated in the sense of this rule. The rank is computed for each individual according to the number of individuals dominating him. If an individual is not dominated by any other individual, he gets the top rank (of course, there may be several non-dominated individuals at the same time). Then come all individuals that are dominated by only one individual, and so on. This is a true multi-objective approach because objectives are considered as independent and there is no trade-off. An individual i that is dominated by j individuals is given $\text{rank}(i) = 1 + j$. Non-dominated individuals have rank 1, so $1 \leq \text{rank}(i) \leq N$. At the end, the best individuals are all the non-dominated individuals over all generations, leading to the computed Pareto front, supposed to represent the real one. The use of this rule allows one to classify the individuals of the population and therefore to calculate the corresponding fitness values, Eq. (2.4).

2.2.3 Evolutionary Algorithm for Multi-objective Problems

Three groups are defined in the population, two for the EA generations and one for storing the non-dominated configurations:

- *Elite.* The currently non-dominated individuals.
- *Parents.* Individuals that may reproduce.
- *Offsprings.* Individuals of next generation, built from parents.

An individual can belong at the same time to several groups. To generate an offspring, one or two parents are selected using their fitness values. The selection of the parents relies on the roulette wheel method. Each roulette wheel slot receives a current individual from the population. An individual with better fitness value is associated with a larger roulette wheel slot size (Fig. 2.3). A larger size of a roulette wheel slot corresponds to a better chance to survive or to reproduce than the others. The new population is produced spinning the roulette wheel N times where N represents the total size of the population. This favors individuals with a higher fitness while leaving a chance for the worst individuals to take part in the reproduction process, hence keeping diversity through the generations. Individuals with an equal rank get the same fitness value and have thus the same probability to survive or to reproduce.

Once the parents have been selected with this method, the offspring can be generated. The genes of the offsprings can be computed using the values of the genes of the parents (Fig. 2.4). To prescribe the properties of the offsprings, we use randomly one of the three following methods:

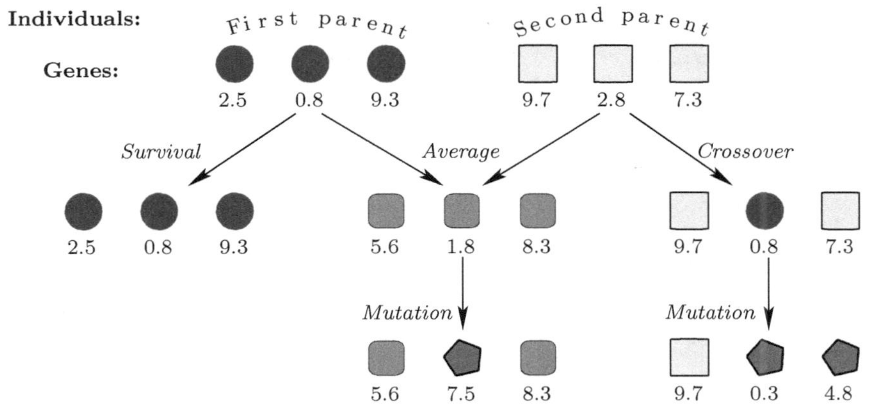

Fig. 2.4 Principle of the EA after selection showing survival, average, crossover and mutation. The individuals have here 3 genes, all of them are between 0 and 10

- *Survival.* Only one individual is selected and the offspring will be the same as this parent without any change.
- *Average.* Two parents are chosen and the genes of the offspring are the average of the corresponding genes of the two parents.
- *Crossover.* The crossover can be used to increase the diversity among the population. The genes are represented by floating-point numbers. In the present studies, only uniform crossover is applied where the genes are selected randomly from either one of the parents during the crossover process. In this way, randomly selected genes from both parents will be kept in the future generations by being associated with the corresponding offspring.

To introduce diversity, the offsprings further go through a mutation step. A mutation operator is needed because important genetic information may occasionally be lost or missing. In many cases, mutation is a key search operator for the domain exploration so that the mutation probability must be 1 or close to 1. A mutation probability of 1 means that all individual genes obtained by averaging or crossover will be modified by mutation. This mutation is performed after defining the offsprings, to randomly modify the genes of the offsprings. Figure 2.4 represents a simple example with three genes based on the floating-point representation between 0 and 10 illustrating the procedure.

Multi-objective methods attempt to localize the Pareto front, which is the set of all non-dominated configurations according to the definition given above. Thus, the multi-objective optimization problem aims at finding a discrete approximation of the Pareto front (sometimes also denoted Pareto Optimal Frontier, POF) which is the set of all non-dominated parameters. As will be shown later, the POF is clearly visible as soon as enough non-dominated configurations have been identified and plotted.

All parameters of the EA procedure used in this work are listed later in corresponding sections (Table 2.2, 2.4 and 2.6). Parameters usually chosen in the literature as well as further parameter sets have been tested extensively in previous works [6], showing that the values retained in these tables are well-suited for the problems considered here. The applied mutation probability may seem quite high. It should be pointed out, that the mutation magnitude is continuously decreasing at each generation to stabilize the population. A high initial mutation probability is needed to obtain a fine-grain resolution of the POF.

2.3 The Optimal Position of the Tubes in a Heat Exchanger (Case A)

2.3.1 Tube Bank Heat Exchanger

A two-dimensional model of a cross-flow tube bank heat exchanger is considered first. One possible configuration can be seen in Fig. 2.1. Air enters the domain at $T_{\text{inlet}} = 293$ K and is warmed up by passing between the tubes in which warm fluid flows in the corresponding practical application. The tubes are supposed to have a constant outer wall temperature, $T_{\text{wall}} = 353$ K. The outlet is at atmospheric pressure.

The optimization problem consists of finding the best locations of the tubes to increase heat exchange while at the same time to limit the pressure loss. The two corresponding numerical parameters to optimize are the average temperature difference ΔT and pressure difference ΔP between inflow and outflow.

The domain bounded by a black line in Fig. 2.1 is simulated in this study. The Reynolds number is equal to 41 defined using the tube diameter $D = 2$ cm and the uniform velocity at the inlet, $v_{\text{inlet}} = 0.03$ m/s. The length of the domain has been chosen to prevent any influence of the inflow or outflow boundary conditions on the inter-blade flow. Corresponding tests have, in particular, demonstrated the importance of extending the computational domain well beyond the last tube in order to avoid the influence of the outflow boundary conditions. The full extent of the numerical domain and a typical numerical grid can be seen in Fig. 2.5.

The stability analysis in Barkley and Henderson [5] proved that three-dimensional effects first appear at a Reynolds number around 188 in the flow around a cylinder. Furthermore, the flow around a single cylinder is steady up to a Reynolds number of 46 [4]. In the present investigation, several cylinders are simulated and will interact with each other, but due to the low

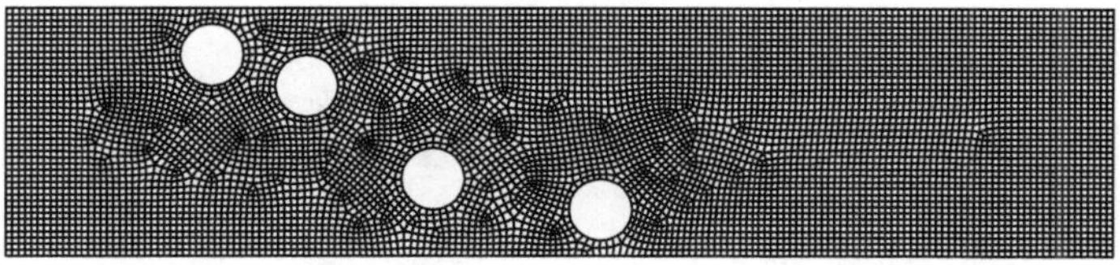

Fig. 2.5 Typical computational grid in Case A

Reynolds number, a steady two-dimensional flow can still be assumed as a good approximation of the true physics.

2.3.2 Problem Parameters

The numerical simulation of a physical problem can be performed using various geometries and/or boundary conditions. Here, for the different simulations, the boundary and inlet conditions are the same, only the computational geometries differ.

The outer dimensions of the computational domain, as well as the number of the tubes in the domain, are fixed and only the positions of the tubes inside the computational region are varied. The positions of all four tubes are always changed simultaneously, but their locations are kept within a predefined range to avoid crossing the boundaries. Furthermore, the positions are constrained to prevent overlapping and direct contacts between cylinders. Since all other properties and boundary conditions are constant, these position parameters are the only input parameters of the optimization algorithm. After defining the computational geometry and obtaining a corresponding mesh, the numerical simulation can be performed. In this study, the industrial CFD program Fluent 6.3 [30] solves the governing equations of the fluid flow phenomena including the energy equation. The two-dimensional fields of pressure and temperature are obtained in this way and provide the two objective parameters, the average temperature and pressure differences: ΔT, ΔP between inflow and outflow.

The set of coupled numerical tools used to solve the multi-objective optimization problem are now described and schematically shown in Fig. 2.6.

2.3.3 Opal (OPtimization ALgorithms) Package

Opal [6, 38] is an object-oriented C++ code for Unix/Linux systems using a Tcl script interpreter [78]. A Tcl script is used for coupling Opal with other

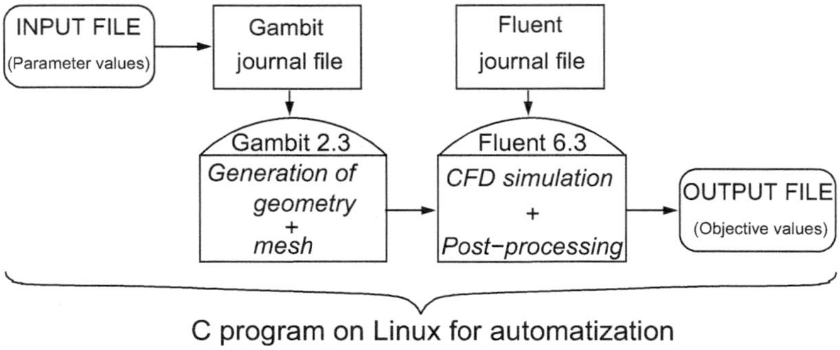

Fig. 2.6 Flow chart showing the numerical solution procedure

computer codes and is employed in our case to call a C interfacing program responsible for the evaluation of the objective functions.

2.3.4 Evaluation of the Objectives for Case A

In the present case, this evaluation relies on the commercial software Gambit [29] for geometry and mesh generation, and Fluent [30] for solving the flow and energy equations. Therefore, the evaluation of an individual set of parameters requires four steps:

1. the generation of the computational geometry using the position variables;
2. the generation of an appropriate mesh for the obtained geometry;
3. the CFD simulation, i.e., the solution of the governing coupled equations for the flow variable and the energy on the mesh generated in the previous step;
4. the post-processing of the obtained results to extract the values of the objective functions for these specific design variables.

Steps 1 and 2 are performed using the commercial software Gambit 2.3 [29], step 3 using the CFD code Fluent 6.3 [30] and step 4 takes place in the in-house interfacing code.

2.3.4.1 Step 1: Computational Geometry of the Heat Exchanger Configuration with 4 Tubes

The geometrical constraints are prescribed in terms of lower and upper limits on the parameters. The positions of the middle points of the four circular tubes of the heat exchanger are given with their two-dimensional coordinates. The domain is separated in six non-overlapping zones in x-direction, the first is the inlet part and the last one is the outlet part. Positioning the tubes are not permitted here to prevent any interaction with the boundary conditions. The four middle zones contain the tubes. Overlapping and direct contacts are not allowed. The locations of the tubes are also varied in the y-direction, but without crossing the periodic boundaries.

2.3.4.2 Step 2: Mesh Generation

After having defined the geometry, the mesh is produced in an automatic manner using the commercial software Gambit 2.3 [29]. This is easily done by modifying the so-called journal file containing the important geometrical parameters as variables. Knowing the middle point coordinates of the tubes $(x_1, y_1), (x_2, y_2), (x_3, y_3)$ and (x_4, y_4) is sufficient to fully define the geometry and therefore to generate the mesh in an automatic manner and export it to Fluent. The script checks that the mesh generation has been successful before going on with the CFD computation, which has always been the case for this simple geometry but might pose a problem for complex three-dimensional configurations. A typical example of the resulting grid is shown in Fig. 2.5. The internal fluid region is meshed using quadrilateral cell elements with the "pave" algorithm of Gambit. The computational nodes are uniformly spaced along the boundaries. The typical number of computational cells in a mesh lies around $11,000 - 12,000$. A systematic grid-independence study has been performed as presented next.

2.3.4.3 Step 3: CFD Simulation

The in-house interfacing code now starts Fluent [30] in an automatic manner using again a journal file. Only the geometry and the mesh change between two CFD computations.

The inlet (left side in Fig. 2.1) boundary is considered as a velocity inlet with imposed conditions for the velocity, set to $v_{\text{inlet}} = 0.03$ m/s, and the temperature $T_{\text{inlet}} = 293$ K. Wall boundary conditions with a constant temperature $T_{\text{wall}} = 353$ K are prescribed on all four tube surfaces. Periodic condition is applied on the top and bottom boundaries. On the right, a pressure outlet condition relaxing to atmospheric pressure is imposed.

The discretized governing equations are solved iteratively using a finite-volume description with the new pressure based solver (PBS) in Fluent 6.3 [30]. To improve the accuracy, second-order upwind discretization is systematically used for all variables, along with a double-precision computation. The normalized residuals are computed at every iteration by Fluent. As soon as all of these residuals fall below a prescribed value, convergence is considered to be reached. In our case, the fixed prescribed value is 10^{-4} for the flow equations and 10^{-6} for the temperature equation, providing a sufficient

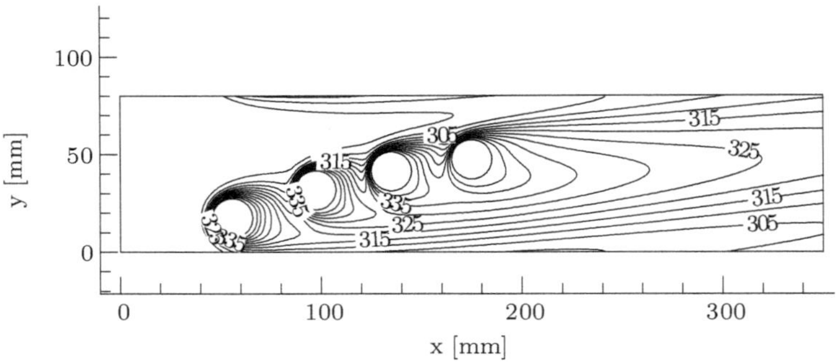

Fig. 2.7 A typical CFD result, showing the obtained temperature field in Kelvin of one of the optimum solutions (see also Fig. 2.8)

accuracy for an acceptable CPU time. This point has been checked for one of the solutions identified as optimal, by further decreasing these thresholds. A completely negligible influence has been observed for the two objective variables.

In the same way, a systematic grid-independence study has been carried out for this selected non-dominated case. By refining several times the grid in a uniform manner, the relative pressure and temperature differences do not change by more than 1.1% and 0.05%, respectively, demonstrating that the initial grid is sufficient to obtain quantitative estimations. Restarting as an unsteady simulation leads only to completely negligible variations in the objective values, confirming that the steady assumption is appropriate for such a low Reynolds-number flow.

The velocity-pressure coupling is treated with the standard SIMPLE pressure-correction method. In most cases, the convergence is achieved in 300 to 500 iteration steps. If the convergence is not reached within 900 iteration steps, the simulation is considered as not converging and is dismissed. This has been observed only for less than 5% of the evaluations.

2.3.4.4 Step 4: Post-processing

After convergence, the temperature difference between the inlet (uniform constant value) and the averaged value along the outlet is computed. The pressure difference between the inlet and outlet averaged pressure values is also computed. These two differences are the two objectives of the optimization problem. The resulting temperature and pressure fields of one of the optimum solutions are presented as an example in Figs. 2.7 and 2.8, respectively.

In the present case, the configuration is two-dimensional and can be easily optimized on a single PC with a reasonable computing time. Nevertheless,

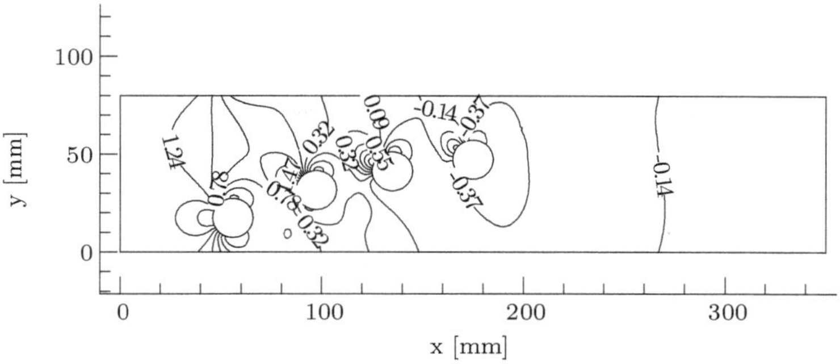

Fig. 2.8 A typical CFD result, showing the obtained relative pressure field in Millipascal of one of the optimum solutions (same solution as in Fig. 2.7)

this will not always be the case. Therefore, a parallel version of this optimization procedure has been developed.

2.3.5 Parallelization

As already mentioned, an important issue related with EAs is the high computational effort needed to perform the necessary evaluations of the objectives associated with the population. The evaluation of a large number of individuals for a large number of generations can lead to unaffordable CPU times in practical engineering cases. On the other hand, the EAs have the advantage to be easily and efficiently parallelizable. At each generation, all N individuals can be evaluated independently on different processors since the central algorithm only needs the values of the objectives to iterate. In the current case, the parallelization has been implemented using the C interfacing program (see Fig. 2.6) responsible for the evaluation of the $N_{\text{obj}} = 2$ objective values associated with the N configurations. A farmer/worker paradigm has been retained [34] using the portable MPICH implementation [36] of the Message Passing Interface (MPI) [58] communication routines. All the evaluations are carried out on a Linux PC cluster running under Red Hat 9. Figure 2.9 shows a principle description of this multi-processor optimization procedure. In the present case, each worker PC performs independently its own CFD simulation.

Alternatively, if one CFD simulation requires a lot of computational effort (memory and time), each evaluation could be performed in parallel. In this situation, all the PC represented in Fig. 2.9 would simultaneously collaborate on one single CFD evaluation. Any optimization method (Simplex, gradient-based, EA,...etc.) can be used in this case for the optimization without impacting parallelization.

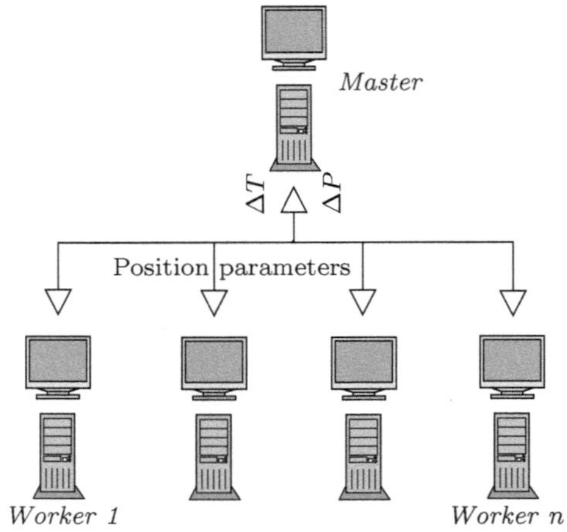

Fig. 2.9 Schematic flow chart showing the multi-objective CFD optimization problem running on a multi-node PC cluster

An EA optimization procedure can easily be carried out in parallel to reduce the complete simulation time. The individuals (here the CFD simulations) are completely independent from each other in one generation. In this way, several evaluations can be performed at the same time on several processors (Fig. 2.9). A single simulation for the presented heat exchanger problem requires only some minutes of CPU time.

The parallel computation can be limited by the required computational resources, i.e., the number of free computers in the system or the available licenses when using commercial codes. Most cluster systems are supervised by a so-called batch system to control and organize the computational work and the available licenses. Depending on the load of the cluster, it is possible that at one time no simulation is running or that several computations are processed at the same time. In this case, the load will not be uniform and the resulting total wall-clock time may not be representative to measure the speed-up of the parallel optimization.

Typical results leading to Pareto fronts can be extracted from these calculations and are presented and discussed next.

2.3.6 Computational Results

2.3.6.1 Pareto Fronts

After several tests, EA optimization was carried out using 40 individuals at every generation. The applied probabilities are given in Table 2.2.

Table 2.2 Parameters of the EA for the heat exchanger optimization, Case A

Parameter	Value
Population size, N	40
Generations	20
Survival probability	25%
Average probability	25%
Crossover probability	50%
Mutation probability	100%
Mutation magnitude	50%[a] (i.e., ±25%)

[a]This value is multiplied by 0.8 at each generation. For example, the mutation magnitude is 6.7% (±3.35%) after 10 generations or 0.72% (±0.36%) after 20 generations. Mutation magnitude must be decreased during the optimization process to stabilize the population.

The parents in the first generation are generated using the quasi-random method proposed by Sobol' [72] and modified by Antonov and Saleev [2]. This leads to an optimal initial coverage of the parameter space, as recommended by the method called Design of Experiments (DOE). The non-dominated individuals at each generation belong to the elite group. At each iteration, 10 individuals can survive for the next generation and 30 new offsprings are added, 10 using averaging and 20 using crossover, leading again to 40 individuals. Since there are 40 parents and only 10 survive for the next generation, the 30 others disappear from the reproduction cycle but are kept in the elite if they are non-dominated. Nevertheless, in this case, they cannot generate offsprings anymore. If more than 30 non-dominated individuals exist in the present population, the integration within the next generation is again based on the roulette wheel method described in Sect. 2.2.2.

The resulting temperature and pressure differences for all points of the elite group evaluated over 20 generations are plotted in Fig. 2.10. The two objectives, temperature and pressure difference are plotted on the x and y axes, respectively. The points already generate a clearly visible POF in the middle part of the figure. In Fig. 2.10, the concept of the Pareto front is also illustrated. The POF is the boundary between infeasible configurations and possible, but non-optimal solutions.

Five different optimal placements obtained by the optimization procedure are presented in Fig. 2.11. The flow direction is from left to right. The first two figures correspond to a low pressure loss, but the average temperature difference between the inlet and the outlet is small. Higher temperature differences can be achieved using the placements shown for example in Fig. 2.11(e) but in this case, the pressure losses increase.

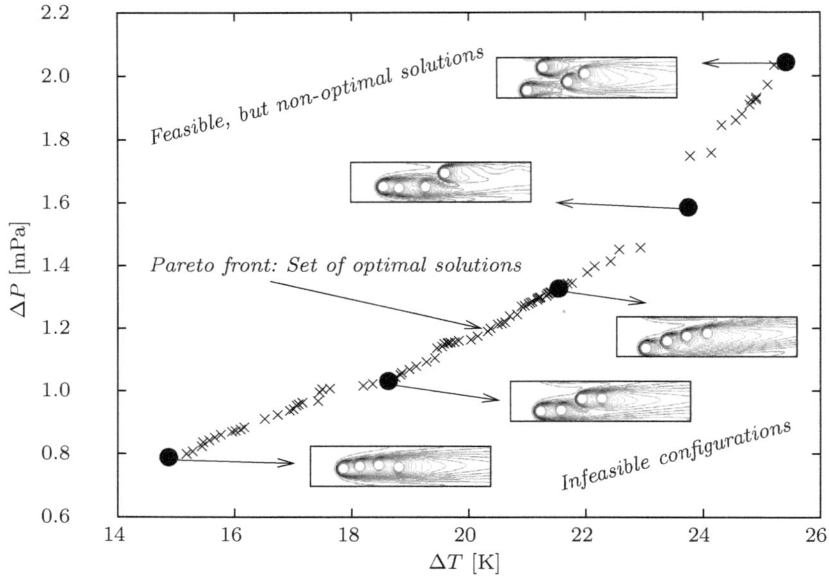

Fig. 2.10 Results of the best evaluations after 20 generations of the EA: Pareto front (POF), feasible and infeasible configurations

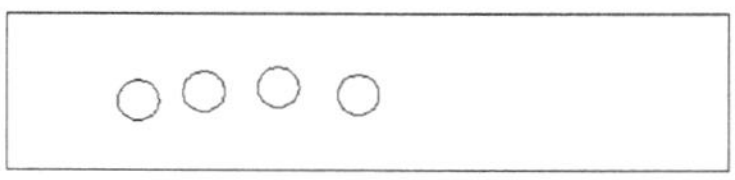

(a) $\Delta T = 14.9$ K, $\Delta P = 0.79$ mPa.

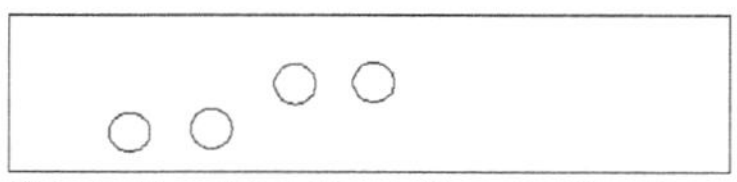

(b) $\Delta T = 18.6$ K, $\Delta P = 1.03$ mPa.

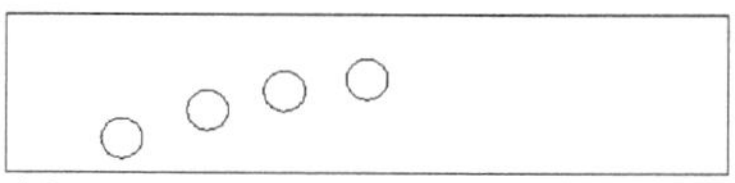

(c) $\Delta T = 21.5$ K, $\Delta P = 1.33$ mPa.

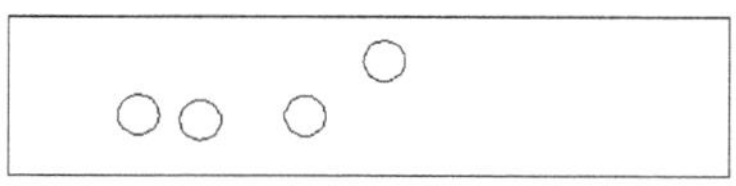

(d) $\Delta T = 23.7$ K, $\Delta P = 1.59$ mPa.

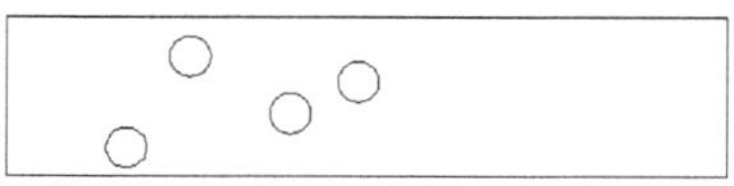

(e) $\Delta T = 25.4$ K, $\Delta P = 2.04$ mPa.

Fig. 2.11 Example of the resulting placement for 5 individuals belonging to the POF. The flow direction is from left to right

2.3.6.2 EA Parameters

The objective values obtained from the simulations are shown as a function of the number of generations in Fig. 2.12, in which only the non-dominated configurations are represented. We observe that the POF can be recognized

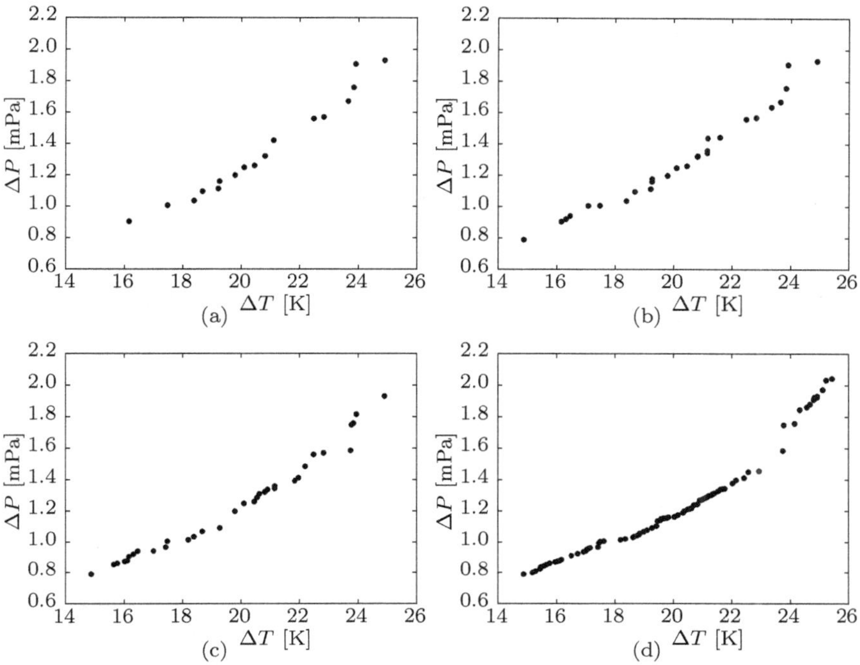

Fig. 2.12 The non-dominated individuals for the heat-exchanger problem (elite). (**a**) The non-dominated individuals after one generation. (**b**) The non-dominated individuals after two generations. (**c**) The non-dominated individuals after five generations. (**d**) The non-dominated individuals after 20 generations

very early, i.e., already after 2 or 3 generations. Nevertheless, the quality increases with the number of generations and the POF becomes more refined. There are two reasons for this progressive improvement: first, we have more and more individuals in the elite group. Secondly, Opal favors individuals that are very different. For example, if we have 5 non-dominated individuals with the same fitness value in the group of the parent, with 4 of them very close to each other in parameter space and the last one quite different, this last individual will be favored for reproduction to enhance diversity, which improves the spatial extension of the POF.

2.3.6.3 Speed-up Obtained Through Parallelization

In order to reduce the waiting time for the user, the process has been parallelized and carried out on a multi-node Linux PC cluster with 16 worker PCs. Each node is a 2.6 GHz/2 GB-RAM Pentium-IV Linux PC. The communications are performed via a Fast Ethernet (1 Gb/s) network connection.

Table 2.3 Speed-up obtained using the parallel optimization method (Case A)

Number of processors	Wall-clock time	Speed-up
1	1280 min	1.00
2	661 min	1.94
5	294 min	4.35
10	153 min	8.37

Table 2.3 shows the resulting CPU times needed for the evaluation of 20 generations consisting of 40 individuals, performed with an increasing number of processors on the PC-cluster. The speed-up is defined here as the ratio between the wall-clock time obtained when using N_{proc} processors compared to the one needed with a single processor. The theoretical optimal value of the speed-up using N_{proc} processors is N_{proc}. In practical cases, the communication times between processors and load-imbalance reduce the speed-up value below the theoretical maximum.

In the present case, the obtained parallel speed-up is nearly optimal. This is not really a surprise since the quantity of information transferred between the so-called farmer and workers is very small: four real parameters in one direction, two in the reverse direction. The communication times are therefore almost negligible compared to the CPU times required for the evaluation of the objectives on each processor. Deviation from the optimal speed-up is thus mainly due to boundary effects at the end of each optimization iteration when some worker PCs become inactive for a short time waiting for the next iteration to start.

2.4 Multi-objective Optimization of a Laminar Burner (Case B)

2.4.1 Governing Equations

Laminar flows involving chemical reactions are considered in this section. In the presented application Mach numbers M are very low. It is observed that pressure variations through laminar flames at low Mach numbers are always of the order of magnitude of a few Pascal and stem mainly from hydrodynamical and not from compressibility effects. Stated differently, density variations only result from heat release due to chemical reactions and from changes in the mixture composition, but not from local fluid compression. Temperature and density vary in opposite directions, such that their effects within the ideal gas law compensate. These physical observations motivate the decomposition of pressure into a bulk background uniform thermodynamic pressure p_u and a hydrodynamic fluctuation term $\tilde{p}$ [33, 53]:

$$p\left(\boldsymbol{x},t\right)=p_u\left(t\right)+\tilde{p}\left(\boldsymbol{x},t\right) \ . \tag{2.5}$$

For the presented application (a domestic gas burner), the numerical domain is considered open, hence p_u equals the atmospheric pressure and is constant. If acoustic waves may propagate in the gas mixture, then an additional acoustic pressure term has to be considered. Since acoustic-flame interactions are not addressed here, it is assumed from now on that acoustic waves are either nonexistent or of negligible effect on the flame structure and on the flow. Readers particularly interested in acoustic-flame interactions may refer to [35, 47] for further specific details.

Within the low-Mach number approximation, the density appears as a function of temperature T and mean molar weight W. When the numerical domain is opened to the atmosphere, the influence of the hydrodynamic pressure p_u on the density must be neglected. The full problem is then described by the following set of balance equations, written in conservative form for mass, momentum, mass fractions and enthalpy, solved in the present case:

$$\frac{\partial \rho}{\partial t}+\boldsymbol{\nabla}\cdot(\rho\boldsymbol{v})=0 \tag{2.6}$$

$$\frac{\partial(\rho\boldsymbol{v})}{\partial t}+\boldsymbol{\nabla}\cdot(\rho\boldsymbol{v}\boldsymbol{v})=-\boldsymbol{\nabla}\tilde{p}+\boldsymbol{\nabla}\cdot\left\{\mu\left(\boldsymbol{\nabla}\boldsymbol{v}+(\boldsymbol{\nabla}\boldsymbol{v})^T\right)\right\} \tag{2.7}$$

$$\frac{\partial(\rho Y_k)}{\partial t}+\boldsymbol{\nabla}\cdot(\rho Y_k\boldsymbol{v})=-\boldsymbol{\nabla}\cdot(\rho Y_k\boldsymbol{V}_k)+W_k\dot{\omega}_k \quad , \quad 1\le k\le K-1 \tag{2.8}$$

$$\frac{\partial(\rho h)}{\partial t}+\boldsymbol{\nabla}\cdot(\rho h\boldsymbol{v})=\boldsymbol{\nabla}\cdot\left(\lambda\boldsymbol{\nabla}T-\rho\sum_{k=1}^{K}h_kY_k\boldsymbol{V}_k\right) \ . \tag{2.9}$$

The notations used here are standard in the combustion community and are explained for example in [39]. Using a unity Lewis number assumption $\mathrm{Le}=\lambda/(\rho c_p D)=1$ for the molecular diffusion model, Eqs. (2.8) and (2.9) become much simpler:

$$\frac{\partial(\rho Y_k)}{\partial t}+\boldsymbol{\nabla}\cdot(\rho Y_k\boldsymbol{v})=\boldsymbol{\nabla}\cdot\left(\frac{\lambda}{c_p}\boldsymbol{\nabla}Y_k\right)+W_k\dot{\omega}_k \tag{2.10}$$

$$\frac{\partial(\rho h)}{\partial t}+\boldsymbol{\nabla}\cdot(\rho h\boldsymbol{v})=\boldsymbol{\nabla}\cdot\left(\frac{\lambda}{c_p}\boldsymbol{\nabla}h\right) \ . \tag{2.11}$$

Equation (2.10) is solved for $N-1$ species, because the last species (the nitrogen) is a non-reacting species (dilutant) simply determined using:

$$Y_{N_2}=1-\sum_{k=1}^{K-1}Y_k \ . \tag{2.12}$$

For all computations presented in this paper, the complete set of chemical species and elementary reactions, with their Arrhenius coefficients A_i (pre-exponential factor), β (temperature exponent) and E_i (activation energy), is taken for methane/air flames from [50]. This detailed reaction scheme involves 29 species and 141 elementary reactions. It thus involves 28 supplementary transport equations in the form of Eq. (2.8) plus Eq. (2.12), leading to a tremendous increase of the requested computing time compared to Case A.

2.4.2 Numerical Solution

The numerical simulation is performed using *UGC*$^+$. This code has been optimized for the computation of steady laminar low-Mach number flows with chemical reactions [7, 68]. It is designed as an application of the multi-purpose *UG* library [8]. *UG* is a modular numerical toolbox originally aimed at investigations of multigrid methods on various model problems described by sets of partial differential equations.

UGC$^+$is based on two main modules: a low-Mach Navier-Stokes solver and a thermo-reactive solver. A joint module has been developed to achieve the full coupling of the two sub-modules into one single Partial Differential Equations (PDE) system. The two solvers are in charge of their own diagonal block of the Jacobian matrix and there is an information interchange between them (mass fluxes, density and viscosity).

The optimization procedure varies the mass flows of the fuel and air at the two inlets. The corresponding velocity values are used at the inlets for the Navier-Stokes equations. On both sides of the numerical domain, symmetry conditions are applied. The temperature of the fresh gas and walls at the inlets is 298 K. The total quantity of methane and air injected in the system (primary + secondary inlet) is kept constant, so that in principle the same energy is always available. At the outlet atmospheric pressure is imposed.

The *UGC*$^+$code attempts to find steady solutions through time-marching. Time discretization is of first-order implicit type. The value of the time-step can be adapted at each iteration, according to convergence or any user-defined condition. The unsteady equations are solved by fixed-point or approximate-Newton iterations and the user can freely specify how often the Jacobian matrix has to be assembled. The linearized equations are then solved by a Bi-CGSTAB [77] algorithm, preconditioned by multigrid V- or W-cycles with an ILU smoother [69]. A dynamic adaptive grid is used to increase resolution for such multi-scale problems (thin reaction zones, large geometries).

The numerical simulation of a physical problem can be performed using various geometries and/or boundary conditions. For the present burner, the computational geometry is fixed for all computations, but the boundary conditions (composition of the mixture for the primary and the secondary inlets) are varied during the optimization procedure.

In previous works [43, 44, 74], the optimization problem was simple (single objective) and the Simplex method [28, 63] has been used to speed-up the computational procedure. Here a multi-objective problem is considered and an EA is employed to investigate the optimization domain for the retained input parameters. For the present problem, one evaluation is extremely costly in CPU time.

One possibility to speed-up the evaluation is to perform every numerical simulation on parallel computers [42] or to use simplified methods to describe the chemical processes (e.g., the flamelet or tabulated chemistry approach [11, 25, 26, 27, 41, 52, 65]). In the presented application, the Navier-Stokes equations are solved for a two-dimensional flow. In the case of detailed chemistry, 29 species conservation equations are solved additionally. Tabulated chemistry involves only three supplementary transport equations. In *UGC*$^+$, both the detailed reaction mechanism and the tabulated chemistry (FPI, for Flame-Prolongation of ILDM) are implemented and available for the computations [41]. In the first case, all chemical parameters are computed using the standard software CHEMKIN [45] while FPI computations rely on tabulated values. The computation using detailed chemistry can be performed either with simple molecular transport using the unity Lewis number assumption or using detailed transport modeling [39]. The presented investigation is based on both detailed chemistry and detailed transport computation. This high level of physical accuracy is important to reduce the modeling uncertainty associated with the CFD solutions, especially for cases with minor differences in the objective functions. However, the stability and the speed of the evaluations are greatly enhanced by starting the simulation using the simplified method (FPI, see [41]) during 30 time-steps. The corresponding results are converted into an initial, detailed chemistry solution based on the tabulated values. The detailed chemistry computation is then continued for another 40 time-steps, leading to the "final" result of the evaluation, used to evaluate the objective functions after the necessary post-processing. Unfortunately, the detailed chemistry computations are extreme stiff. Therefore, in many cases, the residual values will still be very high after this fixed number of iterations, indicating convergence problems. Such simulations are marked as non-feasible results and dismissed from the optimization. "Non-feasible" means also that either no converged solution can be found or that the CPU time required until convergence would be unacceptably long.

All CFD evaluations rely on adaptive time-steps in order to stabilize the solution procedure. A starting value of 0.1 s has been always used here for the restart of the detailed computation from the tabulated chemistry results. Further investigations are needed to check if other time-step values would improve the number of valid evaluations. Although the detailed computations considerably decrease the number of feasible solutions, they deliver also more realistic and accurate results.

The smallest grid spacing employed in the present computations is 62.5 μm. This resolution is needed for very stiff intermediate radicals like HCO and

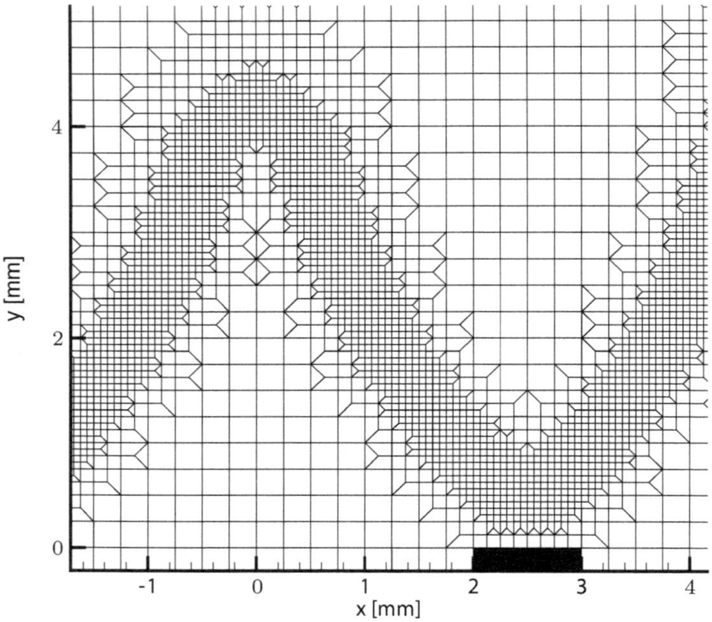

Fig. 2.13 Zoom on the part of the numerical grid close to the flame front

C_2H_2. The numerical grid is refined adaptively according to a predefined criterion, here the mass fraction of HCO. During the computations, the numerical grid contains up to 8,000 finite-volume cells after reaching the fourth level of grid refinement (see Fig. 2.13). Computations using detailed chemistry with more than 9,000 grid elements are not possible due to memory limitation (2 GB physical memory on a single PC in our system). When using detailed chemistry, a higher resolution or more complex configurations can only be computed on parallel supercomputers [42].

The optimization procedure is carried out on five Pentium-IV PC with 2.6 GHz/2 GB-RAM running under Red Hat 9 Linux. Every computation is performed on a single computer. The computation based on tabulated chemistry used as starting condition for the detailed one takes roughly 90 minutes. Typical computation times for one CFD evaluation with detailed chemistry vary between 9 and 22 hours, depending on the inlet conditions.

2.4.3 Optimization of the Laminar Burner

For computing the burner, both the adaptive mesh generation and the flow computations are carried out inside the in-house CFD-software *UGC*$^+$. To illustrate the full computational procedure, the evaluation of an individual set of parameters requires four steps:

1. the computation of the composition of the mixture in the primary and secondary inlet, knowing the specific design variables;
2. the simplified CFD simulation, i.e., the resolution of the governing coupled equations for the flow variables, the energy and the species conservation equations, using first tabulated chemistry (FPI);
3. the high-quality CFD simulation using detailed chemistry and transport models, restarted from the previously obtained FPI solution;
4. the post-processing of the obtained results to extract the values of the objective functions (average mass flow rate of CO along a horizontal cut and maximal variation of the temperature profile along this cut, respectively) for these design variables.

For the case considered here, the configuration is two-dimensional and can be optimized on a single PC with a high but acceptable computing time, provided a CFD software as efficient as *UGC*$^+$is employed. Nevertheless, this will not always be the case. Therefore, a parallel version of this optimization procedure has been developed, as already described for another optimization problem in Sect. 2.3.5. As a whole, 254 configurations have been evaluated during the optimization procedure and 64 feasible cases have been found. The total computational time requested by the optimization using 5 PCs is roughly 3 weeks.

The optimization procedure can now be started with the objectives of reducing as much as possible the average mass flow rate of CO and reducing the temperature variations along a horizontal cut through the solution at $y = 0.025$ m, near the outlet.

The optimization space is defined, as explained before, by two parameters as shown in Figs. 2.14 and 2.15, where the objective functions are represented as function of the input parameters using a contour plot representation. The darker values correspond to a higher mass flow rate of CO and a higher temperature variation, respectively. The top left corners of these figures contain no results. Since the corresponding input parameters would yield a configuration in contradiction with the concept of a central *primary* inlet with a surrounding *secondary* inlet, they are not considered in the procedure.

It can be seen that both objectives improve roughly in the same direction in parameter space. Differences are only observed when looking at the details of these evolutions, so that we observe a posteriori that these objectives are in fact not really concurrent, but approach a nearly identical optimum. The differences between both optimal solutions are so small (less than 2 K for the temperature variation) that they probably approach solution accuracy, even using detailed chemistry.

All EA parameters are listed in Table 2.4.

The two objective values of the feasible cases are shown in Fig. 2.16, in which 20% of the evaluation results are not shown to improve clarity. This

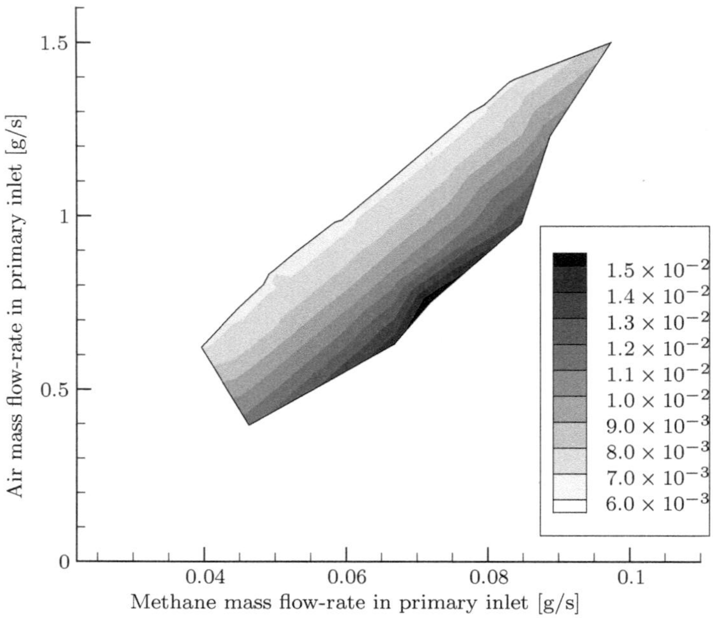

Fig. 2.14 Contour plot of the average mass fraction of CO (in g/s) as function of the input parameters

Table 2.4 Parameters of the EA for the laminar burner optimization (Case B)

Parameter	Value
Population size, N	20
Generations	14
Survival probability	10%
Average probability	40%
Crossover probability	50%
Mutation probability	100%
Mutation magnitude	50%[a] (i.e., ±25%)

[a]This value is multiplied by 0.85 at each generation. For example, the mutation magnitude is 11.6% (±5.8%) after 10 generations. Mutation magnitude must be decreased during the optimization process to stabilize the population.

figure shows again that both objectives improve roughly in the same direction in parameter space and converge to an almost identical optimal solution.

Figure 2.16 does not contain any information on the corresponding input parameters. The relation between the input parameters and the objectives are demonstrated in Fig. 2.17 using parallel coordinates. In this figure, only good configurations are shown to improve readability. The optimal configurations correspond to two almost identical fully premixed and nearly stoichiometric

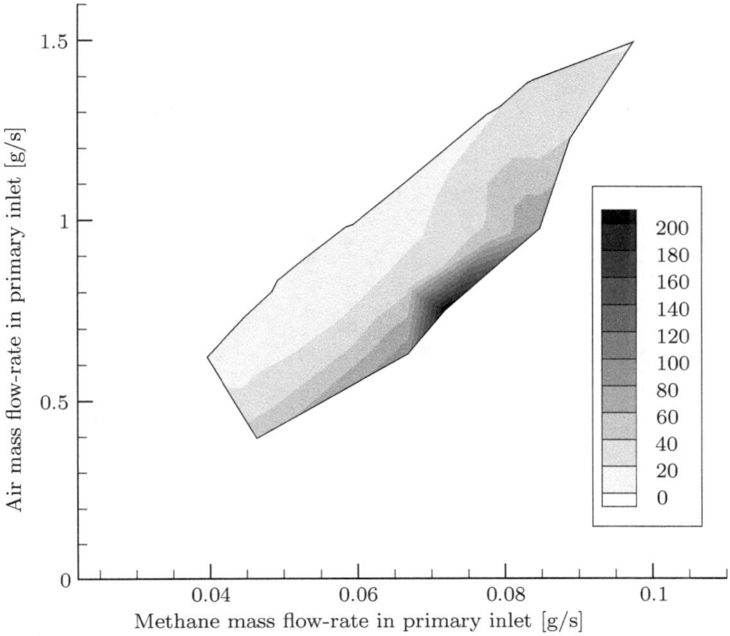

Fig. 2.15 Contour plot of the temperature variation (in K) as function of the input parameters

Table 2.5 Primary inlet flow-rates for the selected simulations

Figure number	Primary injector: Methane mass flow rate [g/s]	Primary injector: Air mass flow rate [g/s]
Fig. 2.18	7.19×10^{-2}	7.49×10^{-1}
Fig. 2.19	4.92×10^{-2}	8.31×10^{-1}

mixtures, injected through the primary and the secondary inlet and leading to the shortest possible flame.

The mass fraction field of CO obtained for the two-dimensional burner considered here at two chosen compositions of the two inlets are presented in Figs. 2.18 and 2.19. In both figures, the mass fraction of the radical HCO is also shown on the right side to visualize the shape and position of the active reaction zone. Figure 2.18 corresponds to a non-optimal solution of the EA. Compared with the optimal solution in Fig. 2.19, a large CO mass fraction and thus CO mass-flow rate is found at the outlet of the combustion chamber for this set of compositions. These two figures employ the same gray-scale values to facilitate comparisons.

The values chosen by the optimization procedure for the primary inlet for the cases corresponding to Figs. 2.18-2.19 are listed in Table 2.5. Consid-

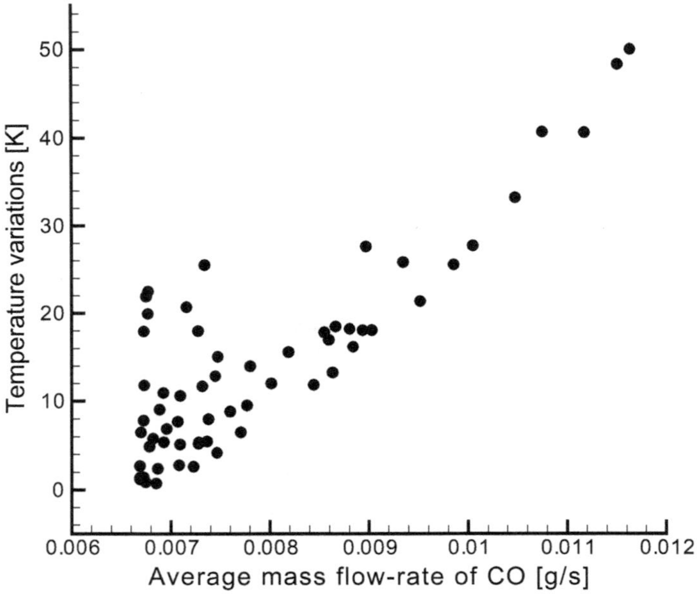

Fig. 2.16 Feasible configurations for Case B. To improve clarity only a part of the results are shown here

erable variations have been tested by the optimization procedure, exploring the whole parameter space before identifying the best solution. Such large variations would certainly not have been considered by a human being carrying out a manual optimization. The automatic optimization requires in this case a considerable computing time but leads to a reduction of the pollutant emission (CO) by a factor 2.5 and of the temperature variation by a factor exceeding 20, compared to the worst feasible configuration (Fig. 2.18).

2.5 Optimization of the Standard k–ω Turbulence Model Parameters (Case C)

Numerical simulations based on RANS are widely used for engineering problems and complex geometries, due to a high computational efficiency. The determination of the closure constants for RANS turbulence models is based on dimensional analysis, theoretical observations or experiments for some special cases like channel or pipe flows. Such experiments often consider two-dimensional flows, so that most model developments are originally suited for two-dimensional situations. Nevertheless, these models are mostly used in quite different – usually complex three-dimensional – configurations so that

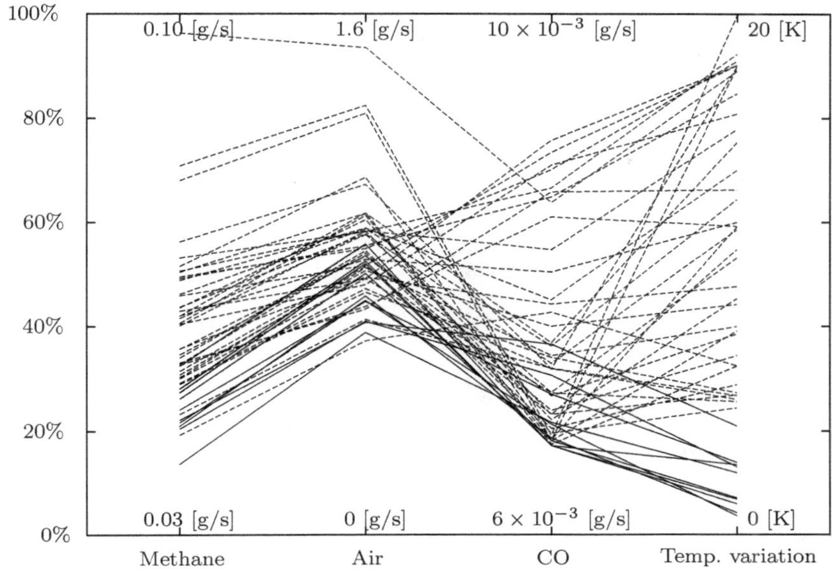

Fig. 2.17 Input parameters and objectives of the EA optimization for good configurations in the laminar burner case

they may fail or lead to inaccurate results. There are well-known issues like swirl, secondary flow, large pressure gradients, strong streamline curvature, etc., where model predictions often become poor. In such cases, some modifications and/or additional terms are needed in the model. Different authors often propose slightly different parameter values when introducing such modifications.

The determination of the model constants for engineering turbulence models is indeed a difficult task. The values are often considered as some ad-hoc values. Changing one parameter in order to observe consequences concerning, for instance, the time-averaged turbulent velocity distribution or the shear-stress distribution is easy. But the simultaneous modification of several parameters of a turbulence model in order to increase accuracy rapidly becomes a formidable issue. If all the model parameters are changed in small steps, then the number of possible combinations would yield an enormous – and probably unnecessary – computational effort to explore the whole domain. In that case, numerical optimization techniques may help to speed-up the search procedure in order to find the best possible combination of the model constants with a minimum computational load, since optimization is much more efficient than a simple trial and error manual procedure.

The objective in this case is to optimize the prediction of the time-averaged turbulent velocity distribution in channel flows. Direct numerical simulation (DNS) data [40, 60] for four different Reynolds numbers are chosen as ref-

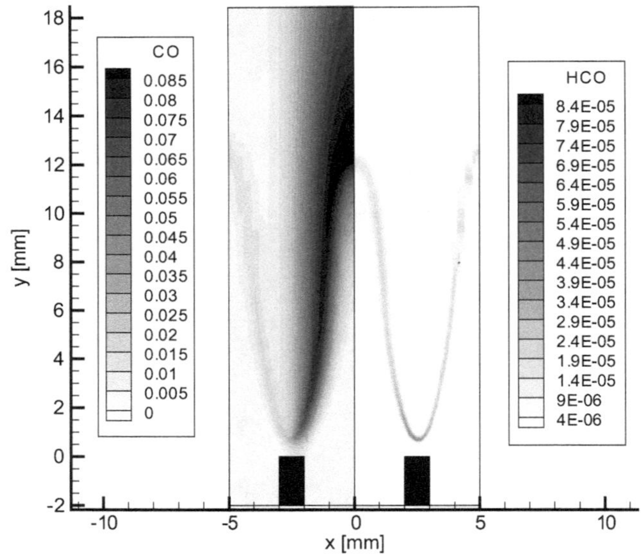

Fig. 2.18 Mass fraction field of CO (left) and HCO (right) in the worst feasible case for the EA optimization

erence. In this way, the optimization problem involves several concurrent objectives that must be fulfilled simultaneously. Generic and robust search methods, such as EA, can be used for such problems as demonstrated here. The present optimization problem consists of finding the best group of values for the model constants of the k–ω turbulence model [79] leading to the most accurate prediction of the velocity distribution, in agreement with the DNS data. The k–ω model is widely used in CFD, demonstrating the importance of this issue. In a first investigation, the shear-stress profiles have been also examined, but the differences were quite small compared with DNS. It was observed that the constants producing a good velocity profile automatically lead to an accurate shear-stress distribution as well, which is not unexpected, since the shear-stress directly depends on the velocity gradients. Therefore, the shear-stress profiles are dropped from the objectives and the number of reference cases is increased, involving different Reynolds numbers.

2.5.1 Governing Equations

The Reynolds-averaged governing equations that describe turbulent flow motion are summarized here following Wilcox [79]. An incompressible flow with constant thermodynamic properties is now assumed, unlike Case B.

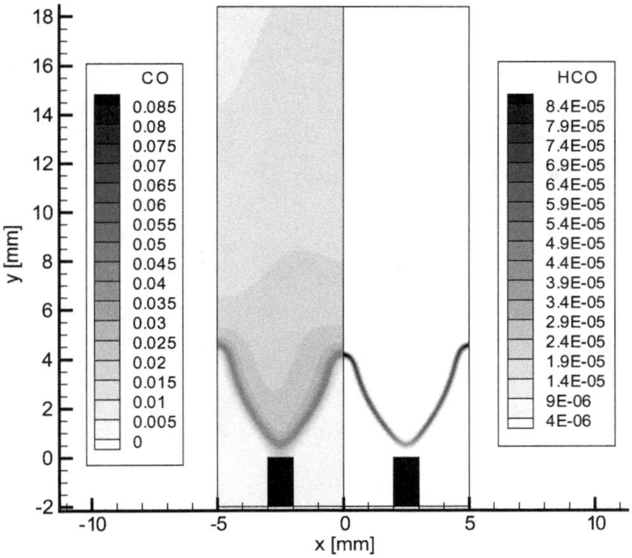

Fig. 2.19 Mass fraction field of CO (left) and HCO (right) at the optimal point

The Reynolds-averaged turbulent flow variables can be separated as:

$$u_i\left(\boldsymbol{x},t\right) = U_i\left(\boldsymbol{x},t\right) + u_i'\left(\boldsymbol{x},t\right) \tag{2.13}$$

where the vector U_i is the time-averaged velocity and u_i' is the fluctuation term. The time-averaged mass conservation equation reads:

$$\frac{\partial U_i}{\partial x_i} = 0 \tag{2.14}$$

and the momentum equations can be written as:

$$\rho\frac{\partial U_i}{\partial t} + \rho U_i\frac{\partial U_i}{\partial x_j} = -\frac{\partial P}{\partial x_i} + \frac{\partial}{\partial x_j}\left[\mu\left(\frac{\partial U_i}{\partial x_j} + \frac{\partial U_j}{\partial x_i}\right)\right] + \frac{\partial}{\partial x_j}(-\rho\overline{u_i'u_j'}) \ . \tag{2.15}$$

According to the Boussinesq hypothesis, the Reynolds stress-tensor (corresponding to the last term in Eq .(2.15)) can be modeled as:

$$-\overline{u_i'u_j'} = \nu_t\left(\frac{\partial U_i}{\partial x_j} + \frac{\partial U_j}{\partial x_i}\right) - \frac{2}{3}k\delta_{ij} \tag{2.16}$$

where ν_t is usually called turbulent eddy viscosity, expressed as:

$$\nu_t = \frac{k}{\omega} \ . \tag{2.17}$$

The modified k–ω turbulence model proposed by Wilcox [79] in 1998 is employed in this study. The transport equations for k and ω can be written as, respectively:

$$\frac{\partial k}{\partial t} + U_j \frac{\partial k}{\partial x_j} = \tau_{ij} \frac{\partial U_i}{\partial x_j} - \beta^* k\omega + \frac{\partial}{\partial x_j} \left[(\nu + \sigma^* \nu_T) \frac{\partial k}{\partial x_j} \right] , \tag{2.18}$$

$$\frac{\partial \omega}{\partial t} + U_j \frac{\partial \omega}{\partial x_j} = \alpha \frac{\omega}{k} \tau_{ij} \frac{\partial U_i}{\partial x_j} - \beta\omega^2 + \frac{\partial}{\partial x_j} \left[(\nu + \sigma \nu_T) \frac{\partial \omega}{\partial x_j} \right] . \tag{2.19}$$

The model includes several auxiliary relations and closure coefficients[1]. In this study, five parameters of the model are selected for the optimization, listed here with their standard values:

$$\alpha = 13/25 \ , \ \beta_0 = 9/125 \ , \ \beta_0^* = 9/100 \ , \ \sigma = 1/2 \ , \ \sigma^* = 1/2 \ . \tag{2.20}$$

A more detailed description of this turbulence model as well as the extension of this model to compressible flows is extensively discussed in Wilcox [79].

2.5.2 Numerical Results

After calculating one set of values for the four selected Reynolds numbers, the time-averaged turbulent velocity profiles are compared with the DNS results. The differences between the four corresponding profiles are measured using the area enclosing the curves. Due to the fine grid, this area can be approximated quite well using a simple numerical integration based on the rectangle rule. The four corresponding numerical parameters to optimize (minimize) are these area values for the four different Reynolds numbers.

The main drawback associated to EAs in general remains their cost in terms of computing time because they require a large number of evaluations on different configurations. Here, a fast computational procedure is chosen in this investigation to speed-up the optimization. The numerical simulations for a fully developed channel flow are performed using the simplified computational code from Wilcox [79]. This program is based on a simple numerical integration and one single simulation takes only a few seconds on a standard PC.

Five values have been selected as the input parameters of the optimization procedure. The model parameters α, β_0, β_0^*, σ and σ^* may freely vary between 0 and 1. A graphical representation in five dimensions is difficult. Therefore, parallel coordinates are used to show the connection between the

[1] In this work, these closure coefficients are referred to as model parameters. Calling them "model constant" is confusing, since these values are probably not constant and possibly problem-dependent.

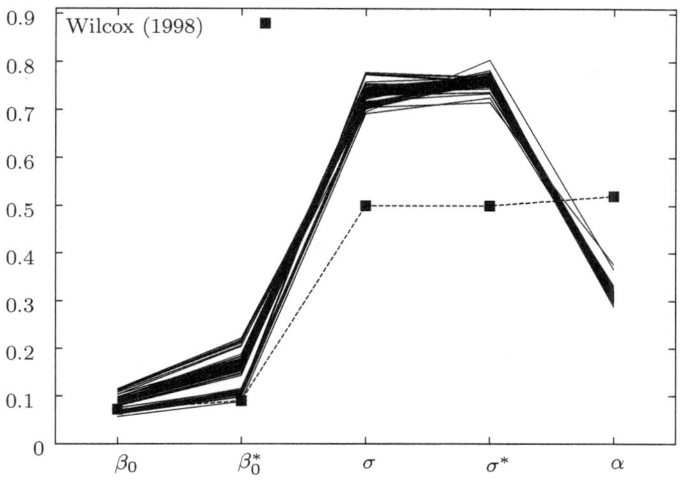

Fig. 2.20 The five input parameters of the optimization represented using parallel coordinates

corresponding values in Fig. 2.20. The original values proposed by Wilcox [79] are also represented in this figure with symbols.

Figure 2.20 also shows that considerable modifications of the model parameters have been tested by the optimization procedure, compared to the original reference values proposed by Wilcox [79].

A total of 5,050 evaluations have been performed during the optimization procedure. Plotting two arbitrary objectives (i.e., the errors for two different Reynolds numbers) against each other, a strong linear dependence can be observed (Fig. 2.21). This is satisfactory, since it demonstrates that, globally, a good parameter set will be valid for different Reynolds numbers, which is essential for practical purposes. However, zooming on the best results (lower left corner) in order to see the details, a Pareto front corresponding to concurrent objectives is detected (Fig. 2.22). This proves that one single, optimal parameter set cannot be found that would be equally valid for all Reynolds numbers.

The computed velocity profiles using the model constants in one optimum configuration are shown in Fig. 2.23. These profiles are compared to DNS data and to the results computed with the original parameter values. The proposed modifications lead to a considerably better agreement with the DNS.

Figure 2.24 finally shows the four objectives of the optimization compared with results obtained by the original model ([79], shown with symbols). Parallel coordinates are used to represent the connection between the corresponding values. It can be clearly seen that, in all four cases, better results are obtained with an optimal parameter set compared with the values proposed by Wilcox in 1998 for channel flows.

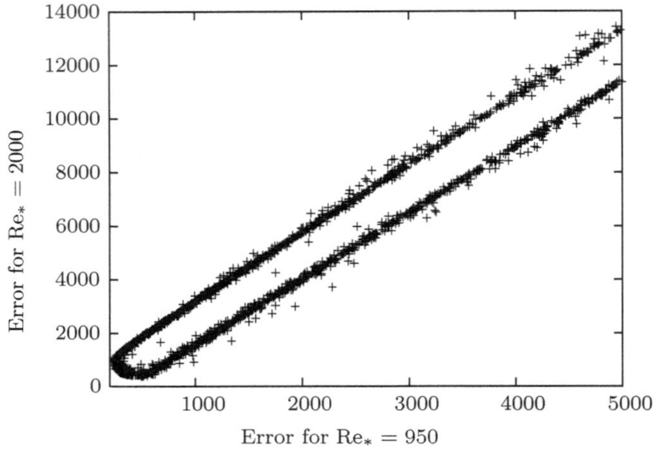

Fig. 2.21 Two objectives in Case C showing a dominantly linear correlation

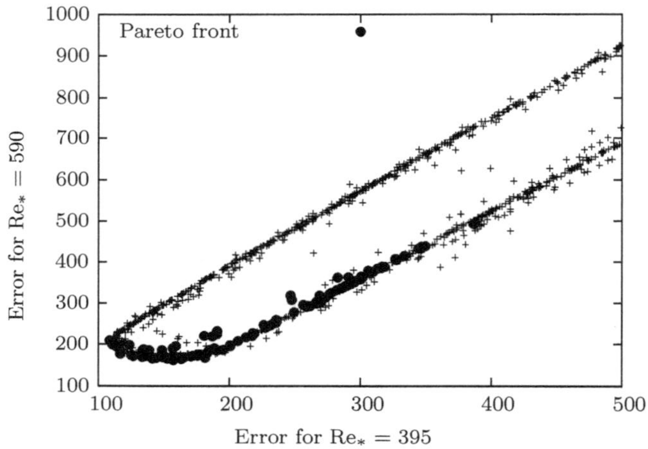

Fig. 2.22 Two objectives in Case C showing the best results and the resulting Pareto front

All EA parameters of the optimization procedure are listed in Table 2.6.

2.6 Conclusions

In this study, EAs have first been applied to a multi-objective shape design optimization problem concerning a heat exchanger configuration close

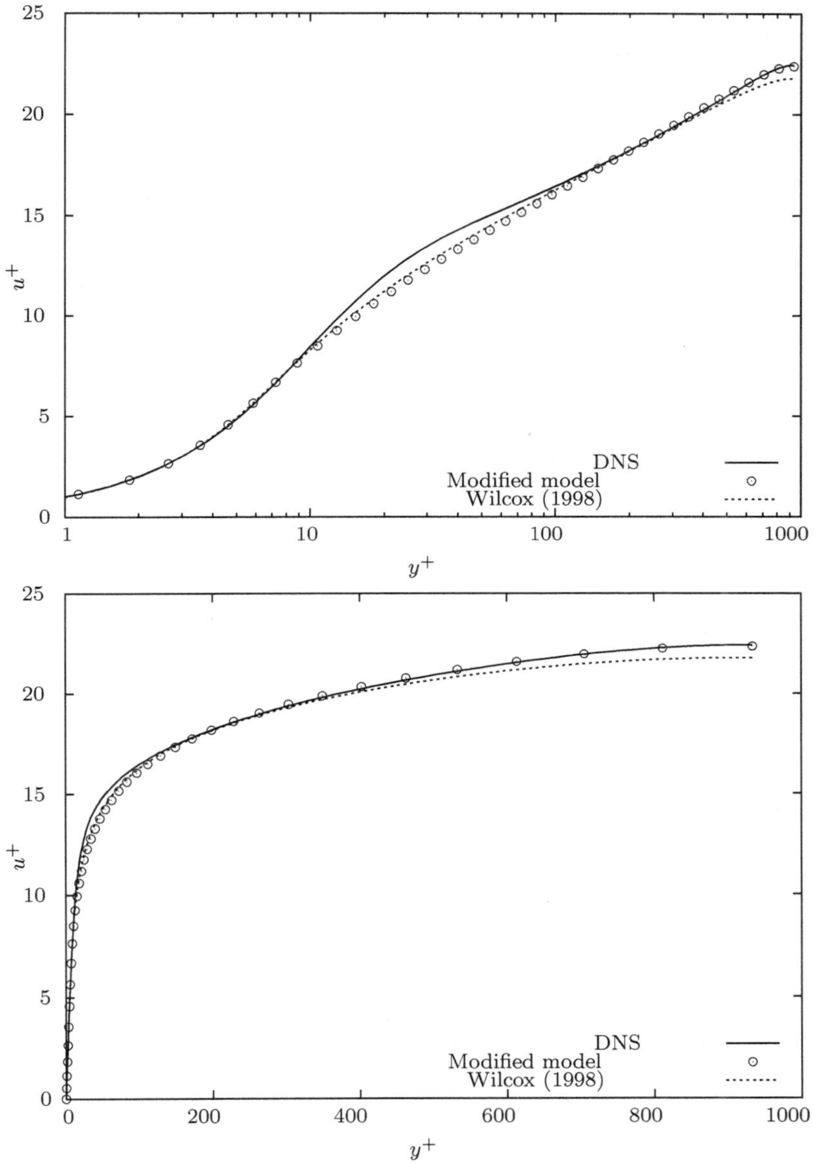

Fig. 2.23 Velocity profiles in a semi-logarithmic plot (top) and in a standard plot (bottom)

to practical applications. The characteristic Pareto front associated with this problem has been obtained within a very reasonable computational time. The heat exchanger model problem is based on a description of the tube positions

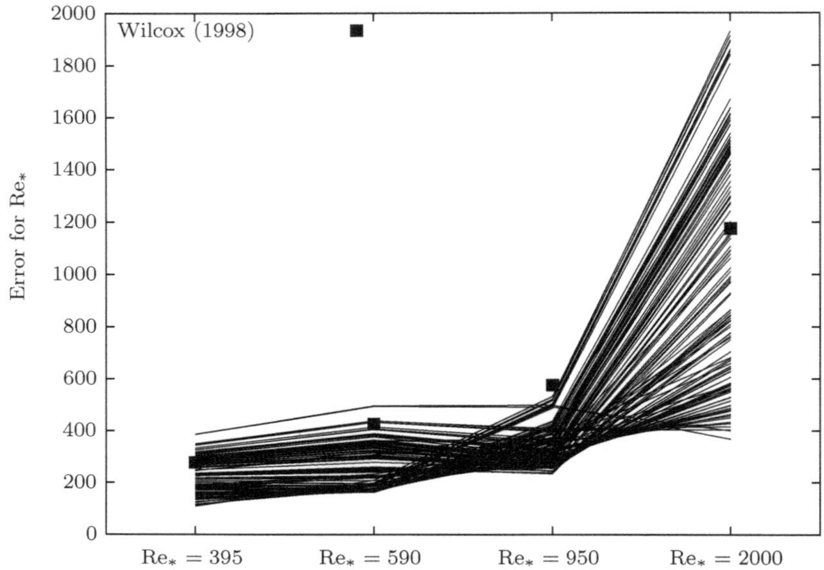

Fig. 2.24 The four objectives of the optimization represented using parallel coordinates for the individuals belonging to the POF

Table 2.6 Parameters of the EA for the turbulence model optimization (Case C)

Parameter	Value
Population size, N	100
Generations	100
Survival probability	50%
Average probability	30%
Crossover probability	20%
Mutation probability	100%
Mutation magnitude	50%[a] (i.e., ±25%)

[a]This value is multiplied by 0.95 at each generation. For example, the mutation magnitude is 31.5% (±1.75%) after 10 generations or 0.3% (±0.15%) after 100 generations. Mutation magnitude must be decreased during the optimization process to stabilize the population.

using eight parameters. This is a simple description that could be refined. Computing times can be further reduced by using parallelization, as demonstrated in this chapter. More complex, practical industrial cases are solvable using on one side appropriate modeling and simplification of the problem and on the other side parallelization.

Furthermore, we have demonstrated that optimization of complex flows involving heat transfer and complex chemical reactions is possible, provided very efficient numerical methods are used for the optimization process (here, the in-house Opal library) as well as for the CFD procedure (here, the in-house *UGC*$^+$code). The optimization domain has been explored using EA, but with a very high computational effort, leading to specific difficulties.

Finally, we have improved the model parameters of a well-known engineering turbulence model using optimization. The proposed new values predict more accurately the time-averaged velocity profiles in channel flows, as shown by a comparison with DNS.

As a whole, this paper has demonstrated that CFD-O can be used for a variety of complex engineering problems, but is still associated with major issues. Some of them will be discussed in more detail in further chapters of this book:

- For complex optimization problems, parallel computations will be absolutely necessary to reduce user waiting time. CFD-O is indeed very well suited for parallel computers. The CFD evaluations can be performed in parallel, or the independent individuals in the optimization can be evaluated in parallel, as presented in this chapter (Cases A and B). It is of course possible to combine both and again speed-up the procedure.
- Evolutionary Algorithms require a large number of evaluations (for example Case C, with more than 5,000 evaluations) to obtain a refined description of the Pareto front. This is a major problem when the evaluation is computationally expensive like in Case B. Therefore, the best way to speed-up optimization is indeed to improve the computational efficiency of the CFD software!
- If an evaluation is computationally very demanding, as in Case B (burner computation using detailed chemistry and transport), it is essential to reduce as far as possible the number and the cost of the evaluations. For this purpose, concepts like Design of Experiments or approximate evaluations based, for example, on Artificial Neural Networks appear promising. As a complement, simplified physical models might be used to get a first approximation as demonstrated in the present chapter.
- When considering many parameters and many concurrent objectives, the visualization of the results becomes increasingly difficult. Graphical and post-processing tools dedicated to optimization would greatly facilitate the analysis.

Acknowledgements Interesting discussions with D. Thévenin, R. Hilbert and R. Baron are gratefully acknowledged. This research project has been initially started at the EM2C Laboratory (École Centrale Paris, France). The library Opal has been first developed by R. Baron during his Ph.D., with the financial support of CETIAT and ADEME. The development of the program *UGC*$^+$ has been mostly carried out by S. Paxion and R. Baron, financed by different French organizations and companies (DGA, ADEME and CETIAT). Concerning *UGC*$^+$ the author acknowledges the essential contributions from G. Wittum, A. Gordner, N. Simus-Paxion, N. Neuss, V. Reichenberger, S. Lang, P. Bastian, B. Fiorina, R. Hilbert and O. Gicquel.

References

1. Ali, N., Behdinan, K.: Optimal geometrical design of aircraft using genetic algorithms. Transactions of the Canadian Society for Mechanical Engineering **26**(4), 373–388 (2003)
2. Antonov, I.A., Saleev, V.M.: An economic method of computing LP_τ-sequences. U.S.S.R. Computational Mathematics and Mathematical Physics **19**, 252–256 (1979)
3. Balagangadhar, D., Roy, S.: Design sensitivity analysis of steady fluid-thermal systems. Computer Methods in Applied Mechanics and Engineering **190**, 5465–5479 (2001)
4. Barkley, D.: Linear analysis of the cylinder wake mean flow. Europhysics Letters **75**(5), 750–756 (2006)
5. Barkley, D., Henderson, R.D.: Three-dimensional Floquet stability analysis of the wake of a circular cylinder. Journal of Fluid Mechanics **322**, 215–241 (1996)
6. Baron, R.: Calcul et optimisation de brûleurs laminaires industriels. Ph.D. thesis, École Centrale Paris, 2002-37 (2002)
7. Baron, R., Paxion, S., Gicquel, O., Simous, N., Bastian, P., Thévenin, D.: Development of a 3D parallel multigrid solver for fast and accurate laminar steady flame computations. In: E.H. Hirschel (ed.) Numerical Flow Simulation III, pp. 115–128. Springer-Verlag (2003)
8. Bastian, P., Birken, K., Johannsen, K., Lang, S., Neuß, N., Rentz-Reichert, H., Wieners, C.: UG - A flexible software toolbox for solving partial differential equations. Computing and Visualization in Science **1**, 27–40 (1997)
9. Bejan, A.: Entropy Generation Minimization. CRC Press, Boca Raton, Florida (1996)
10. Bello-Ochende, T., Bejan, A.: Constructal multi-scale cylinders in cross-flow. International Journal of Heat and Mass Transfer **48**(7), 1373–1383 (2005)
11. Bongers, H., van Oijen, J.A., Somers, L.M.T., de Goey, L.P.H.: The flamelet-generated manifold method applied to steady planar partially premixed counterflow flames. Combustion Science and Technology **177**(12), 2373–2393 (2005)
12. Bonjour, J., Rocha, L.A.O., Bejan, A., Meunier, F.: Dendritic fins optimization for a coaxial two-stream heat exchanger. International Journal of Heat and Mass Transfer **47**(1), 111–124 (2004)
13. Büche, D., Stoll, P., Dornberger, R., Koumoutsakos, P.: Multiobjective evolutionary algorithm for the optimization of noisy combustion processes. IEEE Transactions on Systems, Man and Cybernetics Part C: Applications and Reviews **32**(4), 460–473 (2002)
14. Büche, D., Stoll, P., Koumoutsakos, P.: An evolutionary algorithm for multi-objective optimization of combustion processes. Center for Turbulence Research Annual Research Briefs 2001 pp. 231–239 (2001)
15. Cheng, C.H., Chang, M.H.: Shape design for a cylinder with uniform temperature distribution on the outer surface by inverse heat transfer method. International Journal of Heat and Mass Transfer **46**(1), 101–111 (2003)
16. Deb, K.: Multi-Objective Optimization using Evolutionary Algorithms. Wiley, London (2001)
17. Dias Jr., T., Milanez, L.: Optimal location of heat sources on a vertical wall with natural convection through genetic algorithms. International Journal of Heat and Mass Transfer **49**(13-14), 2090–2096 (2006)
18. Edwards, K., Edgar, T., Manousiouthakis, V.: Kinetic model reduction using genetic algorithms. Computers and Chemical Engineering **22**(1-2), 239–246 (1998)
19. Elliott, L., Ingham, D., Kyne, A., Mera, N., Pourkashanian, M., Wilson, C.: Multiobjective genetic algorithm optimization for calculating the reaction rate coefficients for hydrogen combustion. Industrial and Engineering Chemistry Research **42**(6), 1215–1224 (2003)
20. Elliott, L., Ingham, D., Kyne, A., Mera, N., Pourkashanian, M., Wilson, C.: Reaction mechanism reduction and optimization using genetic algorithms. Industrial and Engineering Chemistry Research **44**(4), 658–667 (2005)

21. Fabbri, G.: Heat transfer optimization in internally finned tubes under laminar flow conditions. International Journal of Heat and Mass Transfer **41**(10), 1243–1253 (1998)
22. Fabbri, G.: Heat transfer optimization in corrugated wall channels. International Journal of Heat and Mass Transfer **43**(23), 4299–4310 (2000)
23. Fabbri, G.: Effect of viscous dissipation on the optimization of the heat transfer in internally finned tubes. International Journal of Heat and Mass Transfer **47**(14–16), 3003–3015 (2004)
24. Falco, I.D.: An introduction to Evolutionary Algorithms and their application to the aerofoil design problem – Part I: the Algorithms. von Kármán Lecture Series on Fluid Dynamics, Bruxelles, Belgium, April 1997. (1997)
25. Fiorina, B., Baron, R., Gicquel, O., Thévenin, D., Carpentier, S., Darabiha, N.: Modelling non-adiabatic partially premixed flames using flame-prolongation of ILDM. Combustion Theory and Modelling **7**(3), 449–470 (2003)
26. Fiorina, B., Gicquel, O., Carpentier, S., Darabiha, N.: Validation of the FPI chemistry reduction method for diluted nonadiabatic premixed flames. Combustion Science and Technology **176**(5-6), 785–797 (2004)
27. Fiorina, B., Gicquel, O., Vervisch, L., Carpentier, S., Darabiha, N.: Approximating the chemical structure of partially premixed and diffusion counterflow flames using FPI flamelet tabulation. Combustion and Flame **140**(3), 147–160 (2005)
28. Fletcher, R.: Practical Methods of Optimization. John Wiley & Sons, New York (1987)
29. Fluent Inc.: GAMBIT 2.3 User's Guide. Lebanon, New Hampshire (2005)
30. Fluent Inc.: FLUENT 6.3 User's Guide. Lebanon, New Hampshire (2006)
31. Foli, K., Okabe, T., Olhofer, M., Jin, Y., Sendhoff, B.: Optimization of micro heat exchanger: CFD, analytical approach and multi-objective evolutionary algorithms. International Journal of Heat and Mass Transfer **49**(5-6), 1090–1099 (2006)
32. Fonseca, C.M., Fleming, P.J.: Genetic algorithms for multiobjective optimization: formulation, discussion and generalization. In: S. Forest (ed.) Genetic Algorithms: Proceedings of the Fifth International Conference, pp. 416–423. Morgan Kaufmann, San Mateo, California (1993)
33. Giovangigli, V.: Multicomponent flow modeling. Birkhäuser, Boston (1999)
34. Goldberg, D.E.: Genetic Algorithms in Search, Optimization and Machine Learning. Addison-Wesley, Reading, Massachusetts (1989)
35. Gordner, A.: Numerische Simulation nichtlinearer Aeroakustik bei kleinen Machzahlen. Ph.D. thesis, Universität Heidelberg, Germany (2005)
36. Gropp, W., Lusk, E.: User's guide for MPICH, a portable implementation of MPI. Tech. Rep. ANL-96/6, Argonne National Laboratory (1994)
37. Guo, Y.Y., He, G., Hsu, A.: Application of genetic algorithms to the development of a variable Schmidt number model for jet-in-crossflows. International Journal of Numerical Methods for Heat & Fluid Flow **11**, 744–760 (2001)
38. Hilbert, R., Janiga, G., Baron, R., Thévenin, D.: Multiobjective shape optimization of a heat exchanger using parallel genetic algorithms. International Journal of Heat and Mass Transfer **49**(15-16), 2567–2577 (2006)
39. Hilbert, R., Tap, F., El-Rabii, H., Thévenin, D.: Impact of detailed chemistry and transport models on turbulent combustion simulations. Progress in Energy and Combustion Science **30**, 61–117 (2004)
40. Hoyas, S., Jiménez, J.: Scaling of the velocity fluctuations in turbulent channels up to Re_τ=2003. Physics of Fluids **18**, 011,702 (2006)
41. Janiga, G., Gicquel, O., Thévenin, D.: High-resolution simulation of three-dimensional laminar burners using tabulated chemistry on parallel computers. In: 2nd ECCOMAS Thematic Conference on Computational Combustion, pp. 1–15. Delft, The Netherlands (2007)
42. Janiga, G., Gordner, A., Shalaby, H., Thévenin, D.: Simulation of laminar burners using detailed chemistry on parallel computers. In: P. Wesseling, E. Onate, J. Périaux (eds.) European Conference on Computational Fluid Dynamics, ECCOMAS CFD 2006, pp. 210/1–210/14. Egmond aan Zee, The Netherlands (2006)

43. Janiga, G., Thévenin, D.: Numerical optimisation of a laminar burner to reduce CO emissions. In: T. Lajos, J. Vad (eds.) Proc. Conf. Modelling Fluid Flow, pp. 109–116. Budapest, Hungary (2006)
44. Janiga, G., Thévenin, D.: Reducing the CO emissions in a laminar burner using different numerical optimization methods. Proceedings of the Institution of Mechanical Engineers, Part A: Journal of Power and Energy **221**(5), 647–655 (2007)
45. Kee, R.J., Rupley, F.M., Miller, J.A.: Chemkin-II: A Fortran chemical kinetics package for the analysis of gas-phase chemical kinetics. Tech. Rep. SAND89-8009, Sandia National Laboratories, Livermore, California (1989)
46. Kock, F., Herwig, H.: Entropy production calculation for turbulent shear flows and their implementation in CFD codes. International Journal of Heat and Fluid Flow **26**, 672–680 (2005)
47. Laverdant, A., Thévenin, D.: Interaction of a gaussian acoustic wave with a turbulent premixed flame. Combustion and Flame **134**, 11–19 (2003)
48. Lee, J., Hajela, P.: Parallel genetic algorithm implementation in multidisciplinary rotor blade design. Journal of Aircraft **33**(5), 962–969 (1996)
49. Lee, K.S., Kim, W.S., Si, J.M.: Optimal shape and arrangement of staggered pins in the channel of a plate heat exchanger. International Journal of Heat and Mass Transfer **44**(17), 3223–3231 (2001)
50. Lindstedt, P.: Modelling of the chemical complexities of flames. Proceedings of the Combustion Institute **27**, 269–285 (1998)
51. Liu, Y., Phan-Thien, N.: An optimum spacing problem for three chips mounted on a vertical substrate in an enclosure. Numerical Heat Transfer, Part A: Applications **37**, 613–630 (2000)
52. Maas, U., Pope, S.: Simplifying chemical kinetics: Intrinsic low-dimensional manifolds in composition space. Combustion and Flame **88**(3-4), 239–264 (1992)
53. Majda, A., Sethian, J.: The derivation and numerical solution of the equations for zero Mach number combustion. Combustion Science and Technology **42**, 185–205 (1985)
54. Mäkinen, R., Neittaanmäki, P., Périaux, J., Toivanen, J.: A genetic algorithm for multiobjective design optimization in aerodynamics and electromagnetics. In: K.D. Papailiou, D. Tsahalis, J. Périaux, D. Knörzer (eds.) Computational Fluid Dynamics '98, Proceedings of the ECCOMAS 98 Conference, vol. 2, pp. 418–422. Wiley, Athens, Greece (1998)
55. Mäkinen, R.A.E., Périaux, J., Toivanen, J.: Multidisciplinary shape optimization in aerodynamics and electromagnetics using genetic algorithms. International Journal for Numerical Methods in Fluids **30**(2), 149–159 (1999)
56. Matos, R., Laursen, T., Vargas, J., Bejan, A.: Three-dimensional optimization of staggered finned circular and elliptic tubes in forced convection. International Journal of Thermal Sciences **43**(5), 477–487 (2004)
57. Matos, R.S., Vargas, J.V.C., Laursen, T.A., Bejan, A.: Optimally staggered finned circular and elliptic tubes in forced convection. International Journal of Heat and Mass Transfer **47**(6–7), 1347–1359 (2004)
58. Message Passing Interface Forum: MPI: A message-passing interface standard. International Journal of Supercomputer Applications **8**(3/4) (1994)
59. Michalewicz, Z.: Genetic Algorithms+Data Structures=Evolution Programs. Springer-Verlag, Berlin, Heidelberg, New York (1996)
60. Moser, R.D., J., K., Mansour, N.N.: DNS of turbulent channel flow up to Re_τ=590. Physics of Fluids **11**, 943–945 (1999)
61. Muyl, F., Dumas, L., Herbert, V.: Hybrid method for aerodynamic shape optimization in automotive industry. Computers & Fluids **33**(5–6), 849–858 (2004)
62. Nejati, V., Matsuuchi, K.: Aerodynamics design and genetic algorithms for optimization of airship bodies. JSME International Journal Series B-Fluids and Thermal Engineering **46**(4), 610–617 (2003)
63. Nelder, J.A., Mead, R.: A simplex method for function minimization. Computer Journal **7**, 308–313 (1965)

64. Obayashi, S., Tsukahara, T., Nakamura, T.: Multiobjective genetic algorithm applied to aerodynamic design of cascade airfoils. IEEE Transactions on Industrial Electronics **47**(1), 211–216 (2000)
65. van Oijen, J.A., de Goey, L.P.H.: Modelling of premixed laminar flames using flamelet-generated manifolds. Combustion Science and Technology **161**, 113–137 (2000)
66. Okabe, T., Foli, K., Olhofer, M., Jin, Y., Sendhoff, B.: Comparative studies on micro heat exchanger optimization. In: Proceedings of the IEEE Congress on Evolutionary Computation, pp. 647–654 (2003)
67. Paschereit, C., Schuermans, B., Büche, D.: Combustion process optimization using evolutionary algorithm. In: American Society of Mechanical Engineers, International Gas Turbine Institute, Turbo Expo (Publication) IGTI, vol. 2, pp. 281–291 (2003)
68. Paxion, S.: Développement d'un solveur multigrille non-structuré parallèle pour la simulation de flammes laminaires en chimie et transport complexes. Phd thesis, École Centrale Paris, France, 1999-40 (1999)
69. Reusken, A.: The smoothing property for regular splittings. In: W. Hackbusch, G. Wittum (eds.) Incomplete Decompositions (ILU)-Algorithms, theory and applications, pp. 130–138. Vieweg, Braunschweig, Germany (1993)
70. da Silva, A., Lorente, S., Bejan, A.: Optimal distribution of discrete heat sources on a plate with laminar forced convection. International Journal of Heat and Mass Transfer **47**(10-11), 2139–2148 (2004)
71. da Silva, A., Lorente, S., Bejan, A.: Optimal distribution of discrete heat sources on a wall with natural convection. International Journal of Heat and Mass Transfer **47**(2), 203–214 (2004)
72. Sobol', I.M.: Distribution of points in a cube and approximate evaluation of integrals. U.S.S.R. Computational Mathematics and Mathematical Physics **7**, 86–112 (1967)
73. Srinivas, N., Deb, K.: Multiobjective optimization using non-dominated sorting in genetic algorithm. Evolutionary Computing **2**(3), 221–248 (1995)
74. Thévenin, D., Zähringer, K., Janiga, G.: Automatic optimization of two-dimensional burners. In: Proceedings of the European Combustion Meeting ECM05, pp. 240/1–240/6. Louvain-la-Neuve, Belgium (2005)
75. Tiwari, S., Maurya, D., Biswas, G., Eswaran, V.: Heat transfer enhancement in cross-flow heat exchangers using oval tubes and multiple delta winglets. International Journal of Heat and Mass Transfer **46**(15), 2841–2856 (2003)
76. Visser, J.A., de Kock, D.J.: Optimization of heat sink mass using the DYNAMIC-Q numerical optimization method. Communications in Numerical Methods in Engineering **18**(10), 721–727 (2002)
77. van der Vorst, H.A.: BI-CGSTAB: A fast and smoothly converging variant of BI-CG for the solution of nonsymmetric linear systems. SIAM Journal on Statistical and Scientific Computing **13**(2), 631–644 (1992)
78. Welch, B., Jones, K., Hobbs, J.: Practical Programming in Tcl and Tk. Prentice Hall (2003)
79. Wilcox, D.C.: Turbulence modeling for CFD. DCW Industries, Inc., La Cañada, California (1998)

Chapter 3
Mathematical Aspects of CFD-based Optimization

Hans Georg Bock and Volker Schulz

Abstract There exist several computational strategies of different efficiency for the solution of model-based optimization problems – particularly, in the case of models based on challenging CFD problems. Applied mathematics provides means for their analysis and for advice on their proper usage.

In this chapter, methods are mainly analyzed based on the explicit treatment of the underlying CFD-problem as a constraint of a nonlinear optimization problem, thus providing the potential for high computational efficiency. Methods of this form are termed optimization boundary value problem methods, simultaneous optimization methods or one-shot optimization methods. The necessary conditions of optimality play a key structural role in devising those strategies. Special attention is given to the following issues: modular sequential quadratic programming with approximate linear solvers, preconditioning of the Karush-Kuhn-Tucker (KKT) system and multigrid optimization in the case of stationary problems. In the case of unsteady problems, we will concentrate on time-domain decomposition such as by multiple shooting, and on algorithmic developments for real-time optimization. The aim of the presentation is to give a survey on advanced and fast methods for optimization within a CFD framework. For details, the reader is referred to the relevant literature.

Hans Georg Bock
Interdisciplinary Center for Scientific Computing (IWR),
University of Heidelberg, Germany
(e-mail: `Bock@iwr.uni-heidelberg.de`)

Volker Schulz
Department of Mathematics, University of Trier, Germany
(e-mail: `Volker.Schulz@uni-trier.de`)

3.1 Introduction

Computational fluid dynamics (CFD) models can be written in the general abstract form

$$F(\dot{y}(t), y(t), u(t), p, t) = 0\,, \quad t \in [0, T] \tag{3.1}$$

where we use the following nomenclature:

t	time within horizon $[0, T]$;
$y(t)$	state vector (dependent variables);
$\dot{y}(t)$	velocity (time derivative) of state vector (dependent variables);
$u(t)$	control vector (independent variables);
$p \in \mathbb{R}^{n_p}$	finite-dimensional parameter vector (independent variables).

The variables $y(t)$ and $\dot{y}$ are functions defined on a spatial domain, the control vector $u(t)$ are usually functions on (a subset of) the spatial domain or its boundary. F is a suitable differential operator that includes transport and diffusion as well as source terms and boundary conditions. We assume that Eq. (3.1) is an initial-boundary value problem that defines the states $y(t)$ uniquely, if $u(t)$, p and initial data $y(0)$ are given. In CFD-model based optimization (CFD-O), we want to determine p and the control function $u(t)$ in such a way that a scalar objective function such as

$$J(y, u, p) := \int_0^T j(y(t), u(t), p, t)dt \tag{3.2}$$

is minimized. The objective function can have different forms in process control, shape optimization, inverse modeling/parameter estimation or optimum experimental design. Usually, there arise additional constraints for the state and control functions in terms of inequalities

$$r(y(t), u(t), p, t) \geq 0 \tag{3.3}$$

modeling technical restrictions. For a sizeable part of the discussion in this paper, it is convenient to choose an even more abstract formulation of the optimization problem above.

If we consider stationary problems (i.e., $\partial F/\partial \dot{y} = 0$ in Eq. (3.1)) and choose a spatial discretization for the steady state y and the steady control u, we can reformulate the problem as

$$\min_{y,p} f(y, p) \tag{3.4}$$

$$\text{s.t. } c(y, p) = 0 \tag{3.5}$$

$$h(y, p) \geq 0\,. \tag{3.6}$$

Here, $y \in \mathbb{R}^n$ is the discretized state vector, the influence vector $p \in \mathbb{R}^{n_p}$ now summarizes the discretized control u and the formerly denoted parameter vector p, $f : \mathbb{R}^n \times \mathbb{R}^{n_p} \rightarrow \mathbb{R}$ denotes the objective function, Eq. (3.2), $c : \mathbb{R}^n \times \mathbb{R}^{n_p} \rightarrow \mathbb{R}^n$ the discretized CFD-model, Eq. (3.1) and $h : \mathbb{R}^n \times \mathbb{R}^{n_p} \rightarrow \mathbb{R}^m$ formulates the restrictions, Eq. (3.3). We arrive at the same formulation also by use of a full space-time discretization of the unsteady CFD-model, Eq. (3.1). Therefore, we use the formulation Eqs. (3.4)-(3.6) for general discussions on CFD-model based optimization in Sect. 3.2 with emphasis on stationary optimization problems and discuss special issues exploiting the structure of unsteady problems in Sect. 3.3.

Again, we assume that the states y are uniquely determined by the state equation (3.5) if the influence vector p is given, which means that the Jacobian $\partial c/\partial y$ is nonsingular. The implicit function theorem ensures the existence of a function ϕ such that $y = \phi(p)$ and one can reformulate the problem described by Eqs. (3.4)-(3.6) in a so-called black-box fashion

$$\min_p f(\phi(p),p) \tag{3.7}$$

$$\text{s.t. } h(\phi(p),p) \geq 0 \ . \tag{3.8}$$

Formulation (3.7, 3.8) is chosen in straightforward implementations of standard optimization techniques like genetic algorithms, the Nelder-Mead-Simplex or simulated annealing. These techniques require a relatively small amount of implementation effort, lead to a modular coupling of the optimization task with the simulation task and are often robust with respect to roughness in the objective function. However, since each evaluation of the implicit function ϕ requires a full CFD simulation, these methods lead to an overall computational effort which is several orders of magnitude higher than a forward simulation of the model (3.1). Furthermore, the inequality conditions (3.8), which are an integral and essential part of all realistic problem formulations, still pose challenges for these techniques. Therefore, we discuss simultaneous techniques for the direct solution of the constrained optimization problem (3.4-3.6), which have the potential for high efficiency, leading to an overall effort of only a small multiple of the forward simulation effort.

3.2 Simultaneous Model-based Optimization

3.2.1 Sequential Quadratic Programming (SQP)

For ease of presentation, we drop the inequalities in problem (3.4-3.6) and lump together the variables as $z := (y,p)$ in order to investigate the generic problem

$$\min_z f(z) \tag{3.9}$$
$$\text{s.t. } c(z) = 0 \ . \tag{3.10}$$

We nevertheless keep in mind that all or some of the equality constraints might later represent so-called active inequality constraints. Optimization theory provides the following necessary conditions in terms of the Lagrangian ($\mathcal{L}$), which hold at an optimal solution to this problem, provided the functions involved are sufficiently smooth

$$\partial\mathcal{L}(z,\lambda)/\partial z = 0\,, \quad \text{where } \mathcal{L}(z,\lambda) := f(z) + \lambda^\top c(z) \ . \tag{3.11}$$

The principle of SQP methods starts at Newton's method for the necessary conditions (3.11) together with the constraints (3.10). This method iterates over z and the adjoint variables λ in the form $(z^{k+1}, \lambda^{k+1}) = (z^k, \lambda^k) + (\Delta z^{k+1}, \Delta\lambda^{k+1})$, where the increments solve the linear system

$$\begin{bmatrix} H & c_z^\top \\ c_z & 0 \end{bmatrix} \begin{pmatrix} \Delta z \\ \Delta\lambda \end{pmatrix} = \begin{pmatrix} -\nabla f^k - c_z(z^k)^\top \lambda^k \\ -c(z^k) \end{pmatrix} . \tag{3.12}$$

Here, H denotes the Hessian of the Lagrangian $\mathcal{L}$ with respect to z, i.e., $H = \mathcal{L}_{zz}(z^k, \lambda^k)$ and subscripts denote respective derivatives.

Because the derivative generation needed to establish the matrix on the left hand side of this equation often turns out to be prohibitively expensive, one often uses approximations instead, i.e., $G :\approx H$ and $A :\approx c_z$ (of course, these approximations may also change from iteration to iteration). This substitution will deteriorate the convergence behavior of this approximate Newton method. However, the fixed points of the iteration are not changed as long as the matrix $\begin{bmatrix} G & A^\top \\ A & 0 \end{bmatrix}$ is nonsingular. If the approximations G and A are sufficiently accurate, one may expect at least linear local convergence of the iteration method. In the implementation, one will terminate the iteration as soon as $(\Delta z^{k+1}, \Delta\lambda^{k+1})$ is sufficiently close to zero in a suitable norm. If an estimate of the convergence rate is known an a priori estimate for the distance to the limit point can be given.

In order to arrive at a generalization of this approach to inequality constrained problems, we first reformulate the system of equations

$$\begin{bmatrix} G & A^\top \\ A & 0 \end{bmatrix} \begin{pmatrix} \Delta z \\ \Delta\lambda \end{pmatrix} = \begin{pmatrix} -\nabla f^k - c_z(z^k)^\top \lambda^k \\ -c(z^k) \end{pmatrix} \tag{3.13}$$

in the form of an equivalent linear-quadratic problem. Setting $\Delta\lambda = \lambda^{k+1} - \lambda^k$, we obtain the formulation

$$\begin{bmatrix} G & A^\top \\ A & 0 \end{bmatrix} \begin{pmatrix} \Delta z \\ \lambda^{k+1} \end{pmatrix} = \begin{pmatrix} -\nabla f^k + (A - c_z(z^k))^\top \lambda^k \\ -c(z^k) \end{pmatrix} .$$

If g is symmetric and positive definite on the null space of A, this is equivalent to the linear quadratic problem

$$\min 1/2 \Delta z^\top G \Delta z + \left(\nabla f^k - (A - c_z(z^k))^\top \lambda^k\right)^\top \Delta z \tag{3.14}$$
$$\text{s.t.} \quad A\Delta z + c(z^k) = 0 \tag{3.15}$$

where the adjoint variable for Eq. (3.15) is λ^{k+1}. In this way, accurate evaluations of the matrices c_z or H can be avoided. Nevertheless, the matrix c_z appears in the objective of the quadratic program (QP), however, only in the form of a matrix-vector product with λ^k, which can be efficiently realized such as in the adjoint mode of automatic differentiation [17].

This formulation is generalizable to inequality constraints, cf. [6, 7], e.g., of the form $c(z) \geq 0$ in (3.10), which leads to linear-quadratic sub-problems of the form

$$\min 1/2 \Delta z^\top G \Delta z + \left(\nabla f^k - (A - c_z(z^k))^\top \lambda^k\right)^\top \Delta z \tag{3.16}$$
$$\text{s.t.} \quad A\Delta z + c(z^k) \geq 0\,. \tag{3.17}$$

The adjoint variables of Eq. (3.17) converge to the adjoint variables of the inequality constraints at the solution. Therefore, they provide a proper decision criterion within an active-set strategy. This formulation of SQP methods allowing for approximations of derivatives, forms also the basis for the real-time investigations presented later (see also [14]).

3.2.2 Modular SQP Methods

Here, we discuss again the separability framework, i.e., problems of the type (3.4-3.6) – first without inequality constraints

$$\min_{y,p} f(y,p) \tag{3.18}$$
$$\text{s.t. } c(y,p) = 0\,. \tag{3.19}$$

The essential feature is the nonsingularity of c_y, which means that we can look at the problem as an unconstrained problem of the form

$$\min_p f(y(p),p)\,. \tag{3.20}$$

When applied to the necessary optimality condition $\nabla_p f(y(p),p) = 0$, Newton's method, or its variants, yield good local convergence properties. Every iteration consists of two steps

(1) solve $B\Delta p = -\nabla_p f(y(p^k),p^k) = 0$, where $B \approx {\nabla_p}^2 f(y(p^k),p^k)$
(2) update $p^{k+1} = p^k + \tau \cdot \Delta p$, where τ is an appropriate step length

The approximation B of the Hessian of f is often performed by Quasi-Newton update formulas as discussed in [29]. Step (1) of this algorithm involves two costly operations which, however, can be performed in a highly modular way. First, $y(p^k)$ has to be computed, which means basically a full nonlinear solution of the flow problem abbreviated by equation (3.19). Furthermore, the corresponding gradient, $\nabla_p f(y(p^k), p^k)$, has to be determined. One of the most efficient methods for this purpose is the adjoint method which means the solution of the linear adjoint problem, since for $y^k = y(p^k)$ one obtains

$$\nabla_p f(y(p^k), p^k) = f_p(y^k, p^k)^\top + c_p(y^k, p^k)^\top \lambda$$

where λ solves the adjoint problem

$$c_y(y^k, p^k)^\top \lambda = -f_y(y^k, p^k)^\top \,.$$

Assuming that the CFD-model (3.19) is solved by Newton's method, one might wonder whether it may be enough to perform only one Newton step per optimization iteration in the algorithm above. This results in the algorithmic steps

(1) solve $\begin{bmatrix} 0 & 0 & c_x^\top \\ 0 & B & c_p^\top \\ c_x & c_p & 0 \end{bmatrix} \begin{pmatrix} \Delta y \\ \Delta p \\ \lambda \end{pmatrix} = \begin{pmatrix} -f_y^\top \\ -f_p^\top \\ -c \end{pmatrix}$

(2) update $(y^{k+1}, p^{k+1}) = (y^k, p^k) + \tau \cdot (\Delta y^{k+1}, \Delta p^{k+1})$

This algorithm is called a reduced SQP algorithm. The local convergence can be again of quadratic, super-linear or linear type [26, 34], depending on how well B approximates the so-called reduced Hessian of the Lagrangian

$$B \approx \mathcal{L}_{pp} - \mathcal{L}_{py} c_y^{-1} c_p - (\mathcal{L}_{py} c_y^{-1} c_p)^\top + c_p^\top c_y^{-\top} \mathcal{L}_{yy} c_y^{-1} c_p \,.$$

The vector λ produced in each step of the reduced SQP algorithm converges to the adjoint variable vector of problem (3.18, 3.19) at the solution. This iteration again can be written in the form of a Newton-type method

$$\begin{bmatrix} 0 & 0 & c_x^\top \\ 0 & B & c_p^\top \\ c_x & c_p & 0 \end{bmatrix} \begin{pmatrix} \Delta y \\ \Delta p \\ \Delta\lambda \end{pmatrix} = \begin{pmatrix} -\mathcal{L}_y^\top \\ -\mathcal{L}_p^\top \\ -c \end{pmatrix}, \quad \begin{pmatrix} y^{k+1} \\ p^{k+1} \\ \lambda^{k+1} \end{pmatrix} = \begin{pmatrix} y^k \\ p^k \\ \lambda^k \end{pmatrix} + \tau \cdot \begin{pmatrix} \Delta y \\ \Delta p \\ \Delta\lambda \end{pmatrix} . \tag{3.21}$$

This iteration can be generalized to inexact linear solves with an approximate matrix $A \approx c_x$ such that the approximate reduced SQP iteration reads as

$$\begin{bmatrix} 0 & 0 & A^\top \\ 0 & B & c_p^\top \\ A & c_p & 0 \end{bmatrix} \begin{pmatrix} \Delta y \\ \Delta p \\ \Delta\lambda \end{pmatrix} = \begin{pmatrix} -\mathcal{L}_y^\top \\ -\mathcal{L}_p^\top \\ -c \end{pmatrix}, \quad \begin{pmatrix} y^{k+1} \\ p^{k+1} \\ \lambda^{k+1} \end{pmatrix} = \begin{pmatrix} y^k \\ p^k \\ \lambda^k \end{pmatrix} + \tau \cdot \begin{pmatrix} \Delta y \\ \Delta p \\ \Delta\lambda \end{pmatrix} . \tag{3.22}$$

It is shown in [23] that in this case, the use of an approximation of the consistently reduced Hessian, i.e.,

$$B \approx \mathcal{L}_{pp} - \mathcal{L}_{py} A^{-1} c_p - (\mathcal{L}_{py} A^{-1} c_p)^\top + c_p^\top A^{-\top} \mathcal{L}_{yy} A^{-1} c_p$$

is recommended. Let us compare this with the SQP-formulation of equation (3.12) in the separability framework

$$\begin{bmatrix} \mathcal{L}_{yy} & \mathcal{L}_{yp} & c_x^\top \\ \mathcal{L}_{py} & \mathcal{L}_{pp} & c_p^\top \\ c_x & c_p & 0 \end{bmatrix} \begin{pmatrix} \Delta y \\ \Delta p \\ \Delta\lambda \end{pmatrix} = \begin{pmatrix} -\mathcal{L}_y^\top \\ -\mathcal{L}_p^\top \\ -c \end{pmatrix}, \quad \begin{pmatrix} y^{k+1} \\ p^{k+1} \\ \lambda^{k+1} \end{pmatrix} = \begin{pmatrix} y^k \\ p^k \\ \lambda^k \end{pmatrix} + \tau \cdot \begin{pmatrix} \Delta y \\ \Delta p \\ \Delta\lambda \end{pmatrix}. \quad (3.23)$$

Because of the triangular structure of the system matrix in Eqs. (3.21) or (3.22), the reduced SQP formulation is much more modular than the full SQP formulation (3.23). This is the reason why the matrices in Eqs. (3.21) or (3.22) are used as pre-conditioner in large scale linear-quadratic optimal control problems [2] or as pre-conditioners in Lagrange-Newton-Krylov methods as discussed in [3, 4]. Indeed, one observes for the (iteration) matrix

$$M = I - \begin{bmatrix} 0 & 0 & c_x^\top \\ 0 & B & c_p^\top \\ c_x & c_p & 0 \end{bmatrix}^{-1} \begin{bmatrix} \mathcal{L}_{yy} & \mathcal{L}_{yp} & c_x^\top \\ \mathcal{L}_{py} & \mathcal{L}_{pp} & c_p^\top \\ c_x & c_p & 0 \end{bmatrix}$$

the fact that $M \neq 0$ in general, but $M^3 = 0$, which is the basis of the convergence considerations in [23].

Now, let us discuss CFD-optimization in the context of one-shot aerodynamic shape optimization as in [20, 21]. The typical problem formulation there is

$$\min_{y,p} f(y,p) \qquad \text{(drag)} \qquad (3.24)$$

$$\text{s.t. } c(y,p) = 0 \qquad \text{(CFD-model)} \qquad (3.25)$$

$$h(y,p) \geq 0 \qquad \text{(lift constraint)} \qquad (3.26)$$

where y collects the state variables of an Euler or Navier-Stokes flow model and p is a finite-dimensional vector parameterizing the shape of a part of an aircraft by means of a suitable spline family. The lift constraint h is a scalar valued function. Additionally, one often also has to treat a pitching moment constraint, which is also a scalar valued function. Engineering knowledge tells us that the lift constraint will be active at the solution. Therefore, we can formulate the constraint right from the beginning in the form of an equality constraint. In this context, a full SQP-approach as in equation (3.23) reads

$$\begin{bmatrix} \mathcal{L}_{yy} & \mathcal{L}_{yp} & h_x^\top & c_x^\top \\ \mathcal{L}_{py} & \mathcal{L}_{pp} & h_p^\top & c_p^\top \\ h_x & h_p & 0 & 0 \\ c_x & c_p & 0 & 0 \end{bmatrix} \begin{pmatrix} \Delta y \\ \Delta p \\ \Delta\mu \\ \Delta\lambda \end{pmatrix} = \begin{pmatrix} -\mathcal{L}_y^\top \\ -\mathcal{L}_p^\top \\ -h \\ -c \end{pmatrix}, \quad \begin{pmatrix} y^{k+1} \\ p^{k+1} \\ \mu^{k+1} \\ \lambda^{k+1} \end{pmatrix} = \begin{pmatrix} y^k \\ p^k \\ \mu^k \\ \lambda^k \end{pmatrix} + \tau \cdot \begin{pmatrix} \Delta y \\ \Delta p \\ \Delta\mu \\ \Delta\lambda \end{pmatrix}. \quad (3.27)$$

This approach is not implementable in general because one usually starts out with a flow solver for $c(y,p) = 0$ and seeks a modular coupling with an optimization approach, which does not necessarily change the whole code structure, as would be the case with formulation (3.23). A modular but nevertheless efficient alternative is an approximate reduced SQP approach as in Eq. (3.22), which is adapted to the case of the additional lift (or pitching) constraint, as established in [16].

$$\begin{bmatrix} 0 & 0 & 0 & A^\top \\ 0 & B & \gamma & c_p^\top \\ 0 & \gamma^\top & 0 & 0 \\ A & c_p & 0 & 0 \end{bmatrix} \begin{pmatrix} \Delta y \\ \Delta p \\ \Delta \mu \\ \Delta \lambda \end{pmatrix} = \begin{pmatrix} -\mathcal{L}_y^\top \\ -\mathcal{L}_p^\top \\ -h \\ -c \end{pmatrix}, \quad \begin{pmatrix} y^{k+1} \\ p^{k+1} \\ \mu^{k+1} \\ \lambda^{k+1} \end{pmatrix} = \begin{pmatrix} y^k \\ p^k \\ \mu^k \\ \lambda^k \end{pmatrix} + \tau \cdot \begin{pmatrix} \Delta y \\ \Delta p \\ \Delta \mu \\ \Delta \lambda \end{pmatrix} \quad (3.28)$$

where

$$\gamma = h_p^\top + c_p^\top \alpha\,, \text{ such that } A^\top \alpha = -h_x^\top\,.$$

An algorithmic version of this modular formulation is given by the following steps:

(1) generate λ^k by performing N iterations of an adjoint solver with right hand side $f_y^\top(y^k, p^k)$ starting in λ^k
(2) generate α^k by performing N iterations of an adjoint solver with right hand side $h_y^\top(y^k, p^k)$ starting in α^k
(3) compute approximate reduced gradients

$$g = f_p^\top + c_p^\top \lambda^{k+1}, \quad \gamma = h_p^\top + c_p^\top \alpha^{k+1}$$

(4) generate B_{k+1} as an approximation of the consistently reduced Hessian
(5) solve the QP

$$\begin{bmatrix} B & \gamma \\ \gamma^\top & 0 \end{bmatrix} \begin{pmatrix} \Delta p \\ \mu^{k+1} \end{pmatrix} = \begin{pmatrix} -g \\ -h \end{pmatrix}$$

(6) update $p^{k+1} = p^k + \Delta p$
(7) compute the corresponding shape geometry and adjust the computational mesh
(8) generate y^{k+1} by performing N iterations of the forward state solver starting from an interpolation of y^k at the new mesh.

This highly modular algorithmic approach is not an exact transcription of Eq. (3.28), but is shown in [16] to be asymptotically equivalent and to converge to the same solution. The overall algorithmic effort for this algorithm is typically in the range of factor 7 to 10 compared to a forward stationary simulation.

3.2.3 Multiple Set-point Optimization

Often, it is not enough to compute an optimal solution for one specific problem setting. Rather, one is interested in computing solutions, which optimize the performance of the process averaged over a range of process parameters or scenarios. Alternatively, one might want to optimize the worst case, which leads to a min-max formulation. This goal leads naturally to robust optimization or working range optimization, which we denote in the form of multiple set-point optimization (cf. [8, 9]). There, instead of problem (3.18, 3.19) we formulate the problem

$$\min_{y_1,y_2,y_3,p} \omega_1 f_1(y_1,p)+\omega_2 f_2(y_2,p)+\omega_3 f_3(y_3,p) \tag{3.29}$$

$$\text{s.t. } c_1(y_1,p)=0 \tag{3.30}$$

$$c_2(y_2,p)=0 \tag{3.31}$$

$$c_3(y_3,p)=0\ . \tag{3.32}$$

For the sake of simplicity, we have restricted the formulation above to a problem with three set-points coupled via the objective, which is a weighted sum of all set-point objectives (weights: $\omega_1, \omega_2, \omega_3$), and via the free optimization variables p, which are the same for all set-points. The generalization to more than three set-points and to additional equality and inequality constraints is obvious. The corresponding Lagrangian in our example is

$$\mathcal{L}(y_1,y_2,y_3,p,\lambda_1,\lambda_2,\lambda_3)=\sum_{i=1}^{3}\omega_i f_i(y_i,p)+\sum_{i=1}^{3}\lambda_i^\top c_i(y_i,p)\ . \tag{3.33}$$

The approximate reduced SQP method (3.22) applied to this case can be written in the following form

$$\begin{bmatrix} 0 & 0 & 0 & 0 & A_1^\top & 0 & 0 \\ 0 & 0 & 0 & 0 & 0 & A_2^\top & 0 \\ 0 & 0 & 0 & 0 & 0 & 0 & A_3^\top \\ 0 & 0 & 0 & B & c_{1,p}^\top & c_{2,p}^\top & c_{3,p}^\top \\ A_1 & 0 & 0 & c_{1,p} & 0 & 0 & 0 \\ 0 & A_2 & 0 & c_{2,p} & 0 & 0 & 0 \\ 0 & 0 & A_3 & c_{3,p} & 0 & 0 & 0 \end{bmatrix} \begin{pmatrix} \Delta y_1 \\ \Delta y_2 \\ \Delta y_3 \\ \Delta p \\ \Delta\lambda_1 \\ \Delta\lambda_2 \\ \Delta\lambda_3 \end{pmatrix} = \begin{pmatrix} -\mathcal{L}_{y_1}^\top \\ -\mathcal{L}_{y_2}^\top \\ -\mathcal{L}_{y_3}^\top \\ -\mathcal{L}_p^\top \\ -c_1 \\ -c_2 \\ -c_3 \end{pmatrix} . \tag{3.34}$$

We notice that the linear sub-problems involving matrices $A_i^\top$ are to be solved independently, and therefore trivially in parallel. The information from all these parallel adjoint problems is collected in the reduced gradient

$$g=\sum_{i=1}^{3}\omega_i f_p^\top+\sum_{i=1}^{3}c_p^\top\lambda_i\ .$$

Next, the solution of optimization step $\Delta p = -B^{-1}g$ is distributed to all approximate linearized forward problems

$$A_i \Delta y_i + c_{i,p} \Delta p = -c_i$$

which can then again be solved in parallel.

3.2.4 Multigrid Optimization

The main motivation for multigrid (MG) methods is derived from the observation that the convergence properties of most iterative methods for the solution of systems of equations, or of optimization problems, deteriorate when the discretization mesh is refined. This means that when approaching the continuous problem by the use of successively finer meshes, the effort required for the solution of the resulting discretized (non-)linear systems increases more than linearly with the number of variables.

The promise of MG methods is to provide grid refinement independent convergence rates by performing more work on coarser grids than on the finer ones. This goal of optimal numerical complexity (which grows then only linearly with respect to the number of unknowns) can be achieved by MG methods in many cases, particularly if the problem possesses the feature that coarser grids are able to provide improvements for finer grids. This is the reason why theoretical convergence results always assume that the coarsest grid is already sufficiently fine. The highest advantage can be gained from MG methods, which are optimally adapted to the problem under investigation, in particular with respect to the spatial distribution of variables in so-called geometric MG methods. With general-purpose MG method solvers, e.g., of algebraic type, one can not expect to achieve optimal complexity, but can however expect to obtain high flexibility.

Besides the huge progress in MG methods for simulation problems, the field of multigrid optimization (MG/OPT) has also come to a certain degree of maturity. The review [12] gives an up-to-date survey on the diverse MG/OPT approaches in the literature. In the following, we want to focus on the most simply implementable MG/OPT approach, which is based on an MG structure only with respect to the free variables p in the reduced formulation (3.20).

Numerical experiments, e.g. [28], demonstrate that MG/OPT greatly improves the efficiency of the underlying optimization scheme used as a "smoother", suggesting that the MG/OPT scheme may be beneficial in combination with well known optimization algorithms. This claim appears to be true as long as a line search along the coarse-grid correction is performed. Also in [28], it is reported that MG/OPT without a line search diverges in some cases. Therefore, a line search appears to be necessary for convergence.

Consider the following (locally) convex optimization problem

$$\min_{p_k} g_k(p_k) := f_k(y_k(p_k), p_k) \tag{3.35}$$

where $k = 1, 2, \ldots, L$ is the resolution or discretization parameter, where L denotes the finest resolution, and p_k is the (unconstrained) optimization variable in the space V_k. For the resolution k, one chooses an appropriate discretized state variable y_k and an accordingly resolved objective evaluation f_k. For variables defined on V_k, we introduce the inner product $(\cdot,\cdot)_k$ with associated norm $\|y\|_k = (y,y)_k^{1/2}$. Between spaces V_k, restriction operators $I_k^{k-1} : V_k \to V_{k-1}$ and prolongation operators $I_{k-1}^k : V_{k-1} \to V_k$ are defined. We require that $(I_k^{k-1} y, v)_{k-1} = (y, I_{k-1}^k v)_k$ for all $y \in V_k$ and $v \in V_{k-1}$.

On each space, denote with S_k an optimization algorithm, e.g., a gradient based technique. Given an initial approximation p_k^0 of the solution to (3.35), the application of S_k results in $g_k(S_k(p_k^0)) < g_k(p_k^0)$.

The MG/OPT scheme is an iterative method. One cycle of this method is defined as follows. Let p_k^0 be the starting approximation at resolution k.

Algorithm 1 (MG/OPT (k) Algorithm) *If $k = 1$ (coarsest resolution) solve Eq. (3.35) exactly.*
Else if $k > 1$:

1. *Pre-optimization. Define $p_k^1 = S_k(p_k^0)$.*
2. *Set up and solve a coarse-grid minimization problem. Define $p_{k-1}^1 = I_k^{k-1} p_k^1$ and $\sigma_{k-1} = \nabla g_{k-1}(y_{k-1}^1) - I_k^{k-1} \nabla g_k(y_k^1)$. The coarse-grid minimization problem is given by*

$$\min_{p_{k-1}} \left(g_{k-1}(p_{k-1}) - \sigma_{k-1}^T \, p_{k-1} \right) . \tag{3.36}$$

   Apply one cycle of MG/OPT(k-1) to Eq. (3.36) to obtain p_{k-1}^2.
3. *Line-search and coarse-grid correction. Perform a line search in the $I_{k-1}^k(p_{k-1}^2 - p_{k-1}^1)$ direction to obtain τ_k. The coarse-grid correction is given by*

$$p_k^2 = p_k^1 + \tau_k \, I_{k-1}^k (p_{k-1}^2 - I_k^{k-1} p_k^1) \; .$$

4. *Post-optimization. Define $p_k^3 = S_k(p_k^2)$.*

Roughly speaking, the essential guideline for constructing g_k on coarse levels is that it must sufficiently well approximate the convexity properties of the functional $f(y(\cdot), \cdot)$ at finest resolution. In addition, we have that the gradient of the coarse-grid functional at $p_{k-1}^1 = I_k^{k-1} p_k^1$ equals the restriction of the gradient of the fine-grid functional at p_k^1. In fact, by adding the term $-\sigma_{k-1}^T \, p_{k-1}$ in Step 2, we have that

$$\nabla \left(g_{k-1}(p_{k-1}) - \sigma_{k-1}^T \, p_{k-1} \right) |_{p_{k-1}^1} = I_k^{k-1} \nabla g_k(p_k^1) \; .$$

It is shown in [12] that therefore the coarse grid correction is indeed a secant direction for the optimization problems on the higher refinement levels.

A typical area of application for this multigrid strategy is the inverse problems for distributed parameters and control or shape optimization problems, where the optimization variable is of function type.

3.3 Unsteady Problems

In principle, unsteady optimization problems of the form (3.1, 3.2, 3.3) can be treated by the methods discussed above after a discretization in space and time. However, the adjoint variables to the CFD equation always have the same dimensionality as the solution of the CFD equation, i.e., if y is a function of space and time, then so is λ. When performing simulations of CFD problems, one typically provides only a space discretization of the flow variables and marches forward in time so that during the whole simulation, the main memory is only populated by one (or two, but at most six, depending on the time marching scheme) spatially discretized state variable pertaining to the solution at the current time-step. Unfortunately, the time direction is reversed for the corresponding adjoint problems and, in nonlinear problems, the solution of the (discretized) adjoint problem needs information from the (discretized) primal CFD solution y. That means, a direct application of the methodology sketched above to time-dependent problems requires storage for at least the size of the whole space-time history of the solution y of the CFD problem. If this amount of storage easily fits into the available memory, one can just skip the following comments and continue with the next section, where we discuss special features of the resulting SQP methods. Otherwise, there are mainly four options:

- It is possible to counteract the lack of storage by investing more computing time. In the situation of time-dependent optimization problems, this can be facilitated by the use of so-called check-pointing strategies as described in [18]. The algorithmic complexity then grows only logarithmically in the number of time-steps to be thus virtually stored and so does the actual storage space required. Note that the multiple shooting approach outlined below provides a natural kind of checkpoints.
- Another interesting option is model reduction. In a trivial form this could just mean choosing a coarser space-grid, which is however not feasible in many cases. But the same line of thought leads to proper orthogonal decomposition (POD), where low dimensional ansatz spaces are constructed by the use of snapshots, which mirror already the major properties of the solution of the CFD problem. The result is a low dimensional system of Ordinary Differential Equations (ODE) or Differential Algebraic Equations (DAE), for which the optimization problems can be solved in the fashion of the methods discussed below. A description of the use of POD

for optimization problems together with strategies for the adaptation of the POD-bases can be found in [1, 22, 25].
- If the dimension of the optimization variables is comparatively low, it is possible to apply the boundary value problem techniques described below in a forward sensitivity driven manner as described in [19, 31, 32, 33]. The resulting storage space required is then equivalent to the dimension of the optimization variables times the dimension of one space discretization of y.
- The direct usage of the reduced formulation (3.20) can be considered as an ultima ratio, if the available main memory is not large enough to exploit all other options. Nevertheless, one has to be careful in evaluating derivatives and also be aware that the resulting unconstrained optimization problem is typically much more nonlinear than the original, constrained formulation (3.18,3.19).

Furthermore, an inconsistent adjoint time integration scheme becomes a major and frequent pitfall in CFD-optimization. It is important that the adjoint discretization scheme is indeed adjoint to the discretized primal CFD time integration scheme, according to the principle of internal numerical differentiation (IND). A deviation from this requirement may lead to gradient approximations which are not descent directions and would thus lead to premature stopping of gradient-based algorithms. The requirement of consistency of the adjoint scheme with the forward scheme leads typically to adjoint schemes which cannot be interpreted as integration schemes for the (infinite) adjoint CFD problem. More details can be found in [11].

3.3.1 Time-domain Decomposition by Multiple Shooting

The direct multiple shooting method for optimization of unsteady processes as described below was introduced by Bock and Plitt for optimal control in [10, 30], and for parameter estimation problems in [5]. The basic idea is to decompose the time history of the problem into subdomains, in which the unsteady PDE solutions are uniquely parameterized by their initial data and by the unknown parameters. For this purpose, one divides the time interval $[0, T]$ into subintervals $[t_i, t_{i+1}]$ with

$$0 = t_0 < t_1 < \ldots < t_N = T \ . \tag{3.37}$$

If the problem depends on a control function rather than a parameter, we parameterize this control on each subinterval $[t_i, t_{i+1}]$ by setting

$$u(t) = \phi_i(t, q_i) \tag{3.38}$$

where ϕ_i are given basis functions parameterized by a finite dimensional parameter vector q_i. The functions ϕ_i may be vectors of polynomials; in practice, one often uses piecewise constant controls, i.e.,

$$\phi_i(t, q_i) = q_i \quad \forall t \in [t_i, t_{i+1}] \; . \tag{3.39}$$

Note that for this particular choice of basis functions bounds on the control u transfer immediately to bounds on the parameter vectors q_i and vice versa.

Additional degrees of freedom are introduced by the variables s_i, $i = 0, \ldots, N-1$ which represent the initial values $y_i(t_i)$ of the state variables $y_i(t)$ on each subinterval $[t_i, t_{i+1}]$, and a variable s_N for the final state. We compute the trajectories as the solutions of the corresponding initial value problems (IVP):

$$\dot{y}_i(t) = f(y_i(t), \phi_i(t, q_i)) \quad \forall\, t \in [t_i, t_{i+1}] \,, \tag{3.40a}$$

$$y_i(t_i) = s_i \, . \tag{3.40b}$$

For ease of presentation, we formulate the CFD-problem in the form of an explicit ODE via the method of lines. Note that every trajectory piece $y_i(t)$ is uniquely determined by the initial value s_i and the control parameter vector q_i. To stress the dependence on the initial value and the control parameters we also write $y_i(t; s_i, q_i)$. To ensure continuity of the whole state trajectory $y(t)$ we set up the *continuity* or *matching conditions*

$$s_{i+1} - y_i(t_{i+1}; s_i, q_i) = 0, \quad i = 0, \ldots, N-1 \; . \tag{3.41}$$

The objective function is evaluated on each subinterval independently:

$$L_i(s_i, q_i) = \int_{t_i}^{t_{i+1}} j(y_i(t), \phi_i(t, q_i)) \, dt \, . \tag{3.42}$$

This leads to the following structured nonlinear program (NLP):

$$\min_{s_i, q_i} \quad \textstyle\sum_{i=0}^{N-1} L_i(s_i, q_i) \tag{3.43a}$$

$$\text{s.t. } s_{i+1} = y_i(t_{i+1}; s_i, q_i), \; i = 0, \ldots, N-1, \tag{3.43b}$$

$$s_0 - y_0 = 0 \tag{3.43c}$$

plus additional path and control constraints. Note that the part of the Jacobian corresponding to the continuity conditions has a special block banded structure and is invertible with respect to $(s_1, \ldots, s_N)$. Exploiting this fact, one can reduce the degrees of freedom in the quadratic subproblem to the variables $\Delta\tilde{w} = (\Delta s_0, \Delta q_0, \ldots, \Delta q_{N-1})$ by applying a *condensing* step before solving the subproblem. A further reduction to the variables $\Delta q = (\Delta q_0, \ldots, \Delta q_{N-1})$ can be achieved by using the linearization of the initial value constraint (3.43c). For detailed information about the condensing technique in general and especially in the context of Nonlinear Model Predictive Control (NMPC), the reader is referred to [13, 14].

Within classical multiple shooting, these blockwise matrices, also called Wronskians, are built up in a sensitivity approach, which means that the Jacobian matrix of the constraints are explicitly constructed and completely stored. Matrix factorization techniques dovetailed to the multiple shooting QP are applied in order to solve the quadratic programs. Note, however, that these Jacobians need only be approximated in a coarse manner, when using formulation (3.16, 3.17) such as by secant update techniques. In case of CFD optimization problems they should only capture unstable and slowly decaying modes. In certain cases, the frequent generation of the constraint Jacobian can also be avoided, so that only the very first Jacobian is constructed and factorized by the use of formulation (3.16, 3.17), where the matrix A denotes the (multiple shooting) Jacobian at the data of the first optimization iteration.

3.3.2 Parallel Multiple Shooting

Parallelization of multiple shooting is trivially based on the idea that the initial value solvers on each multiple shooting interval can work completely independently, provided the initial values s_i are available. Parallel realizations of multiple shooting have existed for a long time, cf. [15, 24, 36], and more recently as parareal in [27]. Ulbrich has described the use of this approach within an optimal control context [35] in a formulation similar to Eqs. (3.16)-(3.17). There the initial values are provided by a coarse discretization scheme (often implicit Euler) which operates only on the multiple shooting nodes. It is obvious that this approach works only well with CFD problems, which are dissipative enough so that even coarse time discretizations give some consistent information.

3.3.3 Real-time Optimization and Nonlinear Model Predictive Control

Real-time optimization is not just fast optimization, rather it means a shift in philosophy: in order to minimize the delay time such as for a control response to a perturbation of the process, one tries to compute as much information as possible before data about the values of states and parameters or of scenarios for the real process become available in real-time. In the CFD context this means in particular that matrix factorizations, which are often avoided for computational speed, or the pre-computation of sophisticated pre-conditioners, receive an increased attention, insofar as they can often be computed in advance, and have a potential to speed up the final solution process.

In the special case of NMPC, a sequence of neighboring optimization problems on a receding time horizon has to be solved, which differ only by (an estimate of) the initial value of the states $y(t_0)$ and possibly by a change in the set-point, i.e., parameters and scenarios. Minimal response delays can be achieved by the "real–time iteration" (RTI) in combination with a "perturbation embedding" as developed in [7, 13, 14]. Here, huge savings in computing time are gained by a kind of hot start property because already the first QP solution in a full step inexact SQP method provides a nearly tangential approximation of the exact solution if the problem is initialized with the solution of the previous problem, even in the presence of a change of the active constraints. Furthermore, the calculations can be divided into a *preparation phase* that can be performed without knowledge of y_0, and a much shorter *feedback phase* that allows to make the delay even shorter than the sampling time. This remaining delay is typically orders of magnitude smaller than the sampling time and thus we can consider the feedback to be instantaneous. So the sampling time is practically only needed to prepare the *following* real–time iteration. Additionally, cheap feasibility and/or optimality improving subiterations (FOI) can be performed that reuse Jacobians and Hessians of the previous step and require only one forward and additionally one adjoint solution, respectively.

For a more detailed description of the real-time iteration scheme and its convergence and nominal stability properties, please refer to [14].

3.3.4 Sensitivity Driven Multiple Shooting

In parameter estimation problems for CFD models, it is possible to use a multiple shooting time-domain decomposition and profit from its superior convergence properties, and on the other hand reduce the optimization space only to the parameters to be estimated. In these problems, one uses a generalized Gauss-Newton approach which achieves rather good convergence already with information of first order only. The linearized least-squares subproblem is condensed to the unknown parameters, which can be achieved while only marching forward in time with a conventional time-integration scheme. This idea was originally developed in [33], and has been successfully applied to multiphase problems, e.g., in [19]. Extensions to Newton-type methods for general objectives, e.g., in control problems were developed in [31, 32].

References

1. Arian, E., Fahl, M., Sachs, E.: Trust-region proper orthogonal decomposition for optimal flow control. NASA/CR-2000-210124 (2000)
2. Battermann, A., Sachs, E.: Block preconditioner for KKT systems in PDE-governed optimal control problems. International Series on Numerical Mathematics (ISNM) **138**, 1–18 (2001)
3. Biros, G., Ghattas, O.: Parallel Lagrange-Newton-Krylov-Schur methods for PDE-constrained optimization. Part I: The Krylov-Schur solver. SIAM Journal on Scientific Computing **27**(2), 687–713 (2005)
4. Biros, G., Ghattas, O.: Parallel Lagrange-Newton-Krylov-Schur methods for PDE-constrained optimization. Part II: The Lagrange-Newton solver, and its application to optimal control of steady viscous flows. SIAM Journal on Scientific Computing **27**(2), 714–739 (2005)
5. Bock, H.: Numerical treatment of inverse problems in chemical reaction kinetics. In: Modelling of Chemical Reaction Systems, vol. 18, pp. 102–125. Springer (1981)
6. Bock, H., Diehl, M., Kostina, E.: SQP methods with inexact Jacobians for inequality constrained optimization. IWR-preprint, Universität Heidelberg, Heidelberg, Germany (2004)
7. Bock, H., Diehl, M., Kostina, E., Schlöder, J.: Constrained optimal feedback control for DAE. In: L. Biegler, O. Ghattas, M. Heinkenschloss, D. Keyes, B. van Bloemen Waanders (eds.) Real-Time PDE-Constrained Optimization, pp. 3–24. SIAM (2007)
8. Bock, H., Egartner, W., Kappis, W., Schulz, V.: Practical shape optimization for turbine and compressor blades. Optimization and Engineering **3**, 395–414 (2002)
9. Bock, H., Kostina, E., Schäfer, A., Schlöder, J., Schulz, V.: Multiple set point partially reduced SQP method for optimal control of PDE. In: W. Jäger, R. Rannacher, J. Warnatz (eds.) Reactive Flows, Diffusion and Transport. Springer (2007)
10. Bock, H., Plitt, K.: A multiple shooting algorithm for direct solution of constrained optimal control problems. In: Proceedings 9th IFAC World Congress Automatic Control. Pergamon Press (1984)
11. Bock, H., Schlöder, J., Schulz, V.: Numerik großer Differentiell-Algebraischer Gleichungen – Simulation und Optimierung. In: H. Schuler (ed.) Prozeßsimulation, pp. 35–80. VCH Verlagsgesellschaft mbH, Weinheim (1994)
12. Borzi, A., Schulz, V.: Multigrid methods for PDE optimization. SIAM Review (to appear) (2007)
13. Diehl, M.: Real-time optimization for large scale nonlinear processes. Ph.D. thesis, University of Heidelberg, Germany (2001)
14. Diehl, M., Bock, H., Schlöder, J.: An iteration scheme for nonlinear optimization in optimal feedback control. SIAM Journal on Control and Optimization **43**(5), 1714–1736 (2005)
15. Gallitzendörfer, J., Bock, H.: Parallel algorithms for optimization boundary value problems in DAE. In: H. Langendörfer (ed.) Praxisorientierte Parallelverarbeitung. Hanser, München, Germany (1994)
16. Gherman, I.: Approximate partially reduced SQP approaches for aerodynamic shape optimization problems. Ph.D. thesis, University of Trier (2007)
17. Griewank, A.: Evaluating Derivatives, Principles and Techniques of Algorithmic Differentiation. Frontiers in Applied Mathematics. SIAM (2000)
18. Griewank, A., Walther, A.: Treeverse: an implementation of checkpointing for the reverse or adjoint mode of computational differentiation. Transactions on Mathematical Software (TOMS) **26** (2000)
19. Hazra, S., Schulz, V.: Numerical parameter identification in multiphase flow through porous media. Computing and Visualization in Science **5**, 107–113 (2002)
20. Hazra, S., Schulz, V.: Simultaneous pseudo-timestepping for aerodynamic shape optimization problems with state constraints. SIAM J. Sci. Comput. **28**(3), 1078–1099 (2006)

21. Hazra, S., Schulz, V., Brezillon, J., Gauger, N.: Aerodynamic shape optimization using simultaneous pseudo-timestepping. Journal of Computational Physics **204**(1), 46–64 (2005)
22. Hinze, M., Volkwein, S.: Proper orthogonal decomposition surrogate models for nonlinear dynamical systems: Error estimates and suboptimal control. In: P. Benner, D. Sorensen, V. Mehrmann (eds.) Dimension Reduction of Large-Scale Systems, *Lecture Notes in Computational Science and Engineering*, vol. 45, pp. 261–306. Springer (2006)
23. Ito, K., Kunisch, K., Schulz, V., Gherman, I.: Approximate nullspace iterations for KKT systems in model based optimization. Tech. Rep. 06-5, FB IV Mathematik/Informatik, University of Trier, Germany (submitted to SIAM Journal on Scientific Computing) (2006)
24. Kiehl, M., Mehlhorn, R., Schumann, M.: Parallel multiple shooting for optimal control problems. J. Optimization Methods and Software **4** (1995)
25. Kunisch, K., Volkwein, S., Xie, L.: HJB-POD based feedback design for the optimal control of evolution problems. SIAM Journal on Applied Dynamical Systems **3**, 701–722 (2004)
26. Kupfer, F.S.: An infinite-dimensional convergence theory for reduced SQP methods in Hilbert space. SIAM Journal on Optimization **6**, 126–163 (1996)
27. Maday, Y., Turinic, G.: A parareal in time procedure for the control of partial differential equations. C. R. Math. Acad. Sci. Paris **335**, 387–392 (2002)
28. Nash, S.: A multigrid approach to discretized optimization problems. Optimization Methods and Software **14**, 99–116 (2000)
29. Nocedal, J., Wright, S.: Numerical optimization. Springer (2000)
30. Plitt, K.J.: Ein superlinear konvergentes Mehrzielverfahren zur direkten Berechnung beschränkter optimaler Steuerungen. Master's thesis, University of Bonn, Germany (1981)
31. Schäfer, A.: Efficient reduced Newton-type methods for solution of large-scale structured optimization problems with application to biological and chemical processes. Ph.D. thesis, University of Heidelberg, Germany (2005)
32. Schäfer, A., Kühl, P., Diehl, M., Schlöder, J., Bock, H.: Fast reduced multiple shooting methods for nonlinear model predictive control. Chemical Engineering and Processing **46**(11), 1200–1214 (2007)
33. Schlöder, J.: Numerische Methoden zur Behandlung hochdimensionaler Aufgaben der Parameteridentifizierung. Bonner Mathematische Schriften 187 (1988)
34. Schulz, V.: Solving discretized optimization problems by partially reduced SQP methods. Computing and Visualization in Science **1**(2), 83–96 (1998)
35. Ulbrich, S.: Generalized SQP-methods with "parareal" time-domain decomposition for time-dependent PDE-constrained optimization. In: L. Biegler, O. Ghattas, M. Heinkenschloss, D. Keyes, B. van Bloemen Waanders (eds.) Real-Time PDE-Constrained Optimization, pp. 145–168. SIAM (2007)
36. Ziesse, M., Bock, H., Gallitzendörfer, J., Schlöder, J.: Parameter estimation in multispecies transport reaction systems using parallel algorithms. In: J. Gottlieb, P. DuChateaux (eds.) Parameter Identification and Inverse Problems in Hydrology, Geology and Ecology. Kluwer (1996)

Chapter 4
Adjoint Methods for Shape Optimization

Kyriakos C. Giannakoglou and Dimitrios I. Papadimitriou

Abstract In aerodynamic shape optimization, gradient-based methods often rely on the adjoint approach, which is capable of computing the objective function sensitivities with respect to the design variables. In the literature adjoint approaches are proved to outperform other relevant methods, such as the direct sensitivity analysis, finite differences or the complex variable approach. They appear in two different formulations, namely the continuous and the discrete one, which are both discussed in this chapter.

In the first part, continuous and discrete approaches for the computation of first derivatives are presented. The mathematical background for both approaches is introduced. Based on it, adjoints for either inverse design problems associated with inviscid or viscous flows or for the minimization of viscous losses in internal aerodynamics are developed. The Navier-Stokes equations are used as state equations. The elimination of field integrals expressed in terms of variations in grid metrics leads to a formulation which is independent of the grid type and can thus be employed with either structured or unstructured grids. From the physical point of view, the minimization of viscous losses in ducts or cascades is handled by minimizing either the difference in total pressure between inlet and outlet (the objective function is, then, a boundary integral) or the field integral of entropy generation. The discrete adjoint approach is, practically, used to compare and cross-check the derivatives computed by means of the continuous approach.

In the second part of this chapter, recent theoretical formulations on the computation and use of the Hessian matrix in optimization problems are presented. It is demonstrated that the combined use of the direct sensitivity analysis for the first derivatives followed by the adjoint approach for second derivatives may support the Newton method at the cost of $N+2$ equivalent

Kyriakos C. Giannakoglou · Dimitrios I. Papadimitriou
Parallel CFD & Optimization Unit, Lab. of Thermal Turbomachines, School of Mechanical Engineering, National Technical University of Athens, Greece
(e-mail: kgianna@central.ntua.gr, e-mail: dpapadim@mail.ntua.gr)

flow solutions per optimization cycle. The computation of the exact Hessian is demonstrated using both discrete and continuous approaches.

Test problems are solved using the proposed methods. They are used to compare the so-computed first and second derivatives with those resulting from the use of finite difference schemes. On the other hand, the efficiency of the proposed methods is demonstrated by presenting and comparing convergence plots for each test problem.

4.1 Introduction

Gradient-based methods, often used in aerodynamic optimization, must be supported by a tool to compute the gradient of an objective function which quantifies the efficiency of candidate solutions to the problem. Since this text focuses on aerodynamic shape optimization only, candidate solutions are considered to be 2D or 3D aerodynamic shapes (airfoils, blades, wings, air inlets, etc.). The objective function is expressed in terms of flow variables computed by numerically solving the flow equations in the corresponding domains, with problem-specific boundary conditions. The adjoint approach, based on control theory, is a means to compute the gradient required by a gradient-based optimization method.

Generally, deterministic algorithms driven by the gradient of the objective function converge to the global optimal solution much faster than evolutionary algorithms, provided that the starting solution does not mislead them by trapping the search around local optima. The reason for the superiority of the adjoint method (despite the aforementioned risk) is that the gradient points towards a better solution whereas the evolutionary algorithms are, generally, unable to exploit local information during the solution refinement. On the other hand, evolutionary algorithms have some other advantages [8, 20]. However, the analysis of these algorithms and their performance is beyond the scope of this chapter which is exclusively concerned with adjoint methods.

Aerodynamic problems usually involve a great number of design variables, so the computation of gradients by means of finite difference schemes renders deterministic algorithms very time-consuming. Compared to other methods, such as the complex-variable approach [48], or the direct sensitivity analysis (as it will become clear as this chapter develops, the direct approach is based on the computation and use of the gradient of flow variables with respect to the design variables as intermediate quantities), the adjoint approach is more efficient to compute the gradient of the objective function. The total cost for the gradient computation does not depend on the number of design variables and is approximately equal to that of solving the flow equations.

Two adjoint approaches, namely the continuous and discrete, have been developed. In continuous adjoint, the adjoint Partial Differential Equations (PDE) are formed starting from the corresponding flow PDE's and are then discretized and numerically solved to compute the adjoint variables' field. In discrete adjoint, the equations are obtained directly from the discretized flow PDE's. Concerning their requirements at the development stage, both approaches have their advantages and disadvantages and the interested reader may refer to [55] for more details. However, according to [43, 44] and the personal experience of the authors, both approaches result in sensitivity derivatives of the same accuracy.

In fluid mechanics, the adjoint method has been introduced by Pironneau, [57] and applied to flows governed by elliptic PDE's. Jameson was the first to present the continuous adjoint formulation for the optimal design of aerodynamic shapes [28, 29, 33] in transonic flows. These papers deal with the inverse design of airfoils and wings in inviscid flows, while the extension to viscous flows is found in [32] where the Navier-Stokes equations are used as state equations.

Since 1988, Jameson has presented many publications on discrete and preferentially on continuous adjoint approaches. Among them, we report on the inverse design of wings in subsonic, transonic and supersonic flows using multiblock structured grids [30, 59], minimization of drag and/or maximization of lift in inviscid [35] and viscous flows [36], sonic boom reduction for supersonic flows [1, 46], shock wave reduction in external aerodynamics [25], unsteady aerodynamic design of isolated airfoils and wings [45, 47], aerodynamic design of full aircraft configurations [34], multipoint drag minimization [39] and multidisciplinary optimization [38, 40].

On the other hand, Giles made a considerable improvement in the development of the discrete adjoint approach. The properties of solutions to the adjoint equations are analyzed in [23, 24]. He also presented the analytical solutions of the adjoint equations using Green's functions [56], an exact approach for the solution of the adjoint equations [22], a shock wave reduction method in external aerodynamics [21] and the harmonic approach to turbomachinery steady and unsteady designs [11, 14].

The discrete adjoint approach using unstructured grids was first presented by Peraire in inviscid, [15] and viscous flows [16], for 2D and 3D [17], configurations. Peraire also applied the adjoint approach to multipoint optimization problems [18].

The continuous adjoint approach for unstructured grids has been developed by Anderson [4, 5] for inviscid and viscous flows. The discrete adjoint formulation for turbulent flows with the Spalart-Allmaras turbulence model can be found in [2, 3]. Improvements of the discrete approach concerning the handling of grid sensitivities can be found in [49, 50] where remeshing and mesh movement strategies used at each optimization cycle are discussed.

The convergence of the iterative optimization algorithm, which in each cycle solves the flow and adjoint equations followed by the updating of the design variables, can be significantly improved by the so-called one-shot method [61, 37]. The three systems of equations are solved simultaneously in one shot, and the convergence is further accelerated using multigrid. Improvements of the one-shot method can be found in [27, 26] where the three systems of equations are solved with the same time-step. Another variation of the one-shot method can be found in [12, 13].

A different approach, which under certain circumstances may lower the total computational cost is the incomplete gradient method [42, 60]. Less important terms in the gradient expression are omitted, reducing thus the accuracy of sensitivity derivatives while increasing the overall efficiency.

The continuous adjoint approach may become unable to deal with functionals not expressed in terms of pressure (inadmissible functionals). To handle inadmissible functionals, the adjoint formulation needs to be carefully modified [6, 9].

An adjoint approach which avoids the computation of sensitivities of grid metrics over the flow field has been presented for inviscid [31] and viscous [53] flows. According to this approach, the gradient depends on flow and adjoint variables as well as grid sensitivities over the boundary only, irrespective of whether the objective function is a boundary or field integral. The lack of internal grid sensitivities in the gradient expression reduces the computational cost and increases accuracy by avoiding repetitive grid remeshings at each optimization cycle. Approaches that skip the computation of grid sensitivities in the discrete adjoint approach are presented in [41, 51].

The computation of the exact Hessian matrix using adjoint approaches and its use in shape optimization is rare in the literature. In aerodynamic design, the Hessian matrix in the neighborhood of the optimal solution is analyzed in [7], while a method to compute the Hessian matrix can be found in [62], although the structural optimization is of concern.

This chapter presents a detailed framework for the development of discrete and especially continuous adjoint methods developed by the authors. The development covers inverse design and optimization problems (minimization of viscous losses, expressed as either entropy generation or total pressure losses) in internal aerodynamics. Applications in external aerodynamics (where the method can be extended to other objectives such as drag minimization constrained by the lift, etc.) have been worked out using the present approach but are not included in this chapter in the interest of space. The Navier-Stokes equations with the Spalart-Allmaras turbulence model are used as state equations. The demonstration is restricted to internal aerodynamics (design of a cascade airfoil and a duct). In addition to theory and examples on the computation and use of gradients (in steepest descent, quasi-Newton methods like BFGS, etc.), recent achievements on the formulation of adjoint-based exact Hessian methods in aerodynamic optimization are presented.

The principles of the adjoint approaches, discrete and continuous, are first presented. Since the discrete adjoint is based on the already discretized flow equations, any further discussion would be useful only for a specific discretization scheme. On the other hand, the continuous adjoint formulation is presented in complete detail. Emphasis is laid on the formulation which leads to an expression of the objective function gradient which is free of field integrals.

In structured grid terminology, such a formulation avoids the computation of grid metrics sensitivities, a procedure which would otherwise require repetitive generations of new meshes in domains defined by slightly perturbing one design variable at a time, followed by the computation of metrics and their variations. Extra computational (CPU) cost and ambiguities relevant to the correct implementation of finite differences are implicit in this procedure.

Both discrete and continuous approaches are discussed with regard to the shape optimization methods that are based on the computation and use of the exact Hessian of the objective function. Irrespective of the approach to be used, it is demonstrated that the computation of the Hessian requires not only the gradient of the objective function (computable using the adjoint approach discussed in the previous paragraph) but also the sensitivity derivatives of the flow quantities with respect to the design variables. It is then necessary to investigate any combination of direct and adjoint approaches (either in discrete or continuous form) that produce the Hessian matrix. There are, in fact, four algorithms: direct-direct, direct-adjoint, adjoint-direct, adjoint-adjoint; the first term denotes the method used to compute the gradient and the second describes how the Hessian can be derived starting from a known gradient. From the CPU cost viewpoint, it is shown that the direct-adjoint approach is the least costly among them and the one developed herein, in discrete and continuous mode. Note that it is unnecessary to make any comparison on their accuracy since all of them compute exact Hessians and, therefore, only some unavoidable numerical errors (due to the different operations and computations involved in each one of them) may appear. Such a comparison has been made but the four curves are almost identical, so there is no need to include it in this chapter. If the exact Hessian is available, the Newton method can be used as optimization algorithm. In this chapter and at least for the examined cases, the superiority of the exact Newton method over other optimization methods (such as quasi-Newton methods, steepest descent, BFGS) is demonstrated.

4.2 Principles of the Adjoint Approach

4.2.1 The Discrete Adjoint Approach

In aerodynamic shape optimization, the adjoint method aims at computing the gradient of an objective function F with respect to the design variables $b_i, i = 1, N$. Total derivatives $\frac{dF}{db_i}$ can be computed using [55],

$$\frac{dF}{db_i} = \frac{\partial F}{\partial b_i} + \frac{\partial F}{\partial U_k}\frac{dU_k}{db_i} \tag{4.1}$$

constrained by the state (flow) equations $R_m(\boldsymbol{U}) = 0, m = 1, ..., M$, which must be satisfied over the M grid nodes. The gradient of these constraints is expressed as

$$\frac{dR_m}{db_i} = \frac{\partial R_m}{\partial b_i} + \frac{\partial R_m}{\partial U_k}\frac{dU_k}{db_i} = 0 \ . \tag{4.2}$$

The combination of Eqs. (4.1) and (4.2) allows the elimination of $\frac{dU_k}{db_i}$ and produces the final expression for $\frac{dF}{db_i}$. By introducing the adjoint or costate variables $\Psi_m, m=1, ..., M$, this expression becomes

$$\frac{dF}{db_i} = \frac{\partial F}{\partial b_i} + \Psi_m \frac{\partial R_m}{\partial b_i} \tag{4.3}$$

where the adjoint variables result from the solution of the adjoint equations

$$R_k^{\Psi} = \frac{\partial F}{\partial U_k} + \Psi_m \frac{\partial R_m}{\partial U_k} = 0 \ . \tag{4.4}$$

Since such a development relies on the discrete state equations, Eqs. (4.2), it is usually referred to as the discrete adjoint method to distinguish it from the continuous one (see below). The advantage of the adjoint formulation is that only one adjoint equation has to be solved irrespective of the value of N.

The adjoint approach, which is associated with Eqs. (4.1) to (4.4) for the computation of the gradient of F, can also be applied to the computation of any quantity $F = \boldsymbol{\gamma}^T \boldsymbol{U}$, where $\boldsymbol{U}$ satisfies the linear system $A\boldsymbol{U} = \boldsymbol{\phi}$; $\boldsymbol{\gamma}$ and $\boldsymbol{\phi}$ are known vectors and A is a known square matrix. Here also, the direct computation of F ($F = \boldsymbol{\gamma}^T A^{-1} \boldsymbol{\phi}$) can be replaced by an adjoint approach. The adjoint equation $A^T \boldsymbol{\Psi} = \boldsymbol{\gamma}$ is first solved and then, F results from $F = \boldsymbol{\Psi}^T \boldsymbol{\phi}$, [55].

4.2.2 The Continuous Adjoint Approach

Section 4.2.1 summarizes the discrete adjoint approach as a means to compute the gradient of an objective function F, in discrete form, constrained by the discretized state PDE's. Alternatively, the same problem may be handled by the continuous adjoint approach, where the adjoint PDE's are first produced and then, discretized and solved.

Assume that the sensitivity derivatives of F with respect to b_i can be expressed as

$$\frac{\delta F}{\delta b_i} = \int_{\Omega} \gamma \frac{\delta U}{\delta b_i} d\Omega + \int_{S} \zeta B_1(\frac{\delta U}{\delta b_i}) dS + \frac{\delta F_g}{\delta b_i} \tag{4.5}$$

where γ and ζ are functions of geometrical quantities and state variables, B_1 is a differential operator and $\frac{\delta F_g}{\delta b_i}$ is the gradient of a function depending on the contour and/or grid sensitivity derivatives. Equation (4.5) contains both field (over the flow domain Ω) and boundary (along its boundary $S = \partial\Omega$) integrals. Expressions for $\frac{\delta U}{\delta b_i}$ can be obtained using the derivatives of the state equations and their boundary conditions that may be written as

$$\begin{aligned} L(\frac{\delta U}{\delta b_i}) &= \phi \ , \ \text{over} \ \ \Omega \\ B_2(\frac{\delta U}{\delta b_i}) &= \epsilon \ , \ \text{along} \ \ S \end{aligned} \tag{4.6}$$

where ϕ, ϵ are known functions and L, B_2 are known operators. Using the continuous adjoint approach, $\frac{\delta F}{\delta b_i}$ can be expressed as

$$\frac{\delta F}{\delta b_i} = \int_\Omega \Psi \phi d\Omega + \int_S (B_1^* \Psi) \epsilon dS + \frac{\delta F_g}{\delta b_i} \tag{4.7}$$

where the adjoint field Ψ is computed by discretizing and solving the adjoint PDE's with appropriate boundary conditions, namely

$$\begin{aligned} L^* \Psi &= \gamma \ , \ \text{over} \ \ \Omega \\ B_2^* \Psi &= \zeta \ , \ \text{along} \ \ S \end{aligned} \tag{4.8}$$

where L^* is the adjoint operator to L and B_1^*, B_2^* are boundary operators satisfying the following equation [55],

$$\begin{aligned} \int_\Omega \Psi L(\frac{\delta U}{\delta b_i}) d\Omega - \int_\Omega (L^* \Psi) \frac{\delta U}{\delta b_i} d\Omega &= \int_S (B_2^* \Psi) B_1(\frac{\delta U}{\delta b_i}) dS \\ &\quad - \int_S (B_1^* \Psi) B_2(\frac{\delta U}{\delta b_i}) dS \ . \end{aligned} \tag{4.9}$$

Advantages and disadvantages of the discrete and continuous approaches are discussed in Sect. 4.2.3. Nevertheless, regardless of the form (discrete or continuous), the adjoint approach, as a means to compute the objective function gradient in aerodynamic optimization, is much more inexpensive than the direct approach.

4.2.3 Differences Between Discrete and Continuous Adjoint

A marked difference that distinguishes the discrete from the continuous adjoint approach is the way the discrete adjoint equations are derived, starting from the state PDE's. The two alternative ways to produce the discrete adjoint equations are schematically shown below.

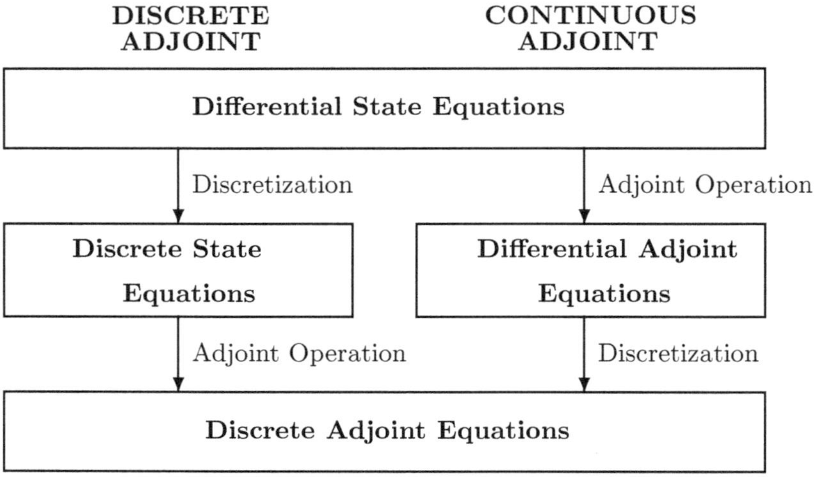

Theoretically, the value of F can be estimated with the same accuracy irrespective of the approach used [43, 44]. However, each approach has its own advantages and disadvantages which influence users' preferences. In the discrete approach, the derivation of the discrete adjoint equations is more cumbersome, unless automatic differentiation is used [58]. Once the discrete adjoint equations have been found, they can be solved using the same solvers as those used to solve the state equations. Using the continuous approach, one might use different discretization schemes. This might become important, since the adjoint equations can be derived and solved without having access to the flow solver source code. However, in practice, it has been proven that the accuracy of the continuous adjoint approach depends on the discretization scheme which must be as close as possible to that used for the discretization of the state equations [4, 5]. Note that differences in the gradient values, computed by the two approaches, may also appear if the computational grid is not adequate enough.

4.3 Inverse Design Using the Euler Equations

The continuous adjoint approach for the inverse design of aerodynamic shapes in inviscid flow is now presented. The Euler equations are used as state equations and these are written in vector form as

$$\frac{\partial \boldsymbol{U}}{\partial t} + \frac{\partial \boldsymbol{f}_k^{inv}}{\partial x_k} = \boldsymbol{0} \tag{4.10}$$

where $\boldsymbol{U}$ is the vector of conservative flow variables, $\boldsymbol{U} = \left[\rho\ ,\ \rho\boldsymbol{v}^T\ ,\ E\right]^T$, $\boldsymbol{f}_k^{inv} = \left[\rho u_k\ ,\ \rho u_k \boldsymbol{v}^T + p\boldsymbol{\delta}_k^T\ ,\ u_k(E+p)\right]^T$ is the vector of the inviscid fluxes, ρ is the local fluid density, u_k are the velocity $\boldsymbol{v}$ components, p is pressure, $E = \rho e + \frac{1}{2}\rho \boldsymbol{v}^2$ is the total energy per unit volume and $\boldsymbol{\delta}_k$ is the vector of Kronecker symbols. For the sake of simplicity, throughout this section which is exclusively concerned with inviscid flows, the inviscid fluxes $\boldsymbol{f}_k^{inv}$ will be denoted by $\boldsymbol{f}_k$.

In inverse design problems, the goal is to compute the aerodynamic shape that produces a given pressure (or any other) distribution over the shape contour. The term "aerodynamic shape" is used to denote isolated or cascade airfoils, ducts, etc. (in 2D) or wings, blades, ducts, etc. (in 3D). The objective function is written as

$$F = \frac{1}{2}\int_{S_w} (p - p_{tar})^2 dS \tag{4.11}$$

where $p_{tar}(S)$ is the target pressure distribution along the solid walls S_w. The gradient of F with respect to the design variables controlling the shape is expressed as

$$\frac{\delta F}{\delta b_i} = \frac{1}{2}\int_{S_w} (p - p_{tar})^2 \frac{\delta (dS)}{\delta b_i} + \int_{S_w} (p - p_{tar})\frac{\delta p}{\delta b_i} dS\ . \tag{4.12}$$

It is straightforward to derive expressions for $\frac{\delta \boldsymbol{U}}{\delta b_i}$ using the so-called direct approach. Starting from the Euler equations, we obtain

$$\frac{\delta}{\delta b_i}\left(\frac{\partial \boldsymbol{f}_k}{\partial x_k}\right) = \frac{\delta}{\delta b_i}\left(A_k \frac{\partial \boldsymbol{U}}{\partial x_k}\right) = \boldsymbol{0}\ . \tag{4.13}$$

Equation (4.13) can be solved for the flow sensitivities $\frac{\delta \boldsymbol{U}}{\delta b_i}$, provided that the sensitivities of spatial derivatives $\frac{\delta}{\delta b_i}\left(\frac{\partial \boldsymbol{U}}{\partial x_k}\right)$ can be transformed to the spatial derivatives of flow sensitivities $\frac{\partial}{\partial x_k}\left(\frac{\delta \boldsymbol{U}}{\delta b_i}\right)$. Since $\frac{\delta p}{\delta b_i}$, which appears in Eq. (4.12), is expressed in terms of $\frac{\delta \boldsymbol{U}}{\delta b_i}$, an expression for $\frac{\delta F}{\delta b_i}$ can be produced. Unfortunately, to compute the grid sensitivities that appear in any possible form of Eq. (4.13) (see, for instance, Eq. (4.17) for grid metrics sensitivities) through finite differences, $2N$ calls to the grid generation software are needed. The detailed development and solution algorithm are omitted.

Alternatively, the partial derivatives of Eq. (4.10) with respect to b_i

$$\frac{\partial}{\partial b_i}\left(\frac{\partial \boldsymbol{f}_k}{\partial x_k}\right) = \frac{\partial}{\partial x_k}\left(A_k \frac{\partial \boldsymbol{U}}{\partial b_i}\right) = \boldsymbol{0} \tag{4.14}$$

can be used to produce expressions for $\frac{\partial \boldsymbol{U}}{\partial b_i}$. The assumption that the Jacobian matrix derivatives can be omitted is made. Once $\frac{\partial \boldsymbol{U}}{\partial b_i}$ are known, $\frac{\delta \boldsymbol{U}}{\delta b_i}$ can be computed using

$$\frac{\delta \boldsymbol{U}}{\delta b_i} = \frac{\partial \boldsymbol{U}}{\partial b_i} + \frac{\partial \boldsymbol{U}}{\partial x_m}\frac{\delta x_m}{\delta b_i} \ . \tag{4.15}$$

This alternative formulation of the direct approach skips the computation of grid sensitivities at the interior, since Eq. (4.12) depends only on $\frac{\delta p}{\delta b_i}$ at the boundaries. However, as stated in the previous section, the CPU cost of the direct approach is high since N solutions of the linearized flow equations are necessary.

To set up the continuous adjoint approach, the augmented objective function F_{aug} is introduced by adding to F the field integral of the adjoint variables multiplied by the flow equations. One may directly express the gradient of F with respect to b_i, by using either the total or the partial sensitivity derivatives of the flow equations. The two alternative expressions are

$$\frac{\delta F_{aug}}{\delta b_i} = \frac{\delta F}{\delta b_i} + \int_{\Omega} \boldsymbol{\Psi}^T \frac{\delta}{\delta b_i}\left(\frac{\partial \boldsymbol{f}_k}{\partial x_k}\right) d\Omega \tag{4.16a}$$

$$\frac{\delta F_{aug}}{\delta b_i} = \frac{\delta F}{\delta b_i} + \int_{\Omega} \boldsymbol{\Psi}^T \frac{\partial}{\partial b_i}\left(\frac{\partial \boldsymbol{f}_k}{\partial x_k}\right) d\Omega \ . \tag{4.16b}$$

Starting from Eq. (4.16a), using the transformation from $\frac{\delta}{\delta b_i}\left(\frac{\partial \boldsymbol{U}}{\partial x_k}\right)$ to $\frac{\partial}{\partial x_k}\left(\frac{\delta \boldsymbol{U}}{\delta b_i}\right)$, [32], based on equation

$$\frac{\delta}{\delta b_i}\left(\frac{\partial \boldsymbol{U}}{\partial x_k}\right) = \frac{\partial}{\partial x_k}\left(\frac{\delta \boldsymbol{U}}{\delta b_i}\right) + \frac{\partial \boldsymbol{U}}{\partial \xi^m}\frac{\delta}{\delta b_i}\left(\frac{\partial \xi^m}{\partial x_k}\right) \tag{4.17}$$

where ξ^j are structured grid metrics and by integrating by parts, $\frac{\delta F}{\delta b_i}$ is expressed as follows

$$\frac{\delta F}{\delta b_i} = \frac{1}{2}\int_{S_w} (p - p_{tar})^2 \frac{\delta(dS)}{\delta b_i} - \int_{\Omega} \boldsymbol{\Psi}^T \frac{\partial f_k}{\partial \xi^m}\frac{\delta}{\delta b_i}\left(\frac{\partial \xi^m}{\partial x_k}\right) d\Omega$$
$$+ \int_{S_w} (\Psi_{k+1} p - \boldsymbol{\Psi}^T \boldsymbol{f}_k)\frac{\delta(n_k dS)}{\delta b_i} \ . \tag{4.18}$$

Adjoint variables are computed by solving the field adjoint equations

$$\frac{\partial \boldsymbol{\Psi}}{\partial t} - {A_k}^T \frac{\partial \boldsymbol{\Psi}}{\partial x_k} = \boldsymbol{0} \tag{4.19}$$

with appropriate boundary conditions along the solid walls

$$(p - p_{tar}) + \Psi_{k+1} n_k = 0 \tag{4.20}$$

and the inlet/outlet boundary

$$\frac{\delta \boldsymbol{U}^T}{\delta b_i}(A_n^T \boldsymbol{\Psi}) = 0 \tag{4.21}$$

$A_n = A_k n_k$ and n_k are the outward normal to the boundary, unit vector components. It is important to note that Eq. (4.18) includes a field integral of grid metric sensitivities $\frac{\delta}{\delta b_i}\left(\frac{\partial \xi^m}{\partial x_k}\right)$, the computation of which increases the CPU cost and becomes a source of inaccuracies. The standard way to compute grid metric sensitivities is through finite differences (after perturbating one design variable at a time, modifying the contour and remeshing). If Eq. (4.16a) or

$$\frac{\delta F_{aug}}{\delta b_i} = \frac{\delta F}{\delta b_i} + \int_{\Omega} \boldsymbol{\Psi}^T \frac{\partial}{\partial x_k}\left(\frac{\partial \boldsymbol{f}_k}{\partial b_i}\right) d\Omega \tag{4.22}$$

is used instead, the adjoint equations and boundary conditions remain exactly the same (Eqs. (4.19), (4.20) and (4.21)) but the gradient expression becomes

$$\begin{aligned} \frac{\delta F}{\delta b_i} = \frac{1}{2}\int_{S_w} (p - p_{tar})^2 \frac{\delta(dS)}{\delta b_i} - \int_{S_w} \frac{\partial \boldsymbol{U}}{\partial x_k}^T {A_n}^T \boldsymbol{\Psi} \frac{\delta x_k}{\delta b_i} dS \\ + \int_{S_w} (\Psi_{k+1} p - \boldsymbol{\Psi}^T \boldsymbol{f}_k) \frac{\delta(n_k dS)}{\delta b_i} \end{aligned} \tag{4.23}$$

which differs from Eq. (4.18) since Eq. (4.23) is free of field integrals. Thus, the gradient values are computed with less computational cost and greater accuracy. In addition, Eq. (4.23) is more general and can be used with either structured or unstructured grids.

Proof of Eq. (4.23):

The integration by parts of the second term on the right-hand side (r.h.s.) of Eq. (4.22) gives

$$\begin{aligned} \int_{\Omega} \boldsymbol{\Psi}^T \frac{\partial}{\partial x_k}\left(\frac{\partial \boldsymbol{f}_k}{\partial b_i}\right) d\Omega = - \int_{\Omega} \frac{\partial \boldsymbol{U}^T}{\partial b_i}\left(A_k^T \frac{\partial \boldsymbol{\Psi}}{\partial x_k}\right) d\Omega \\ + \int_{S_{i,o,w}} \boldsymbol{\Psi}^T \frac{\partial \boldsymbol{f}_k}{\partial b_n} n_k dS \, . \end{aligned} \tag{4.24}$$

Employing the no-penetration condition, the second integral on the r.h.s of Eq. (4.24) is written as

$$\begin{aligned} \int_{S_w} \boldsymbol{\Psi}^T \frac{\partial \boldsymbol{f}_k}{\partial b_i} n_k dS &= \int_{S_w} \boldsymbol{\Psi}^T \frac{\delta \boldsymbol{f}_k}{\delta b_i} n_k dS - \int_{S_w} \boldsymbol{\Psi}^T \frac{\partial \boldsymbol{f}_k}{\partial x_m}\frac{\delta x_m}{\delta b_i} n_k dS \\ &= \int_{S_w} \Psi_{k+1} n_k \frac{\delta p}{\delta b_i} dS + \int_{S_w} (\Psi_{k+1} p - \boldsymbol{\Psi}^T \boldsymbol{f}_k) \frac{\delta(n_k dS)}{\delta b_i} \\ &\quad - \int_{S_w} \frac{\partial \boldsymbol{U}}{\partial x_k}^T {A_n}^T \boldsymbol{\Psi} \frac{\delta x_k}{\delta b_i} dS \, . \end{aligned} \tag{4.25}$$

The gradient of F_{aug} with respect to the design variables is finally given by

$$\frac{\delta F_{aug}}{\delta b_i} = \frac{1}{2}\int_{S_w}(p-p_{tar})^2\frac{\delta(dS)}{\delta b_i} + \underbrace{\int_{S_w}(p-p_{tar})\frac{\delta p}{\delta b_i}dS}_{BCW}$$

$$\underbrace{-\int_{\Omega}\left(\frac{\delta \boldsymbol{U}^T}{\delta b_i} - \frac{\partial \boldsymbol{U}}{\partial x_m}^T\frac{\delta x_m}{\delta b_i}\right)\left(A_k^T\frac{\partial \boldsymbol{\Psi}}{\partial x_k}\right)d\Omega}_{FAE}$$

$$-\int_{S_w}\frac{\partial \boldsymbol{U}}{\partial x_k}^T A_n{}^T\boldsymbol{\Psi}\frac{\delta x_k}{\delta b_i}dS + \underbrace{\int_{S_w}\Psi_{k+1}n_k\frac{\delta p}{\delta b_i}dS}_{BCW}$$

$$+\int_{S_w}(\Psi_{k+1}p - \boldsymbol{\Psi}^T\boldsymbol{f}_k)\frac{\delta(n_k dS)}{\delta b_i} + \underbrace{\int_{S_{i,o}}\frac{\delta \boldsymbol{U}^T}{\delta b_i}(A_n^T\boldsymbol{\Psi})dS}_{BCIO} \ . \qquad (4.26)$$

In Eq. (4.26), the field integral marked with FAE, which depends on the flow variable sensitivities, is eliminated giving rise to the field adjoint equations, Eq. (4.19). Terms marked with BCW and $BCIO$ are also eliminated by satisfying the adjoint boundary conditions (Eqs. (4.20) and (4.21)). The remaining terms determine the sensitivity derivatives of the objective function with respect to the design variables, Eq. (4.23).

4.4 Inverse Design Using the Navier-Stokes Equations

In real flows, the inverse design of optimal aerodynamic shapes is based on the Navier-Stokes equations

$$\frac{\partial \boldsymbol{U}}{\partial t} + \frac{\partial \boldsymbol{f}_k^{inv}}{\partial x_k} - \frac{\partial \boldsymbol{f}_k^{vis}}{\partial x_k} = \boldsymbol{0} \qquad (4.27)$$

where $\boldsymbol{f}_k^{vis} = \left[0 \ , \ \boldsymbol{\tau}_k^T \ , \ u_m\tau_{km} + q_k\right]^T$ is the vector of viscous fluxes, $\boldsymbol{\tau}_k$ are the viscous stresses and $q_k = k\frac{\partial T}{\partial x_k}$ are the thermal heat flux components.

Using the same objective function, the extra work to be done to develop the direct approach for the computation of $\frac{\delta F}{\delta b_i}$ is to account for the sensitivities of the viscous stresses, too. Using the continuous adjoint approach, based on the partial derivatives of the Navier-Stokes equations with respect to b_i, the sensitivity derivatives of F are finally given by [53]

$$\frac{\delta F}{\delta b_i} = \frac{1}{2}\int_{S_w} (p-p_{tar})^2 \frac{\delta(dS)}{\delta b_i} + \int_{S_w} (\Psi_{k+1} p - \boldsymbol{\Psi}^T \boldsymbol{f}_k) \frac{\delta(n_k dS)}{\delta b_i}$$
$$-\int_{S_w} \boldsymbol{\Psi}^T \left(\frac{\partial \boldsymbol{f}_k^{inv}}{\partial x_m} - \frac{\partial \boldsymbol{f}_k^{vis}}{\partial x_m} \right) \frac{\delta x_m}{\delta b_i} n_k dS$$
$$+\int_{S_w} \frac{\Psi_{k+1}}{n_k} \tau_{km} \delta(n_k n_m) dS + \int_{S_w} \Psi_\Lambda q_k \frac{\delta(n_k dS)}{\delta b_i}$$
$$-\int_{S_w} \frac{\partial u_k}{\partial x_l} \left[\mu \left(\frac{\partial \Psi_{m+1}}{\partial x_k} + \frac{\partial \Psi_{k+1}}{\partial x_m} \right) + \lambda \delta_{km} \frac{\partial \Psi_{n+1}}{\partial x_n} \right] \frac{\delta x_l}{\delta b_i} n_m dS \quad (4.28)$$

($\Lambda = 4$ in 2D and $\Lambda = 5$ in 3D). In viscous flows, $\boldsymbol{\Psi}$ should satisfy the field adjoint equations

$$\frac{\partial \boldsymbol{\Psi}}{\partial t} - A_k^T \frac{\partial \boldsymbol{\Psi}}{\partial x_k} - M^{-T} \boldsymbol{K} = \boldsymbol{0} \quad (4.29)$$

where M is the transformation matrix from the non-conservative to the conservative flow variables and $\boldsymbol{K}$ corresponds to the adjoint viscous stresses, where

$$K_1 = -\frac{T}{\rho} \frac{\partial}{\partial x_k} \left(k \frac{\partial \Psi_\Lambda}{\partial x_k} \right) \qquad K_\Lambda = \frac{T}{p} \frac{\partial}{\partial x_k} \left(k \frac{\partial \Psi_\Lambda}{\partial x_k} \right)$$
$$K_{k+1} = \frac{\partial}{\partial x_m} \left[\mu \left(\frac{\partial \Psi_{m+1}}{\partial x_k} + \frac{\partial \Psi_{k+1}}{\partial x_m} \right) + \lambda \delta_{km} \frac{\partial \Psi_{l+1}}{\partial x_l} \right]$$
$$+ \frac{\partial}{\partial x_m} \left[\mu \left(u_m \frac{\partial \Psi_\Lambda}{\partial x_k} + u_k \frac{\partial \Psi_\Lambda}{\partial x_m} \right) + \lambda \delta_{km} u_l \frac{\partial \Psi_\Lambda}{\partial x_l} \right] - \tau_{km} \frac{\partial \Psi_\Lambda}{\partial x_m} \ . \quad (4.30)$$

Along the solid walls, the boundary conditions for the adjoint variables which correspond to the velocity components read

$$\Psi_{k+1} = -(p - p_{tar}) n_k \quad , \quad k = 1, ..., \Lambda \ . \quad (4.31)$$

The boundary condition for Ψ_Λ, which corresponds to the energy equation, is either homogeneous Dirichlet (constant wall temperature) or Neumann (adiabatic walls). The inlet/outlet adjoint boundary conditions are still given by Eq. (4.21), neglecting locally the spatial derivatives of flow variables.

4.5 Viscous Losses Minimization in Internal Flows

In this section, the continuous adjoint approach is adapted to shape optimization problems in internal flows where the minimization of viscous losses is targeted. Viscous losses can be expressed as either total pressure drop or entropy generation. Both are discussed in this section. Concerning the two alternative formulations, a distinguishing feature stems from the (field or boundary) integral used to define the objective function.

4.5.1 Minimization of Total Pressure Losses

In contrast to inverse design problems, F is defined by integrals along the inlet and outlet boundaries of the domain, (S_i, S_o) instead of the parameterized wall boundary. Consequently, F and b_i are "non-collocated". For this reason, the way inlet/outlet boundary conditions for the adjoint variables are imposed is described below [54].

The total pressure losses functional is defined as

$$F = \int_{S_{i,o}} p_t dS = \int_{S_i} p_t dS - \int_{S_o} p_t dS \tag{4.32}$$

and its gradient is simply given by

$$\frac{\delta F}{\delta b_n} = \int_{S_i} \frac{\delta p_t}{\delta b_i} dS - \int_{S_o} \frac{\delta p_t}{\delta b_i} dS \tag{4.33}$$

since the inlet and exit boundaries remain fixed. Using the adjoint approach, the gradient of the objective function can be expressed as follows

$$\begin{aligned}\frac{\delta F}{\delta b_i} =& -\int_{S_w} \left(\boldsymbol{\Psi}^T \boldsymbol{f}_k\right) \frac{\delta(n_k dS)}{\delta b_i} + \int_{S_w} \Psi_\Lambda q_m \frac{\delta(n_m dS)}{\delta b_i} \\ & - \int_{S_w} \frac{\partial u_k}{\partial x_l} \left[\mu \left(\frac{\partial \Psi_{m+1}}{\partial x_k} + \frac{\partial \Psi_{k+1}}{\partial x_m} \right) + \lambda \delta_{km} \frac{\partial \Psi_{n+1}}{\partial x_n} \right] \frac{\delta x_l}{\delta b_i} n_m dS \\ & - \int_{S_w} \boldsymbol{\Psi}^T \left(\frac{\partial \boldsymbol{f}_k^{inv}}{\partial x_m} - \frac{\partial \boldsymbol{f}_k^{vis}}{\partial x_m} \right) \frac{\delta x_m}{\delta b_i} n_k dS \, .\end{aligned} \tag{4.34}$$

The field adjoint equations are given by Eq. (4.29) and are similar to the ones used in viscous inverse designs. The solid wall conditions for the $\boldsymbol{\Psi}$ components that correspond to the velocity components are homogeneous Dirichlet. Dirichlet or Neumann conditions are imposed for Ψ_Λ, depending on the wall condition on temperature.

The adjoint boundary conditions over the outlet of the flow domain are imposed by solving

$$\left[(P\bar{A})^T \boldsymbol{\Psi}\right]^T + \left(\frac{\partial p_t}{\partial \boldsymbol{V}}\right)^T L = 0 \tag{4.35}$$

for the characteristic flow variables directed outwards. $\frac{\partial p_t}{\partial \boldsymbol{V}}$ is the derivative of the total pressure with respect to the nonconservative variables $\boldsymbol{V}$, P is the matrix with the right eigenvectors of the conservative Jacobian matrix $A_k n_k$, L is the matrix with the right eigenvectors of the nonconservative Jacobian matrix and $\bar{\Lambda}$ is the diagonal matrix with the eigenvalues of the Jacobian matrix. The boundary condition at the inlet of the flow domain is imposed by solving

$$\left[(P\bar{\Lambda})^T \boldsymbol{\Psi}\right]^T = 0 \tag{4.36}$$

since p_t is constant at the inlet.

4.5.2 Minimization of Entropy Generation

From a practical point of view, it makes no difference to use either total pressure losses or entropy generation as a measure of viscous losses in internal flows. From the mathematical point of view, however, the use of entropy generation has the additional feature that the objective function is a field integral. Thus, F is defined as

$$F = \int_\Omega \rho u_k \frac{\partial s}{\partial x_k} d\Omega = \int_\Omega \frac{1}{T} \tau_{km} \frac{\partial u_k}{\partial x_m} d\Omega \ . \tag{4.37}$$

The gradient of F can be computed using the direct approach, by expressing the sensitivities of the finite volume $d\Omega$ as, [53],

$$\frac{\delta(d\Omega)}{\delta b_i} = \frac{\partial}{\partial x_k}\left(\frac{\delta x_k}{\delta b_i}\right) d\Omega \ . \tag{4.38}$$

The integral that includes $\frac{\partial}{\partial x_k}\left(\frac{\delta x_k}{\delta b_i}\right)$ is integrated by parts and the gradient of F is, finally, given by

$$\begin{aligned}\frac{\delta F}{\delta b_i} = &- \int_\Omega \frac{1}{T^2} \tau_{km} \frac{\partial u_k}{\partial x_m} \frac{\partial T}{\partial b_i} d\Omega - 2 \int_\Omega \frac{\partial}{\partial x_m}\left(\frac{1}{T}\tau_{km}\right) \frac{\partial u_k}{\partial b_i} d\Omega \\ &- 2 \int_{S_w} \frac{1}{T} \tau_{km} \frac{\partial u_m}{\partial x_l} n_k \frac{\delta x_l}{\delta b_i} dS + \int_{S_w} \frac{1}{T} \tau_{km} \frac{\partial u_k}{\partial x_m} \frac{\delta x_l}{\delta b_i} n_l dS\end{aligned} \tag{4.39}$$

where $\frac{\partial T}{\partial b_i}$ and $\frac{\partial u_k}{\partial b_i}$ are computed using Eq. (4.14) (direct approach), with $\boldsymbol{f}_k = \boldsymbol{f}_k^{inv} - \boldsymbol{f}_k^{vis}$. The adjoint approach, through Eq. (4.38) and integrations by parts, gives the final expression of $\frac{\delta F}{\delta b_i}$ as follows

$$\frac{\delta F}{\delta b_i}=-\int_{S_w}\left(\boldsymbol{\Psi}^T \boldsymbol{f}_k\right)\frac{\delta(n_k dS)}{\delta b_i}-\int_{S_w}\boldsymbol{\Psi}^T\left(\frac{\partial \boldsymbol{f}_k^{inv}}{\partial x_l}-\frac{\partial \boldsymbol{f}_k^{vis}}{\partial x_l}\right)\frac{\delta x_l}{\delta b_i}n_k dS$$
$$+\int_{S_w}\Psi_\Lambda q_m\frac{\delta(n_m dS)}{\delta b_i}-\int_{S_w}\frac{\partial u_k}{\partial x_l}\left[\mu\left(\frac{\partial \Psi_{m+1}}{\partial x_k}+\frac{\partial \Psi_{k+1}}{\partial x_m}\right)+\lambda\delta_{km}\frac{\partial \Psi_{q+1}}{\partial x_q}\right]\frac{\delta x_l}{\delta b_i}n_m dS$$
$$-2\int_{S_w}\tau_{km}\frac{\partial u_m}{\partial x_l}n_k\frac{\delta x_l}{\delta b_i}dS+\int_{S_w}\frac{1}{T}\tau_{km}\frac{\partial u_k}{\partial x_m}\frac{\delta x_l}{\delta b_i}n_l dS \tag{4.40}$$

where $\boldsymbol{\Psi}$ satisfies the field adjoint equations

$$\frac{\partial \boldsymbol{\Psi}}{\partial t}-A_k^T\frac{\partial \boldsymbol{\Psi}}{\partial x_k}-M^{-T}\boldsymbol{K}-M^{-T}\boldsymbol{L}=\boldsymbol{0}\ . \tag{4.41}$$

The vector of source terms $\boldsymbol{L}$ is defined as

$$L_1=-\frac{1}{T^2}\tau_{km}\frac{\partial u_k}{\partial x_m}\frac{p}{\rho^2(\gamma-1)}\ ,\ L_{k+1}=2\frac{\partial}{\partial x_m}\left(\frac{\mu}{T}\tau_{km}\right)\ ,\ L_\Lambda=-\frac{\rho}{p}L_1 \tag{4.42}$$

The boundary conditions over the solid wall are the same to those used for the total pressure losses minimization; at the inlet and outlet boundaries, Eq. (4.21) is imposed.

From Eq. (4.40), it can be stated that the gradient of F does not depend on field integrals, even if F is, in fact, a field integral retaining thus, the advantages of better accuracy and lower computational cost.

4.6 Computation of the Hessian Matrix

Starting from either the direct or the adjoint approach to compute first derivatives, the use of the direct or adjoint approach anew leads to the computation of the Hessian matrix. Consequently, four approaches to compute the exact Hessian matrices have been developed. These approaches (either discrete or continuous) differ with respect to their CPU cost.

Once the $\frac{N(N+1)}{2}$ distinct values of the symmetrical Hessian matrix as well as the gradient of the objective function have been computed, the design variables are updated using the (exact) Newton algorithm

$$\frac{d^2F^\lambda}{db_i db_j}db_i^\lambda=-\frac{dF^\lambda}{db_j} \tag{4.43a}$$
$$b_i^{\lambda+1}=b_i^\lambda+db_i^\lambda\quad ,\quad i=1,...,N \tag{4.43b}$$

where λ is the optimization cycle counter.

Considering that the adjoint approach is much less time-consuming than the direct approach for the gradient computation, one may think that the exclusive use of adjoints would be advantageous for the computation of the

Hessian as well. However, if first-order sensitivity derivatives are computed using the adjoint approach, N pairs of additional systems of direct or adjoint equations must be solved for the computation of the Hessian matrix. So, the total cost to compute all quantities needed by Eq. (4.43) is equal to that of $1+2N$ equivalent flow solutions.

The high cost of the previous approaches is due to the fact that the adjoint approach is used for the first derivative and therefore, $\frac{d\boldsymbol{U}}{db_i}$ values are not available; as it will become clear below, these quantities are necessary in order to proceed to the computation of the Hessian matrix. For this reason, the direct approach to compute the gradient of F must be employed. Given this decision, one should investigate the expected gain from the subsequent use of the adjoint approach to compute the Hessian (direct-adjoint approach) compared to the direct-direct approach, where the computation of the Hessian is based on the repeated application of the direct approach. It can be shown (this development is omitted) that the direct-direct approach requires $N+\frac{N(N+1)}{2}$ equivalent flow solutions at each optimization cycle. On the other hand, the direct-adjoint approach, in which second derivatives are computed using the adjoint approach, is the less time-consuming one since it requires the solution of $1+N$ equivalent flow problems. Note that the previous figures must be augmented by one, so as to account for the cost of solving the flow equations and computing the value of F.

4.6.1 Discrete Direct-adjoint Approach for the Hessian

Starting from Eq. (4.1), the second derivative of F is given by

$$\frac{d^2F}{db_idb_j}=\frac{\partial^2 F}{\partial b_i\partial b_j}+\frac{\partial^2 F}{\partial b_i\partial U_k}\frac{dU_k}{db_j}+\frac{\partial^2 F}{\partial U_k\partial b_j}\frac{dU_k}{db_i}+\frac{\partial^2 F}{\partial U_k\partial U_m}\frac{dU_k}{db_i}\frac{dU_m}{db_j}+\frac{\partial F}{\partial U_k}\frac{d^2U_k}{db_idb_j}\ . \tag{4.44}$$

Equation (4.44) can be used to compute $\frac{d^2F}{db_idb_j}$ provided that $\frac{d\boldsymbol{U}}{db_i}$ (N quantities) and $\frac{d^2\boldsymbol{U}}{db_idb_j}$ ($\frac{N(N+1)}{2}$ quantities) can also be computed. The first derivatives of $\boldsymbol{U}$ are computed at the cost of N equivalent flow solutions, Eq. (4.2), according to the so-called direct approach. For the additional $\frac{N(N+1)}{2}$ quantities (second derivatives of $\boldsymbol{U}$), from Eq. (4.2), we get

$$\frac{d^2R_n}{db_idb_j}=\frac{\partial^2 R_n}{\partial b_i\partial b_j}+\frac{\partial^2 R_n}{\partial b_i\partial U_k}\frac{dU_k}{db_j}+\frac{\partial^2 R_n}{\partial U_k\partial b_j}\frac{dU_k}{db_i}+\frac{\partial^2 R_n}{\partial U_k\partial U_m}\frac{dU_k}{db_i}\frac{dU_m}{db_j}+\frac{\partial R_n}{\partial U_k}\frac{d^2U_k}{db_idb_j}=0\ . \tag{4.45}$$

Equations (4.45) can be solved to provide $\frac{d^2\boldsymbol{U}}{db_i db_j}$ at the cost of $\frac{N(N+1)}{2}$ equivalent flow solutions. Instead, one may form the second derivative of the (new) augmented objective function $\hat{F}$ by introducing (new) adjoint variables $\hat{\Psi}_n$, as follows

$$\frac{d^2\hat{F}}{db_i db_j} = \frac{d^2 F}{db_i db_j} + \hat{\Psi}_n \frac{d^2 R_n}{db_i db_j} \ . \tag{4.46}$$

Substituting Eqs. (4.44) and (4.45) in Eq. (4.46) and rearranging the terms we obtain

$$\begin{aligned}\frac{d^2\hat{F}}{db_i db_j} = & \frac{\partial^2 F}{\partial b_i \partial b_j} + \hat{\Psi}_n \frac{\partial^2 R_n}{\partial b_i \partial b_j} + \frac{\partial^2 F}{\partial U_k \partial U_m} \frac{dU_k}{db_i} \frac{dU_m}{db_j} + \hat{\Psi}_n \frac{\partial^2 R_n}{\partial U_k \partial U_m} \frac{dU_k}{db_i} \frac{dU_m}{db_j} \\ & + \frac{\partial^2 F}{\partial b_i \partial U_k} \frac{dU_k}{db_j} + \hat{\Psi}_n \frac{\partial^2 R_n}{\partial b_i \partial U_k} \frac{dU_k}{db_j} + \frac{\partial^2 F}{\partial U_k \partial b_j} \frac{dU_k}{db_i} + \hat{\Psi}_n \frac{\partial^2 R_n}{\partial U_k \partial b_j} \frac{dU_k}{db_i} \\ & + \left(\frac{\partial F}{\partial U_k} + \hat{\Psi}_n \frac{\partial R_n}{\partial U_k} \right) \frac{d^2 U_k}{db_i db_j} \ . \end{aligned} \tag{4.47}$$

The term depending on the second derivatives of the flow variables (last term) is eliminated by satisfying the adjoint equations

$$\frac{\partial F}{\partial U_k} + \hat{\Psi}_n \frac{\partial R_n}{\partial U_k} = 0 \tag{4.48}$$

at the cost of only one equivalent flow solution. The Hessian matrix is finally expressed as follows

$$\begin{aligned}\frac{d^2\hat{F}}{db_i db_j} = & \frac{\partial^2 F}{\partial b_i \partial b_j} + \hat{\Psi}_n \frac{\partial^2 R_n}{\partial b_i \partial b_j} + \left(\frac{\partial^2 F}{\partial U_k \partial U_m} + \hat{\Psi}_n \frac{\partial^2 R_n}{\partial U_k \partial U_m} \right) \frac{dU_k}{db_i} \frac{dU_m}{db_j} \\ & + \left(\frac{\partial^2 F}{\partial b_i \partial U_k} + \hat{\Psi}_n \frac{\partial^2 R_n}{\partial b_i \partial U_k} \right) \frac{dU_k}{db_j} \\ & + \left(\frac{\partial^2 F}{\partial U_k \partial b_j} + \hat{\Psi}_n \frac{\partial^2 R_n}{\partial U_k \partial b_j} \right) \frac{dU_k}{db_i} \ . \end{aligned} \tag{4.49}$$

Thus, using the so-called direct-adjoint approach, the total CPU cost for a Newton optimization cycle, Eqs. (4.43), is equal to $1+N+1$ equivalent flow solutions (including the solution of the flow equations).

4.6.2 Continuous Direct-adjoint Approach for the Hessian (Inverse Design)

In the continuous approach, the gradient of F is given by Eq. (4.12). Starting from Eq. (4.12), the second derivative of F is expressed as follows

$$\frac{\delta^2 F}{\delta b_i \delta b_j}=\int_{S_w}\frac{\delta p}{\delta b_i}\frac{\delta p}{\delta b_j}dS+\int_{S_w}(p-p_{tar})\frac{\delta^2 p}{\delta b_i \delta b_j}dS+\int_{S_w}(p-p_{tar})\frac{\delta p}{\delta b_i}\frac{\delta(dS)}{\delta b_j}$$
$$+\int_{S_w}(p-p_{tar})\frac{\delta p}{\delta b_j}\frac{\delta(dS)}{\delta b_i}+\frac{1}{2}\int_{S_w}(p-p_{tar})^2\frac{\delta^2(dS)}{\delta b_i \delta b_j}\;. \tag{4.50}$$

As in discrete adjoint, the use of Eq. (4.50) requires the knowledge of $\frac{\delta \boldsymbol{U}}{\delta b_i}$ and $\frac{\delta^2 \boldsymbol{U}}{\delta b_i \delta b_j}$ in terms of which $\frac{\delta p}{\delta b_i}$ and $\frac{\delta^2 p}{\delta b_i \delta b_j}$ can be expressed. First derivatives can be computed according to the direct (continuous) approach. Second derivatives can be computed by solving the equations resulting from the second derivatives of the flow equations ($\frac{N(N+1)}{2}$ equations),

$$\frac{\partial}{\partial x_k}\left(A_k \frac{\partial^2 \boldsymbol{U}}{\partial b_i \partial b_j}\right)=\boldsymbol{0} \tag{4.51}$$

with the usual assumption that the derivatives of the Jacobian matrices can be eliminated. Instead of solving Eqs. (4.51), the direct-adjoint approach requires the solution of the field adjoint Eqs. (4.19), together with the boundary conditions (4.20) and (4.21). The Hessian matrix is finally computed as

$$\frac{\delta^2 F_{aug}}{\delta b_i \delta b_j}=\int_{S_w}\frac{\delta p}{\delta b_i}\frac{\delta p}{\delta b_j}dS+\int_{S_w}(p-p_{tar})\frac{\delta p}{\delta b_i}\frac{\delta(dS)}{\delta b_j}+\int_{S_w}(p-p_{tar})\frac{\delta p}{\delta b_j}\frac{\delta(dS)}{\delta b_i}$$
$$+\frac{1}{2}\int_{S_w}(p-p_{tar})^2\frac{\delta^2(dS)}{\delta b_i \delta b_j}+\int_{S_w}(\Psi_{k+1}p-\boldsymbol{\Psi}^T\boldsymbol{f}_k)\frac{\delta^2 n_k}{\delta b_i \delta b_j}dS$$
$$+\int_{S_w}\left(\Psi_{k+1}\frac{\delta p}{\delta b_i}-\boldsymbol{\Psi}^T\frac{\delta \boldsymbol{f}_k}{\delta b_i}\right)\frac{\delta n_k}{\delta b_j}dS+\int_{S_w}\left(\Psi_{k+1}\frac{\delta p}{\delta b_j}-\boldsymbol{\Psi}^T\frac{\delta \boldsymbol{f}_k}{\delta b_j}\right)\frac{\delta n_k}{\delta b_i}dS$$
$$-\int_{S_w}\boldsymbol{\Psi}^T A_k n_k\left(\frac{\partial^2 \boldsymbol{U}}{\partial b_i \partial x_l}\frac{\delta x_l}{\delta b_j}+\frac{\partial^2 \boldsymbol{U}}{\partial b_j \partial x_l}\frac{\delta x_l}{\delta b_i}\right)dS$$
$$-\int_{S_w}\boldsymbol{\Psi}^T A_k n_k\left(\frac{\partial^2 \boldsymbol{U}}{\partial x_l \partial x_m}\frac{\delta x_l}{\delta b_i}\frac{\delta x_m}{\delta b_j}+\frac{\partial \boldsymbol{U}}{\partial x_l}\frac{\delta^2 x_l}{\delta b_i \delta b_j}\right)dS\;. \tag{4.52}$$

As with the discrete approach, the computational cost of the continuous direct-adjoint approach is equal to 1+N+1 equivalent flow solutions.

4.7 Applications

4.7.1 Gradient and Hessian-based Inverse Design of a 2D Duct

The inverse design of a 2D duct, using the Euler equations as state equations, is first presented. The flow conditions are: $\alpha_1=0°$ and $M_2=0.4$. The upper

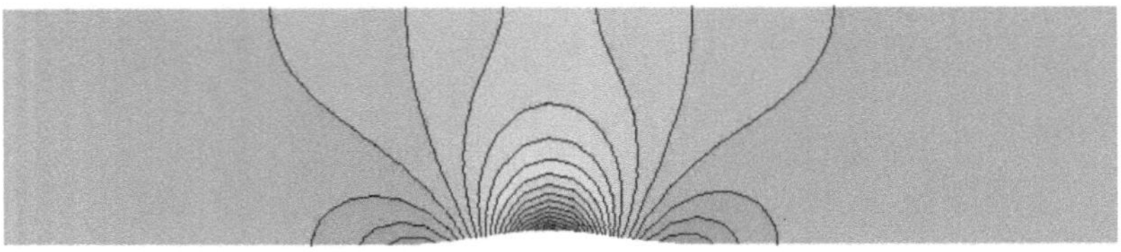

Fig. 4.1 Inverse design of a 2D duct. Mach number distribution over the duct flow domain

wall of the duct is straight and the shape of the symmetrical bump which is located in the middle of the lower side is parameterized using 5 control points. Among them, only the normal-to-the-wall components are allowed to vary. The targeted shape of the duct and the corresponding Mach number contours are illustrated in Fig. 4.1.

Both gradient and Hessian-based optimization algorithms are used and compared with each other in terms of CPU cost and quality of the predicted optimal solutions. The gradient of F with respect to the design variables is computed using the metrics-free continuous adjoint approach, the direct approach and the conventional adjoint approach. In the results section, this last term is used to denote the adjoint approach which leads to the gradient expression of Eq. (4.18), including a field integral of grid metrics and finite differences. All of them are compared in Fig. 4.2(a). From this comparison, it is clear that there is practically no difference in the accuracy of the gradient between these approaches although the CPU cost of the direct approach and finite differences is much higher (N=5 equivalent flow solutions per cycle) than that of the adjoint approaches (a single equivalent flow solution). Also, the CPU cost of the conventional adjoint approach is higher than that of the metrics-free one by the $2N$ additional calls to the grid generation software (central finite differences).

The Hessian matrix values are computed using the direct-adjoint approach and compared with finite differences, Fig. 4.2(b). Both present the same accuracy but the direct-adjoint approach is less costly. Furthermore, there is no need to make the distinction between continuous and discrete approaches since both of them lead to almost identical results.

Using the gradient computed by any of the previous methods, several gradient-based optimization algorithms are used and compared, Fig. 4.3, top-left, showing the superiority of the quasi-Newton (BFGS) algorithm over the other algorithms. At the top-right of Fig. 4.3, the BFGS algorithm is also compared with the (exact) Newton method. The Newton method which is based on the exact Hessian requires a considerably smaller number of cycles to reach the target pressure distribution, almost at machine accuracy. Either BFGS or Newton method are supported by the steepest descent algorithm which is applied during the first ten cycles. At the bottom of Fig. 4.3, BFGS and Newton methods are also compared in terms of required equivalent flow

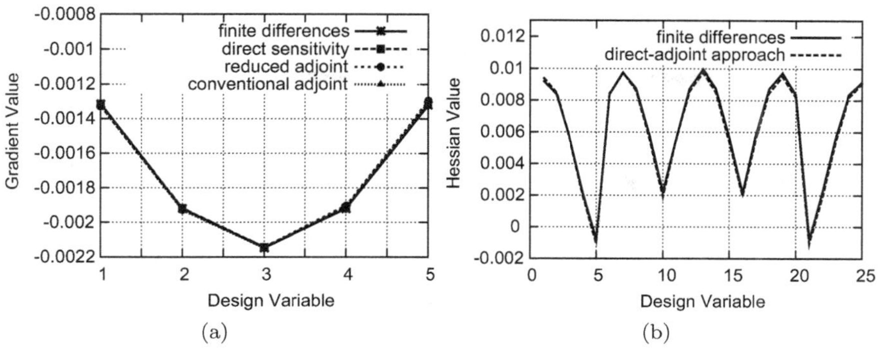

Fig. 4.2 Inverse design of a 2D duct (**a**) objective function gradient values computed using the metrics-free and the conventional adjoint approach, the direct approach and finite differences (**b**) Hessian matrix values using the direct-adjoint approach and finite differences. The first five values correspond to the first row of $\frac{d^2F}{db_i db_j}$ and so forth (5 columns $\times$ 5 rows $=$ 25 values)

solutions which are a direct measure of CPU cost. It is clear that the exact Newton approach converges faster to the optimal solution, at least with this particular number of design variables.

The change in the gradient vector values is shown in Fig. 4.4(a), in semi-log scale. Gradient values are zeroed upon convergence. In Fig. 4.4(b), it is shown that the Hessian matrix values change slightly from cycle to cycle and the greater change occurs during the first optimization cycles.

In Fig. 4.5(a), the initial, reference and optimal bump shapes are compared. The shape computed via the Newton method is shown, although there are practically no apparent differences between optimal solutions obtained by any of the two methods. The corresponding pressure distributions are also compared in the same figure, right. The slow convergence of steepest descent and conjugate gradient algorithms results (after approximately 200 cycles) into computations which were not run to convergence; so there are some discrepancies between the optimal and target distributions (not evident, in the scale of the figures shown herein).

4.7.2 Losses Minimization of a 2D Compressor Cascade

The next application example is concerned with the minimization of viscous losses of the flow in a 2D compressor cascade through the redesign of the airfoil shape. The Navier Stokes equations are solved together with the Spalart-Allmaras turbulence model for the computation of the total pressure

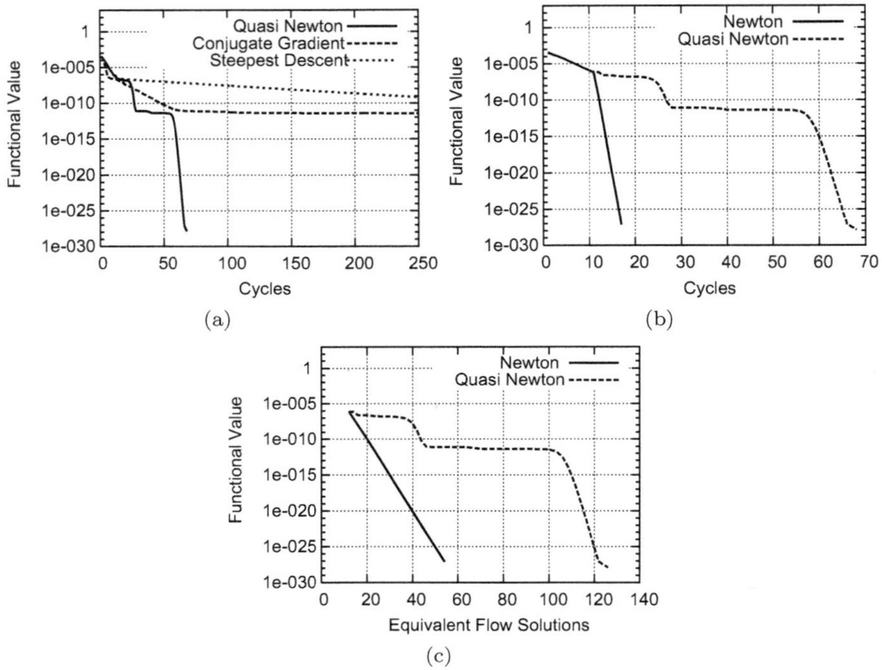

Fig. 4.3 Inverse design of a 2D duct (**a**) convergence rates of the three gradient-based algorithms (steepest descent, conjugate gradient and quasi-Newton) (**b**) convergence of Newton and quasi-Newton algorithms in terms of optimization cycles **c** convergence of Newton and quasi-Newton algorithms in terms of required equivalent flow solutions

losses used as objective function and the corresponding adjoint equations are solved for the computation of the gradient. The sensitivity of the turbulent viscosity is not taken into account in the adjoint formulation. Geometrical constraints are imposed in order to prevent the blade shape from becoming too thin. The gradient of the constraint is added to the gradient of F, using the augmented Lagrange multipliers method [10].

Each blade side is parameterized with 13 control points using Bézier-Bernstein polynomials [19]. Only the blade-to-blade coordinates of the control points are allowed to vary. The first three control points (determining the curvature of the leading edge) and the last one are kept constant on both sides. The H-type structured grid, chosen after a grid-independency study, has $400 \times 200 = 80,000$ nodes. The flow conditions are: $\alpha_1 = 44°$, $M_{2,is} = 0.45$ and the chord-based Reynolds number is $Re_c = 8 \times 10^5$. The grid together with the Mach number distribution over the optimal cascade airfoil is shown in Fig. 4.6.

The convergence history of F (total pressure losses) is shown in Fig. 4.7(a). In Fig. 4.7(b), the convergence history of the sum of violated constraints is

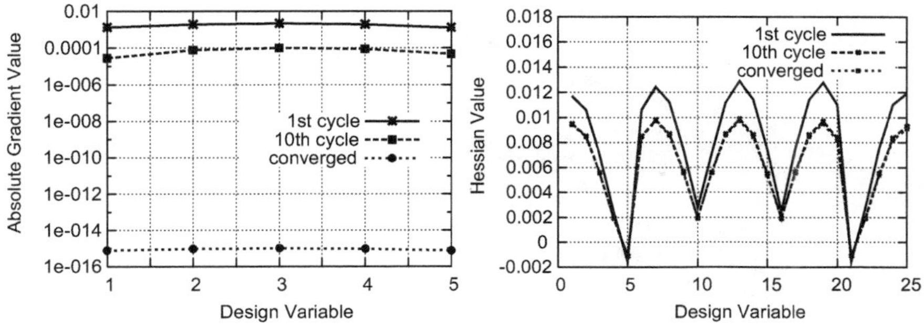

(a) (b)

Fig. 4.4 Inverse design of a 2D duct (**a**) change in the objective function gradient values (absolute values) during the optimization (**b**) corresponding change in Hessian matrix values

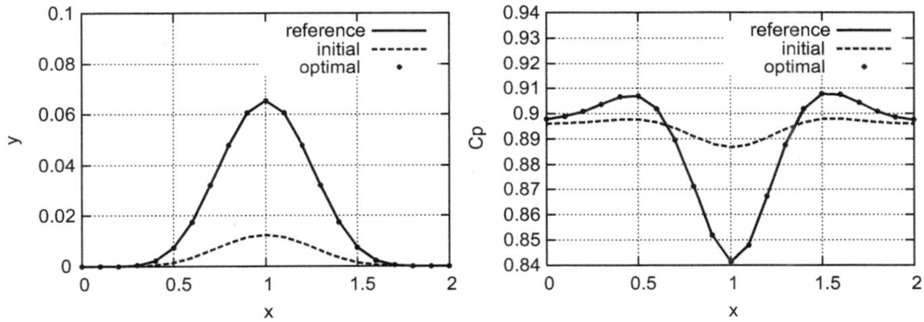

(a) (b)

Fig. 4.5 Inverse design of a 2D duct (**a**) comparison of the initial, optimal and reference airfoil contours (not in scale) (**b**) comparison of the initial, optimal and target pressure distributions

also shown. This curve quantifies the difference (absolute value) in the actual blade airfoil thickness, at specific chord-wise positions, from the corresponding minimum allowed ones. The latter are defined to be equal to the 90% of an existing (reference) airfoil. If any of these local constraints are violated, a penalty value is added to the total penalty value. So, zero penalty values correspond to a non-violated thickness constraint. During the first five cycles, losses are reduced almost by 1.5%, while none of the thickness constraints are violated. The reduction rate in total pressure losses decreased during the subsequent cycles and oscillations became apparent due to occasional constraint violations. Upon convergence (for which around 40 cycles are needed), the thickness constraints are not violated any more or are just slightly violated.

The reduction rate of the gradient values for all design variables are shown in Fig. 4.8. The objective function gradient with respect to the suction side

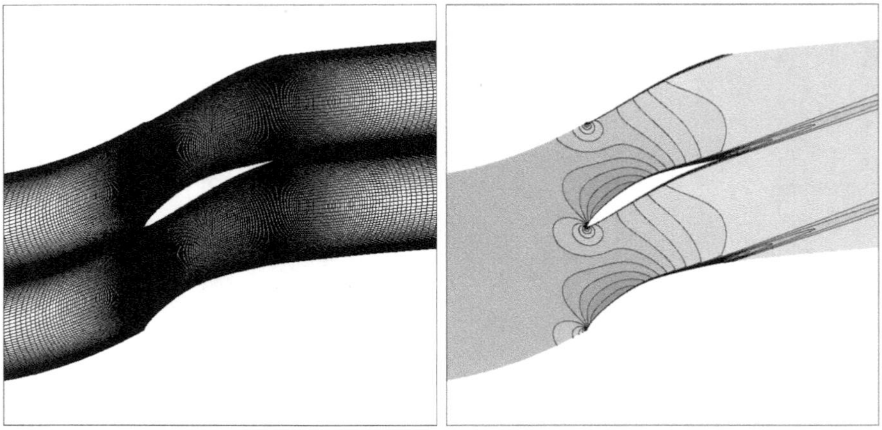

(a) (b)

Fig. 4.6 Total pressure losses minimization in a 2D compressor cascade. Structured computational grid and Mach number distribution over the optimal compressor cascade

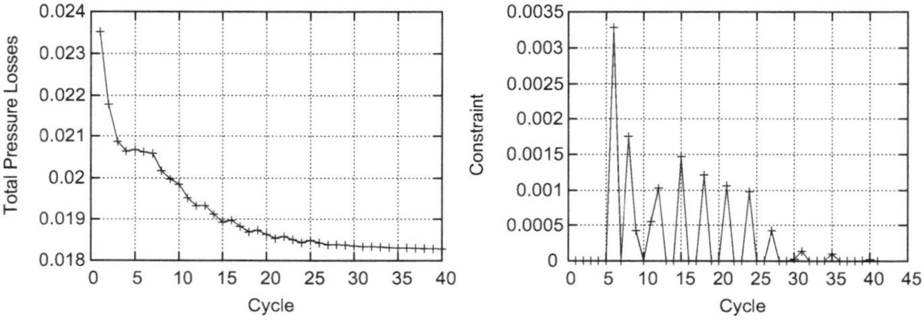

(a) (b)

Fig. 4.7 Total pressure losses minimization in a 2D compressor cascade (**a**) convergence rate of total pressure losses and (**b**) sum of violated geometrical constraints (total penalty value)

control points (first nine values) are greater than the corresponding gradient components for the pressure side points (remaining nine values). Note, also, that the gradient values at the optimal solution are not zeroed due to the thickness constraint imposition.

The initial and optimal set of design variables and the corresponding blade airfoil contours are shown in Fig. 4.9. It is evident that, due to their greater (positive) gradient values, the suction side control points move towards the pressure side and tend to reduce the blade thickness. Once the airfoil becomes too thin, penalization due to the constraint violation is activated. So the

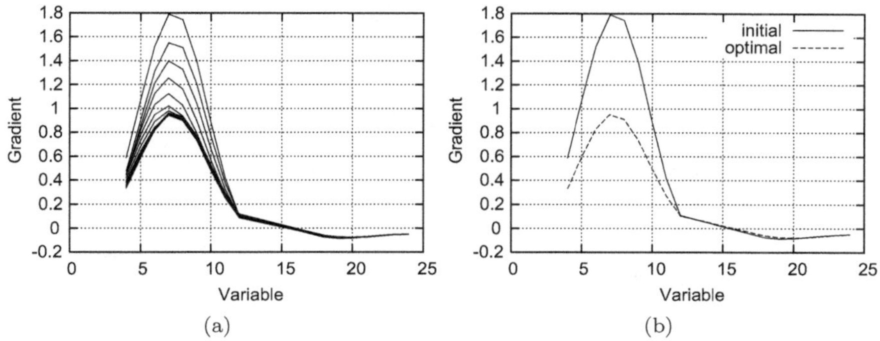

Fig. 4.8 Total pressure losses minimization in a 2D compressor cascade (**a**) convergence history of the objective function gradient values and (**b**) gradient values for the initial and optimal cascade airfoils

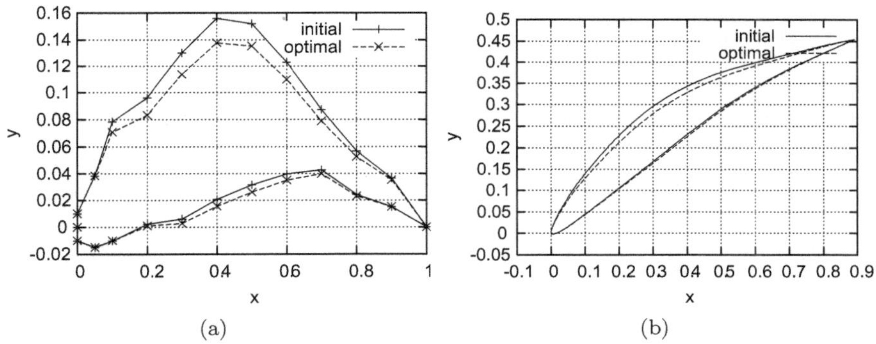

Fig. 4.9 Total pressure losses minimization in a 2D compressor cascade (**a**) initial and optimal set of Bézier control points and (**b**) the corresponding cascade airfoil contours

corrected gradient values "push" both suction and pressure side control points away from each other.

Improvements in aerodynamic characteristics are shown in Figs. 4.10 and 4.11 where the initial and optimal pressure and friction coefficients are shown. A close-up view of the flow developed close to the trailing edge over the suction side is shown. Both figures show that separation on the optimal blade airfoil is kept minimum or even disappears. The static pressure plateau and the negative friction coefficient values are significantly reduced. The improvement of separation is also shown in Fig. 4.12 where the Mach number distribution is shown for the initial and optimal configuration.

Similar results can be obtained using the entropy generation instead of the total pressure loss as objective function [52, 53]. These are omitted in the interest of space.

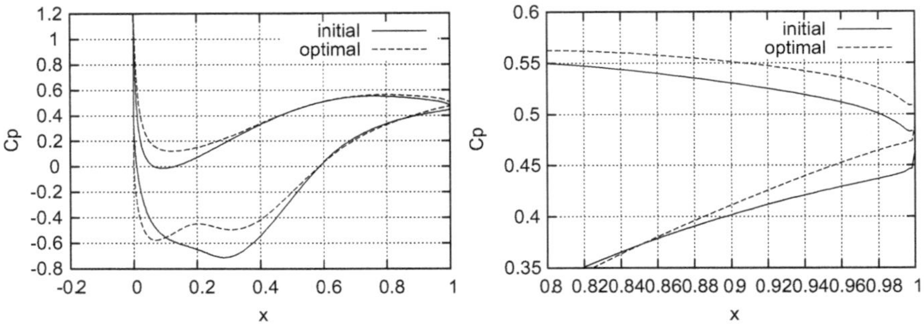

(a) (b)

Fig. 4.10 Total pressure losses minimization in a 2D compressor cascade (**a**) pressure coefficient distribution for the initial and optimal cascade airfoil and (**b**) focus on the separation region

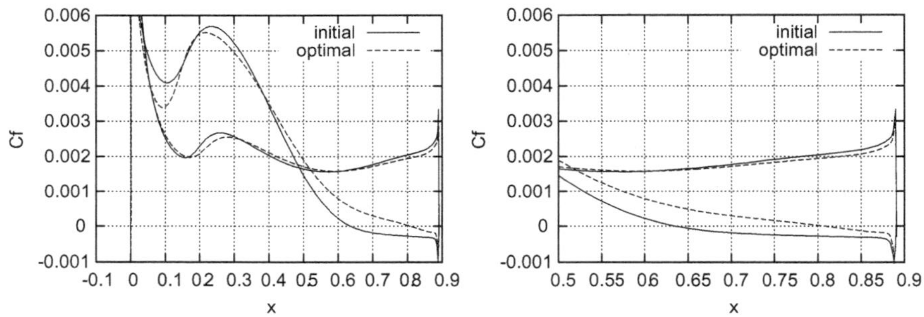

(a) (b)

Fig. 4.11 Total pressure losses minimization in a 2D compressor cascade (**a**) friction coefficient distribution for the initial and optimal cascade airfoil and (**b**) focus on the separation region

4.8 Conclusions

Discrete and continuous adjoint approaches for use in aerodynamic shape optimization problems were presented. These can be used to compute the gradient of objective functions in inviscid or viscous flows. Different objective functions (the standard one, used in inverse shape design problems, or others, which are appropriate when the target is the minimization of entropy generation or total pressure losses in internal flows), were handled. An improved formulation of the adjoint problem avoids the computation of field integrals containing metrics and other geometrical sensitivities and reduces the overall computational cost. The extension of the adjoint formulation for the computation of the Hessian matrix of a functional was then presented in both discrete and continuous forms. The application of the Newton method

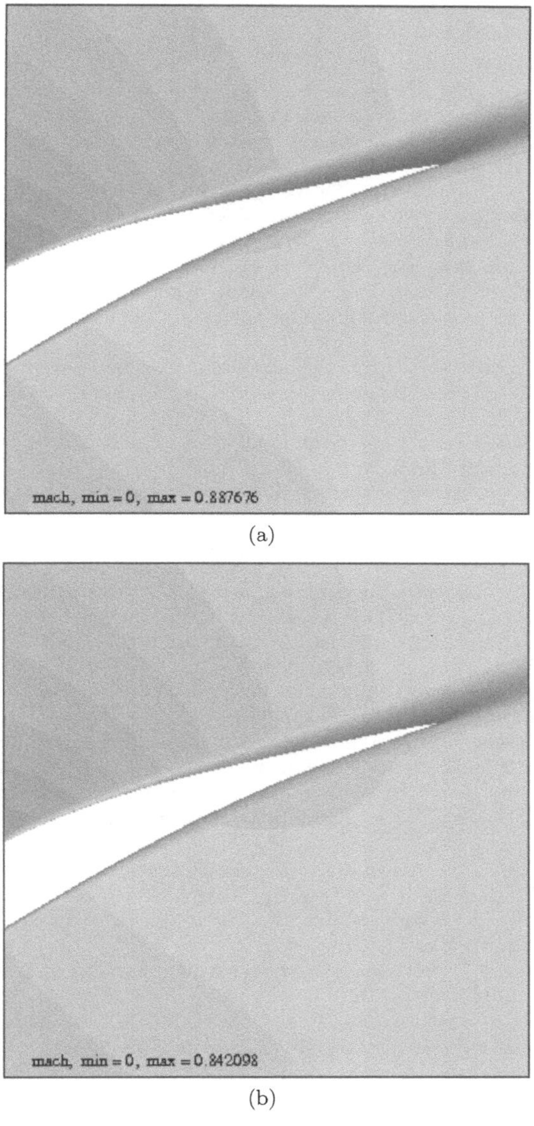

Fig. 4.12 Total pressure losses minimization in a 2D compressor cascade (**a**) Mach number distribution at the separated region for the initial and (**b**) optimal cascade

can lead to fast design optimization cycles when the exact Hessian matrix elements are known. Applications of the adjoint-based computed gradient vector and Hessian matrix were shown, demonstrating that in either case, the benefit from the adjoint approach to the optimization of aerodynamic shapes was noticeable.

References

1. Alonso, J.J., Kroo, I.M., Jameson, A.: Advanced algorithms for design and optimization of quiet supersonic platform. AIAA Paper 2002-0144 (2002)
2. Anderson, W.K., Bonhaus, D.L.: Airfoil design on unstructured grids for turbulent flows. AIAA Journal **37**(2), 185–191 (1999)
3. Anderson, W.K., Bonhaus, D.L., Daryl, L.: Aerodynamic design on unstructured grids for turbulent flows. NASA Technical Memorandum 112867 (1997)
4. Anderson, W.K., Venkatakrishnan, V.: Aerodynamic design optimization on unstructured grids with a continuous adjoint formulation. AIAA Paper 97-0643 (1997)
5. Anderson, W.K., Venkatakrishnan, V.: Aerodynamic design optimization on unstructured grids with a continuous adjoint formulation. Computers and Fluids **28**, 443–480 (1999)
6. Arian, E., Salas, M.D.: Admitting the inadmissible: Adjoint formulation for incomplete cost functionals in aerodynamic optimization. NASA/CR-97-206269, ICASE Report No.97–69 (1997)
7. Arian, E., Taasan, S.: Analysis of the Hessian for aerodynamic optimization: Inviscid flow. Computers and Fluids **28**(7), 853–877 (1999)
8. Bäck, T.: Evolutionary Algorithms in Theory and Practice. Oxford University Press (1996)
9. Baysal, O., Ghayour, K.: Continuous adjoint sensitivities for optimization with general cost functionals on unstructured meshes. AIAA Journal **39**(1), 48–55 (2001)
10. Bertsekas, D.P.: Nonlinear Programming, 2nd edn. Athena Scientific (1999)
11. Campobasso, M.S., Duta, M.C., Giles, M.B.: Adjoint calculation of sensitivities of turbomachinery objective functions. AIAA Journal of Propulsion and Power **19**(4), 693–703 (2003)
12. Dadone, A., Grossman, B.: Progressive optimization of inverse fluid dynamic design problems. Computers and Fluids **29**, 1–32 (2000)
13. Dadone, A., Grossman, B.: Fast convergence of inviscid fluid dynamic design problems. Computers and Fluids **32**, 607–627 (2003)
14. Duta, M.C., Giles, M.B., Campobasso, M.S.: The harmonic adjoint approach to unsteady turbomachinery design. International Journal for Numerical Methods in Fluids **40**(3-4), 323–332 (2002)
15. Elliot, J., Peraire, J.: Aerodynamic design using unstructured meshes. AIAA Paper 96-1941 (1996)
16. Elliot, J., Peraire, J.: Aerodynamic optimization using unstructured meshes with viscous effects. AIAA Paper 97-1849 (1997)
17. Elliot, J., Peraire, J.: Practical 3d aerodynamic design and optimization using unstructured meshes. AIAA Journal **35**(9), 1479–1485 (1997)
18. Elliot, J., Peraire, J.: Constrained, multipoint shape optimisation for complex 3d configurations. Aeronautical Journal **102**(1017), 365–376 (1998)
19. Forest, A.R.: Interactive interpolation and approximation by Bezier polynomials. The Computer Journal **15**(1), 71–79 (1972)
20. Giannakoglou, K.C.: Design of optimal aerodynamic shapes using stochastic optimization methods and computational intelligence. Progress in Aerospace Sciences **38**, 43–76 (2002)
21. Giles, M.B.: Discrete adjoint approximations with shocks. In: T. Hou, E. Tadmor (eds.) Hyperbolic Problems: Theory, Numerics, Applications. Springer-Verlag (2003)
22. Giles, M.B., Duta, M.C., Muller, J.D., Pierce, N.A.: Algorithm developments for discrete adjoint methods. AIAA Journal **41**(2), 198–205 (2003)
23. Giles, M.B., Pierce, N.A.: Adjoint equations in CFD: duality, boundary conditions and solution behaviour. AIAA Paper 97-1850 (1997)
24. Giles, M.B., Pierce, N.A.: On the properties of solutions of the adjoint Euler equations. In: 6th ICFD Conference on Numerical Methods for Fluid Dynamics. Oxford, UK (1998)

25. Harbeck, M., Jameson, A.: Exploring the limits of transonic shock-free airfoil design. AIAA Paper 2005-1041 (2005)
26. Hazra, S., Schulz, V., Brezillon, J., Gauger, N.R.: Aerodynamic shape optimization using simultaneous pseudo-timestepping. Journal of Computational Physics **204**(1), 46–64 (2005)
27. Hazra, S.B., Schulz, V.: Simultaneous pseudo-timestepping for aerodynamic shape optimization problems with state constraints. SIAM Journal of Scientific Computing **28**(3), 1078–1099 (2006)
28. Jameson, A.: Aerodynamic design via control theory. Journal of Scientific Computing **3**, 233–260 (1988)
29. Jameson, A.: Optimum aerodynamic design using CFD and control theory. AIAA Paper 95-1729 (1995)
30. Jameson, A., Alonso, J.J.: Automatic aerodynamic optimization on distributed memory architectures. AIAA Paper 96-0409 (1996)
31. Jameson, A., Kim, S.: Reduction of the adjoint gradient formula in the continous limit. AIAA Paper 2003-0040 (2003)
32. Jameson, A., Pierce, N., Martinelli, L.: Optimum aerodynamic design using the Navier-Stokes equations. Theoretical and Computational Fluid Dynamics **10**, 213–237 (1998)
33. Jameson, A., Reuther, J.: Control theory based airfoil design using the Euler equations. AIAA Paper 94-4272 (1994)
34. Jameson, A., Shankaran, S., Martinelli, L., Haimes, B.: Aerodynamic shape optimization of complete aircraft configuration. AIAA Paper 2004-0533 (2004)
35. Kim, S.K., Alonso, J.J., Jameson, A.: Two-dimensional high-lift aerodynamic optimization using the continuous adjoint method. AIAA Paper 2000-4741 (2000)
36. Kim, S.K., Alonso, J.J., Jameson, A.: Design optimization of high-lift configurations using a viscous continuous adjoint method. AIAA Paper 2002-0844 (2002)
37. Kuruvila, G., Taasan, S., Salas, M.D.: Airfoil design and optimization by the one-shot method. AIAA Paper 95-0478 (1995)
38. Leoviriyakit, K., Jameson, A.: Aero-structural wing planform optimization. AIAA Paper 2004-0029 (2004)
39. Leoviriyakit, K., Jameson, A.: Multi-point wing planform optimization via control theory. AIAA Paper 2005-0450 (2005)
40. Leoviriyakit, K., Kim, S., Jameson, A.: Aero-structural wing planform optimization using the Navier-Stokes equations. AIAA Paper 2004-4479 (2004)
41. Mavriplis, D.J.: A discrete adjoint-based approach for optimization problems on three-dimensional unstructured meshes. AIAA Paper 2006-50 (2006)
42. Mohammadi, B., Pironneau, O.: Shape optimization in fluid mechanics. Annual Review of Fluid Mechanics **36**, 255–279 (2004)
43. Nadarajah, S., Jameson, A.: A comparison of the continuous and discrete adjoint approach to automatic aerodynamic optimization. AIAA Paper 2000-0667 (2000)
44. Nadarajah, S., Jameson, A.: Studies of the continuous and discrete adjoint approaches to viscous automatic aerodynamic shape optimization. AIAA Paper 2001-2530 (2001)
45. Nadarajah, S., Jameson, A.: Optimum shape design for unsteady three-dimensional viscous flows using a non-linear frequency domain method. AIAA Paper 2002-2838 (2002)
46. Nadarajah, S., Jameson, A., Alonso, J.J.: An adjoint method for the calculation of remote sensitivities in supersonic flow. AIAA Paper 2002-0261 (2002)
47. Nadarajah, S., McMullen, M., Jameson, A.: Non-linear frequency domain based optimum shape design for unsteady three-dimensional flow. AIAA Paper 2002-2838 (2002)
48. Newman, J.C., Anderson, W.K., Whitfield, D.L.: Multidisciplinary sensitivity derivatives using complex variables. Tech. Rep. MSSU-COE-ERC-98-08, Mississippi State University (1998)
49. Nielsen, E.J., Anderson, W.K.: Recent improvements in aerodynamic design optimization on unstructured meshes. AIAA Paper 2001-0596 (2001)

50. Nielsen, E.J., Anderson, W.K.: Recent improvements in aerodynamic design optimization on unstructured meshes. AIAA Journal **40**(6), 1155–1163 (2002)
51. Nielsen, E.J., Park, M.A.: Using an adjoint approach to eliminate mesh sensitivities in computational design. AIAA Paper 2005-0491 (2005)
52. Papadimitriou, D.I., Giannakoglou, K.C.: Compressor blade optimization using a continuous adjoint formulation. In: ASME Turbo Expo, GT2006/90466 (2006)
53. Papadimitriou, D.I., Giannakoglou, K.C.: A continuous adjoint method with objective function derivatives based on boundary integrals for inviscid and viscous flows. Computers and Fluids **36**, 325–341 (2007)
54. Papadimitriou, D.I., Giannakoglou, K.C.: Total pressure losses minimization in turbomachinery cascades, using a new continuous adjoint formulation. Journal of Power and Energy: Special Issue on Turbomachinery **221**(6), 865–872 (2007)
55. Pierce, N.A., Giles, M.B.: An introduction to the adjoint approach to design. Flow, Turbulence and Combustion **65**(3-4), 393–415 (2000)
56. Pierce, N.A., Giles, M.B.: Analytic adjoint solutions for the quasi-one-dimensional Euler equations. Journal of Fluid Mechanics **426**, 327–345 (2001)
57. Pironneau, O.: Optimal shape design for elliptic systems. Springer-Verlag, New York (1984)
58. Rall, L.B.: Automatic Differentiation: Techniques and Applications. Springer-Verlag (1981)
59. Reuther, J., Alonso, J.J., Rimlinger, M.J., Jameson, A.: Aerodynamic shape optimization of supersonic aircraft configurations via an adjoint formulation on distributed memory parallel computers. Computers and Fluids **28**, 675–700 (1999)
60. Soto, O., Lohner, R.: On the computation of flow sensitivities from boundary integrals. AIAA Paper 2004-0112 (2004)
61. Taasan, S., Kuruvila, G., Salas, M.D.: Aerodynamic design and optimization in one-shot. AIAA Paper 91-005 (1992)
62. Tortorelli, D., Michaleris, P.: Design sensitivity analysis: Overview and review. Inverse Problems in Engineering **1**, 71–105 (1994)

Part II
Specific Applications of CFD-based Optimization to Engineering Problems

Chapter 5
Efficient Deterministic Approaches for Aerodynamic Shape Optimization

Nicolas R. Gauger

Abstract Because detailed aerodynamic shape optimizations still suffer from high computational costs, efficient optimization strategies are required. Regarding the deterministic optimization methods, the adjoint approach is seen as a promising alternative to the classical finite difference approach. With the adjoint approach, the sensitivities needed for the aerodynamic shape optimization can be efficiently obtained using the adjoint flow equations. Here, one is independent of the number of design variables with respect to the numerical costs for determining the sensitivities. Another advantage of the adjoint approach is that one obtains accurate sensitivities and gets rid of the laborious tuning of the denominator step sizes for the finite differences.

Differentiation between continuous and discrete adjoint approaches is noted. In the continuous case, one formulates the optimality condition first, then derives the adjoint problem and finally does the discretization of the so obtained adjoint flow equations. In the discrete case, one takes the discretized flow equations for the derivation of the discrete adjoint problem. This can be automated by so-called algorithmic differentiation (AD) tools.

The different adjoint approaches will be explained for single disciplinary aerodynamic shape optimization first and then their extension to multidisciplinary design optimization (MDO) problems will be discussed for aero-structure cases. Finally, we will discuss the so-called one-shot methods. Here, one breaks open the simulation loop for optimization.

Nicolas R. Gauger
German Aerospace Center (DLR), Institute of Aerodynamics and Flow Technology, Lilienthalplatz 7, 38108 Braunschweig, Germany
together with
Humboldt University Berlin, Department of Mathematics,
Unter den Linden 6, 10099 Berlin, Germany
(e-mail: Nicolas.Gauger@dlr.de)

Nomenclature

$(x, y) \in \mathrm{IR}^2$	cartesian coordinates	M_∞	Mach number
$(\xi, \eta) \in [0, 1]^2$	body fitted coordinates	$)_\infty$	... at free stream
$D \subset \mathrm{IR}^2$	flow field domain	γ	ratio of specific heats
$\partial D = B \cup C$	flow field boundary	C_{ref}	cord length
$B = \{(\xi, 1)\}$	farfield	C_p	pressure coefficient
$C = \{(\xi, 0)\}$	solid wall	C_D	drag coefficient
$\boldsymbol{n} = \begin{pmatrix} n_x \\ n_y \end{pmatrix} \perp D$	outward pointing normal unit vector	C_L	lift coefficient
α	angle of attack	C_m	pitching moment coefficient
ρ	density	(x_m, y_m)	pitching moment's reference point
$\boldsymbol{v} = \begin{pmatrix} u \\ v \end{pmatrix}$	velocity	I	cost function
p	pressure	$-d(I)$	adjoint boundary condition's RHS on C
E	specific total energy	$X \in \mathrm{IR}^n$	vector of design variables
H	total enthalpy	Z	displacement field

5.1 Introduction

In aerodynamic shape optimization, the task of computing sensitivities is essential for the application of gradient-based optimization strategies. Gradient computations for a given cost function $I(X)$, for a design vector X out of a defined design space, can generally be done with several methods.

One way is the finite difference method (FD), which approximates the components of the gradient by difference quotients of the cost function evaluated for the initial aerodynamic shape as well as the shapes generated by perturbations of the design variables, for a given step size in the design space. Hence, the computational effort for the gradient approximation using finite differences is proportional to the number of design variables. Therefore, problems with this method occur if the computation of the cost function is extremely expensive or if there are many design variables. Again, the finite differences are just approximations and therefore one has to take care of the accuracy. As we will see in Sect. 5.4, step size tuning requires a lot of effort as well.

An alternative are the so-called adjoint methods that include two kinds of approach: continuous and discrete adjoint approaches.

In the continuous case one formulates the optimality condition first, then derives the adjoint problem and finally does the discretization of the so obtained adjoint flow equations. The continuous adjoint method was first used in aerodynamic shape optimization by Jameson in his works on potential equations [19] and later also for the Euler [21] as well as Navier-Stokes [20] equations. Its main advantage over the finite differences is the significant increase in speed since the corresponding numerical effort is independent of the number of design variables. However, the implementation of the continuous adjoint approach may be time-consuming and error-prone.

On the other hand, in the discrete case, one takes the discretized flow equations for the derivation of the discrete adjoint problem (see e.g., [3, 12]). This can be automated by so-called algorithmic, or automatic, differentiation (AD) tools.

AD is a comparatively new field of mathematical sciences. This technique is based on the observation that various elemental operations (like $+, -, \times$) build up the cost function as their concatenation. Therefore, applying the chain rule to this concatenation results in an automated differentiation of the cost function. Depending on the starting point of the differentiation process – either at the beginning or at the end of the respective chain of concatenations – one distinguishes between the *forward* mode and the *reverse* mode of AD. Using the reverse mode of AD, gradients can be computed very accurately at a computational cost that is independent of the number of design variables. This is the reason why this method is also called a discrete adjoint method.

At DLR, a few attempts have been made in AD computations. One attempt is the differentiation of the DLR TAUij code by ADOL-C [14], which is presented in this chapter as part of a differentiated optimization chain.

In the next sections, the different adjoint approaches will be explained for single disciplinary aerodynamic shape optimization first and then their extension to multidisciplinary design optimization (MDO) problems will be discussed for aero-structure cases.

Finally, we will discuss the so-called one-shot methods. Here one breaks open the simulation loop for optimization.

To start out, we explain how one can parameterize aerodynamic shapes by the use of deformation techniques.

5.2 Parameterization by Deformation

In aerodynamic shape optimization, a geometry is either given by a parameterization or can be changed by parameterized deformation. This means that based on these parameters, a shape can be built up or deformed by a design vector. Furthermore, the obtained shape has some aerodynamic properties like the drag coefficient or pressure distribution. Therefore, the task of the aerodynamic shape optimization is to optimize this design vector and its dependent shape for some aerodynamic cost function.

When optimizing, there must be some chain to calculate the cost function value at a given parameterization. This can be done by deforming a static

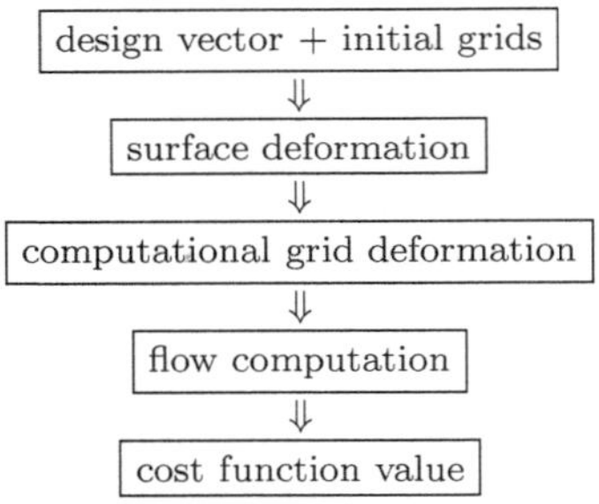

Fig. 5.1 Cost function computation

initial shape or surface mesh and its dependent computational grid based on the parameterization and afterwards evaluating the cost function. A schema of the procedure is illustrated in Fig. 5.1.

5.2.1 Surface Deformation

The basic idea for deforming the surface of an airfoil is to compute functions and then add their values to the upper and lower side of the surface. Therefore, every design parameter is used to scale a specific function which is afterwards added to the shape.

Several kinds of functions are considered for the deformation. The first are the Hicks-Henne functions which are defined as

$$h_{a,b} : [0,1] \to [0,1] : h_{a,b}(x) = \left(\sin(\pi x^{\frac{\log 0,5}{\log a}})\right)^b .$$

These functions are positive, defined and mapped in the interval $[0,1]$ where their peak is at position a. Furthermore, they are analytically smooth at zero and one.

The used parameterization operates with Hicks-Henne functions with a fixed b of 3.0 and a varies from $\frac{3}{n+5}$ to $\frac{n+3}{n+5}$ where n is the number of design parameters.

The second kind of functions considered is transformed cosine functions. These cosine functions are defined for $q \in [0,1]$ as $c_q(x) : [0,1] \to [0,1]$ where

$$c_q(x) = \begin{Bmatrix} \frac{1}{2}(1-\cos(\frac{x}{q}\pi)) & \text{for} & x \leq 2q \\ 0 & \text{for} & x > 2q \end{Bmatrix} \quad \text{for} \quad q \leq \frac{1}{2}$$

and

$$c_q(x) = \begin{Bmatrix} 0 & \text{for} & x < 2q-1 \\ \frac{1}{2}(1-\cos(\frac{x-2q+1}{1-q}\pi)) & \text{for} & x \geq 2q-1 \end{Bmatrix} \quad \text{for} \quad q > \frac{1}{2} .$$

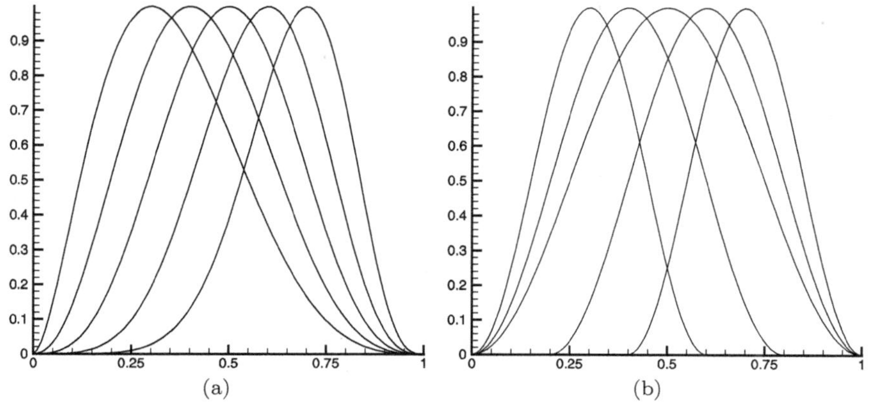

Fig. 5.2 Hicks-Henne functions for (**a**) $b = 3$ and $a = 0.3 - 0.7$ and (**b**) transformed cosine functions for $q = 0.3 - 0.7$ (right)

These functions have the positive property that their impact on the surface deformation is local because their support is $2\min(q, 1-q)$. The used parameterization operates with these functions where q varies from $\frac{3}{n+5}$ to $\frac{n+3}{n+5}$.

In principle, one can use each kind of functions for surface deformations. The extension to 3D is straightforward by adding, e.g., 3D spline functions to the surface.

5.2.2 Grid Deformation

After having deformed the surface, there is a need to deform the computational grid as well. This deformation should be related to the changes of the surface. Within the following work, this is done via the volume spline method by Hounjet et al. (see [18]). This method is a general interpolation approach for n interpolation points (x_i, y_i, z_i) and their values $f_i (1 \leq i \leq n)$ which is given by

$$f(x,y,z) = \alpha_1 + \alpha_2 x + \alpha_3 y + \alpha_4 z + \sum_{i=1}^{n} \beta_i \sqrt{(x-x_i)^2 + (y-y_i)^2 + (z-z_i)^2}. \tag{5.1}$$

The coefficients α_i and β_i can be determined by the condition that the interpolation f should be exact at its n interpolation points

$$f(x_i, y_i, z_i) = f_i \quad (1 \leq i \leq n)$$

and the four additional conditions

$$\sum_{i=1}^{n} \beta_i = 0$$
$$\sum_{i=1}^{n} \beta_i \, x_i = 0$$
$$\sum_{i=1}^{n} \beta_i \, y_i = 0$$
$$\sum_{i=1}^{n} \beta_i \, z_i = 0$$

which can be physically interpreted as equilibrium equations.

This results in solving the following linear system of equations

$$\begin{pmatrix} 0 \\ 0 \\ 0 \\ 0 \\ f_1 \\ f_2 \\ \vdots \\ f_n \end{pmatrix} = \begin{pmatrix} 0 & 0 & 0 & 0 & 1 & 1 & \dots & 1 \\ 0 & 0 & 0 & 0 & x_1 & x_2 & \dots & x_n \\ 0 & 0 & 0 & 0 & y_1 & y_2 & \dots & y_n \\ 0 & 0 & 0 & 0 & z_1 & z_2 & \dots & z_n \\ 1 & x_1 & y_1 & z_1 & 0 & \epsilon_{12} & \dots & \epsilon_{1n} \\ 1 & x_2 & y_2 & z_2 & \epsilon_{21} & 0 & \dots & \epsilon_{2n} \\ \vdots & \vdots & \vdots & \vdots & \vdots & \vdots & \dots & \vdots \\ 1 & x_n & y_n & z_n & \epsilon_{n2} & \epsilon_{n2} & \dots & 0 \end{pmatrix} \cdot \begin{pmatrix} \alpha_1 \\ \alpha_2 \\ \alpha_3 \\ \alpha_4 \\ \beta_1 \\ \beta_2 \\ \vdots \\ \beta_n \end{pmatrix}$$

where $\epsilon_{ij} = \sqrt{(x_i - x_j)^2 + (y_i - y_j)^2 + (z_i - z_j)^2}$ is the Euclidean distance between the interpolation points (x_i, y_i, z_i) and (x_j, y_j, z_j).

After solving this system of equations the interpolation is ready to be used with the given formula (5.1) for arbitrary points (x, y, z).

This general interpolation method is now applied to the differences of the original and deformed surface dx, dy, dz. These functions are each interpolated with the differences of the surfaces as interpolation points. Afterwards, dx, dy, dz are applied to the computational grid and therefore yield a grid deformation.

Therefore, let $(x_{\text{old},i}, y_{\text{old},i}, z_{\text{old},i})$ be the old and $(x_{\text{new},i}, y_{\text{new},i}, z_{\text{new},i})$ be the new surface points $(1 \le i \le n)$. Then the functions dx, dy, dz can be interpolated with the interpolation point values

$$dx_i = x_{\text{new},i} - x_{\text{old},i} \ ,$$
$$dy_i = y_{\text{new},i} - y_{\text{old},i} \ ,$$
$$dz_i = z_{\text{new},i} - z_{\text{old},i}$$

at $(x_{\text{old},i}, y_{\text{old},i}, z_{\text{old},i})$. These obtained functions dx, dy, dz can then be computed at arbitrary points. Now let $(a_{\text{old},j}, b_{\text{old},j}, c_{\text{old},j})$ be the old computational grid points and $(a_{\text{new},j}, b_{\text{new},j}, c_{\text{new},j})$ their corresponding new points $(1 \leq j \leq m)$. Finally, the grid deformation is given by

$$\begin{aligned} a_{\text{new},j} &= a_{\text{old},j} + dx(a_{\text{old},j}, b_{\text{old},j}, c_{\text{old},j}) \ , \\ b_{\text{new},j} &= b_{\text{old},j} + dy(a_{\text{old},j}, b_{\text{old},j}, c_{\text{old},j}) \ , \\ c_{\text{new},j} &= c_{\text{old},j} + dz(a_{\text{old},j}, b_{\text{old},j}, c_{\text{old},j}) \ . \end{aligned}$$

Instead of deforming the computational grid there is also the possibility to generate a new grid at each optimization step. But this adds costs to the computation overhead because grid generation is expensive. Therefore, the present work uses the above explained volume spline interpolation method to deform the grid and save computation time.

5.3 Sensitivity-based Aerodynamic Shape Optimization

For convenience reasons, the following analysis is restricted to the 2D Euler equations. Let $X \in \mathbb{R}^n$ denote the vector of design variables. Then X determines the airfoil $C(X)$ and its physics $w(X)$, where $w = \begin{pmatrix} \rho \\ \rho u \\ \rho v \\ \rho E \end{pmatrix}$ is the vector of the conserved variables. w is assumed to be the solution of the quasi-unsteady Euler equations

$$\frac{\partial w}{\partial t} + \frac{\partial f}{\partial x} + \frac{\partial g}{\partial y} = 0 \qquad \text{in } D \tag{5.2}$$

where $\boldsymbol{n}^\top \boldsymbol{v} = 0$ on $C = C(X)$, with $f = \begin{pmatrix} \rho u \\ \rho u^2 + p \\ \rho uv \\ \rho uH \end{pmatrix}$ and $g = \begin{pmatrix} \rho v \\ \rho vu \\ \rho v^2 + p \\ \rho vH \end{pmatrix}$.
On the far-field, free stream conditions are assumed. For a perfect gas

$$p = (\gamma - 1)\rho(E - \frac{1}{2}(u^2 + v^2)) \tag{5.3}$$

holds for the pressure, and finally C_p, C_D, C_L and C_m are defined as

$$C_p := \frac{2(p - p_\infty)}{\gamma M_\infty^2 p_\infty} \ , \tag{5.4}$$

$$C_D := \frac{1}{C_{\text{ref}}} \int_C C_p(n_x \cos\alpha + n_y \sin\alpha) dl \ , \tag{5.5}$$

$$C_L := \frac{1}{C_{\text{ref}}} \int_C C_p(n_y \cos\alpha - n_x \sin\alpha) dl \;, \tag{5.6}$$

$$C_m := \frac{1}{C_{\text{ref}}^2} \int_C C_p(n_y(x - x_m) - n_x(y - y_m)) \; dl \;. \tag{5.7}$$

If the geometry is now perturbed from $C(X)$ to $C(X + \delta X)$, then via the solution of

$$\begin{aligned} & \frac{\partial(w+\delta w)}{\partial t} + \frac{\partial(f+\delta f)}{\partial x} + \frac{\partial(g+\delta g)}{\partial y} = 0 \\ \Leftrightarrow \quad & \frac{\partial(\delta w)}{\partial t} + \frac{\partial(\delta f)}{\partial x} + \frac{\partial(\delta g)}{\partial y} = 0 \qquad \text{in } D \end{aligned} \tag{5.8}$$

where

$$\boldsymbol{n}^\top \boldsymbol{v} = 0 \;\text{ on }\; C = C(X + \delta X) \;, \tag{5.9}$$

the associated variation of pressure is as follows

$$\delta C_p = \frac{2\delta p}{\gamma M_\infty^2 p_\infty} \approx \frac{2(p(X+\delta X) - p(X))}{\gamma M_\infty^2 p_\infty} \;. \tag{5.10}$$

Finally via

$$\delta n_x \approx n_x(X + \delta X) - n_x(X) \tag{5.11}$$

and

$$\delta n_y \approx n_y(X + \delta X) - n_y(X) \tag{5.12}$$

the variations of C_D, C_L and C_m are obtained as

$$\begin{aligned} \delta C_D = & \frac{2}{\gamma M_\infty^2 p_\infty C_{\text{ref}}} \int_C \delta p(n_x \cos\alpha + n_y \sin\alpha) \; dl \\ & + \frac{1}{C_{\text{ref}}} \int_C C_p(\delta n_x \cos\alpha + \delta n_y \sin\alpha) \; dl \;, \end{aligned} \tag{5.13}$$

$$\begin{aligned} \delta C_L = & \frac{2}{\gamma M_\infty^2 p_\infty C_{\text{ref}}} \int_C \delta p(n_y \cos\alpha - n_x \sin\alpha) \; dl \\ & + \frac{1}{C_{\text{ref}}} \int_C C_p(\delta n_y \cos\alpha - \delta n_x \sin\alpha) \; dl \;, \end{aligned} \tag{5.14}$$

$$\begin{aligned} \delta C_m = & \frac{2}{\gamma M_\infty^2 p_\infty C_{\text{ref}}^2} \int_C \delta p(n_y(x - x_m) - n_x(y - y_m)) \; dl \\ & + \frac{1}{C_{\text{ref}}^2} \int_C C_p \delta(n_y(x - x_m) - n_x(y - y_m)) \; dl \;. \end{aligned} \tag{5.15}$$

Proceeding as described above for the n perturbations $\delta_i X$ in each of the n components of the design vector X, the gradient of the cost function I (e.g., drag, lift or pitching moment coefficients) is obtained as $\nabla_X I = (\delta_i I)_{i=1,\dots,n}$ after $n+1$ flow calculations.

The easiest gradient-based optimization strategy is the steepest descent method. There, a recursive line search in the direction $-\nabla_{X^{(k)}} I$, starting from the point $X^{(k)}$, leads to an optimal geometry

$$X^{(k+1)} = X^{(k)} - \varepsilon^{(k)} \nabla_{X^{(k)}} I \tag{5.16}$$

with respect to the cost function I in that direction. This is repeated until the norm of the gradient of the cost function becomes zero.

In addition to the gradient of the cost function, one can determine gradients of constraints in the same way, e.g., for the task of drag reduction by constant lift. Furthermore, one can make use of second order sensitivity informations, or at least their approximations, in order to speed up the optimization process. Often, the so-called constrained SQP methods are used with the BFGS-updates (BFGS - named after Broyden, Fletcher, Goldfarb and Shanno). The abbreviation SQP stands for sequentially quadratic programming (quadratic - second order sensitivities). A good survey of several possible deterministic optimization strategies is provided in the book by Nocedal and Wright [25].

But one can see that the numerical costs for the determination of the gradient of the cost function or constraints are directly proportional to the number of design variables. This finite difference or brute force approach becomes more and more inefficient as the number of design variables increases.

5.4 Sensitivity Computations

In aerodynamic shape optimization, the task of computing sensitivities is very important in order to have the possibility to use gradient based optimization strategies. Gradient computations for a given cost function $I(X)$ for a design vector X out of a defined design space can generally be done with several methods.

5.4.1 Finite Difference Method

The first is the finite difference (FD) method which approximates the gradient as follows

$$\frac{\partial I}{\partial x_i}(X) \approx \frac{I(X + he_i) - I(X)}{h} \quad (1 \le i \le n) \tag{5.17}$$

where n is the number of design parameters, e_i the ith unit vector, and h is the scalar step size. Problems with this method occur if the computation of the cost function is extremely expensive. As can be seen in the approximation (5.17), this cost function has to be calculated once at point X and further n times at $(X + he_i)$ for $1 \leq i \leq n$. This results in $(n+1)$ cost function evaluations which can take a long time. Another problem may occur if the step size h is not accurately chosen. This is based on the fact that

$$I(X + he_i) = I(X) + \frac{\partial I}{\partial x_i}(X)h + \frac{\partial^2 I}{\partial^2 x_i}(X)h^2 + O(h^3)$$

which can be transformed into

$$\frac{\partial I}{\partial x_i}(X) = \frac{I(X + he_i) - I(X)}{h} - \frac{\partial^2 I}{\partial^2 x_i}(X)h + O(h^2)\ . \tag{5.18}$$

If the chosen h is too large, the first order term on the right side of Eq. (5.18) would have a large impact on the quality of the approximation in (5.17). This means on the one hand that the h chosen has to be relatively small. On the other hand, the h chosen can not be arbitrarily small because of numerical stability. This is based on the division in the approximation term (5.17) which will be error intensive if h is too small and therefore results in numerical noise. Therefore, the step length h has to be manually tuned with respect to the cost function, parameterization and used geometry.

5.4.2 Continuous Adjoint Formulation

The second method to compute sensitivities is the continuous adjoint approach. Again, just for convenience reasons, the following analysis is restricted to the 2D Euler equations. In order to determine the gradient of the cost function independently of the design variables with respect to the numerical costs, one can use the following continuous adjoint formulation. Let $\psi = \begin{pmatrix} \psi_1 \\ \psi_2 \\ \psi_3 \\ \psi_4 \end{pmatrix}$ denote the vector of the adjoint variables. Instead of solving $n+1$ times the quasi-unsteady Euler equations to get the gradient, the Euler equations are solved just once in order to get the transposed Jacobians $\left(\frac{\partial f}{\partial w}\right)^\top$, $\left(\frac{\partial g}{\partial w}\right)^\top$ and then the quasi-unsteady continuous adjoint Euler equations

$$-\frac{\partial \psi}{\partial t} - \left(\frac{\partial f}{\partial w}\right)^\top \frac{\partial \psi}{\partial x} - \left(\frac{\partial g}{\partial w}\right)^\top \frac{\partial \psi}{\partial y} = 0 \qquad \text{in } D \tag{5.19}$$

where

$$n_x \psi_2 + n_y \psi_3 = -d(I) \quad \text{on} \quad C = C(X) \tag{5.20}$$

and

$$\delta x_\xi, \ldots, \delta y_\eta = 0, \;\; \delta w = 0 \;\; \text{on} \;\; B = B(X) \tag{5.21}$$

are also solved just once.

The right hand side $-d(I)$ of the wall boundary condition of the quasi-unsteady adjoint Euler equations is dependent on the cost function I. The adjoint far-field boundary condition just states that the geometrical position of the far-field is fixed and free stream conditions apply there.

Finally, the components of the gradient $\nabla_X I = (\delta_i I)_{i=1,\ldots,n}$ can now be determined via an integration just over the adjoint solution and the metric sensitivities $\delta x_\xi, \ldots, \delta y_\eta$ and

$$\begin{aligned}\delta I = &- \int_C p(-\psi_2 \delta y_\xi + \psi_3 \delta x_\xi)\, dl + K(I) \\ &- \int_D \psi_\xi^\top (\delta y_\eta f - \delta x_\eta g) + \psi_\eta^\top (-\delta y_\xi f + \delta x_\xi g)\, dA\end{aligned} \tag{5.22}$$

is obtained where $K(I)$ is again a term dependent on the cost function I.

For the gradient of the drag, the following right hand side adjoint boundary on C is used

$$d(C_D) = \frac{2}{\gamma M_\infty^2 p_\infty C_{\text{ref}}} (n_x \cos\alpha + n_y \sin\alpha) \tag{5.23}$$

and to get the corresponding gradient, $K(I)$ is

$$K(C_D) = \frac{1}{C_{\text{ref}}} \int_C C_p (\delta n_x \cos\alpha + \delta n_y \sin\alpha)\, dl \tag{5.24}$$

for the gradient of the lift

$$d(C_L) = \frac{2}{\gamma M_\infty^2 p_\infty C_{\text{ref}}} (n_y \cos\alpha - n_x \sin\alpha) \tag{5.25}$$

and

$$K(C_L) = \frac{1}{C_{\text{ref}}} \int_C C_p (\delta n_y \cos\alpha - \delta n_x \sin\alpha)\, dl \tag{5.26}$$

are used, and for the gradient of the pitching moment

$$d(C_m) = \frac{2}{\gamma M_\infty^2 p_\infty C_{\text{ref}}^2} (n_y(x - x_m) - n_x(y - y_m)) \tag{5.27}$$

and

$$K(C_m) = \frac{1}{C_{\text{ref}}^2} \int_C C_p \delta(n_y(x - x_m) - n_x(y - y_m))\, dl \tag{5.28}$$

are used. For more details see [7] or [8].

Table 5.1 An evaluation trace of the simple example

$v_{-1} = x_1$	$= 1.5$		
$v_0 = x_2$	$= 0.5$		
$v_1 = v_{-1}/v_0$	$= 1.5/0.5$	$= 3.0000$
$v_2 = \sin(v_1)$	$= \sin(3.0)$	$= 0.1411$
$v_3 = v_1 + v_2$	$= 3.0 + 0.1411$	$= 3.1411$
$y = v_3$	$= 3.1411$	

5.4.3 Algorithmic Differentiation (AD)

A third method is a discrete approach which is usually called algorithmic or automatic differentiation and depends on a specific implementation of the cost function. This implementation consists of various elemental operations (like $+, -, \times$) which build up the cost function as their concatenation. Therefore, applying the chain rule to this concatenation results in a differentiation of the cost function (after having dealt with possible inconsistencies and other problems which are beyond the scope of this chapter, see [13] for detailed information).

To give the reader a feeling of the principal ideas of AD, we will introduce the two basic concepts with the help of a simple example. Let f be a cost function which depends on two input parameters x_1 and x_2 which is given by

$$f(x_1, x_2) = \sin(x_1/x_2) + x_1/x_2 \ .$$

We now wish to compute the value of $y = f(1.5, 0.5)$ and its derivative by AD. Then a possible evaluation trace is given in Table 5.1.

The first possibility to apply the chain rule is to differentiate every single operation in the order of the evaluation trace. Let us suppose we want to differentiate the output variable y with respect to x_1. Then we associate with every variable v_i of the evaluation trace another variable $\dot{v}_i = \partial v_i/\partial x_1$. Applying the chain rule to each line in the evaluation trace, in order, leads to a numeric value of $\dot{y}$ which is the wanted sensitivity of y with respect to x_1. Clearly, $\dot{v}_{-1} = \partial v_{-1}/\partial x_1 = 1.0$ and $\dot{v}_0 = \partial v_0/\partial x_1 = 0.0$. Augmenting the evaluation trace of Table 5.1 gives the derived trace in Table 5.2. The total floating-point operation count of the added lines to evaluate $\partial y/\partial x_1$ is a small multiple of that for the underlying code to evaluate y.

Exactly the same code can be used to evaluate $\partial y/\partial x_2$ as well; the only change is to set $\dot{x}_1 = 0.0$ and $\dot{x}_2 = 1.0$ at the beginning.

The second possibility is to apply the chain rule in reverse order and is hence called the reverse mode. This concept can be seen as a discrete adjoint approach. Therefore, we associate for every v_i another variable $\overline{v}_i = \partial y/\partial v_i$ called the adjoint variable. By definition $\overline{y} = 1.0$ and since the only ways in which v_1 can affect y are via the definitions $v_2 = \sin(v_1)$ and $v_3 = v_1 + v_2$

Table 5.2 Forward differentiated evaluation trace

$v_{-1} = x_1$	$= 1.5$	
$\dot{v}_{-1} = \dot{x}_1$	$= 1.0$	
$v_0 = x_2$	$= 0.5$	
$\dot{v}_0 = \dot{x}_2$	$= 0.0$	
$v_1 = v_{-1}/v_0$	$= 1.5/0.5$	$= 3.0000$
$\dot{v}_1 = \dot{v}_{-1}/v_0 - v_{-1}\dot{v}_0/v_0/v_0$		
$= (\dot{v}_{-1} - v_1\dot{v}_0)/v_0$	$= (1.0 - 3.0 \times 0.0)/0.5$	$= 2.0000$
$v_2 = \sin(v_1)$	$= \sin(3.0)$	$= 0.1411$
$\dot{v}_2 = \cos(v_1)\dot{v}_1$	$= (-0.99) \times 2.0$	$= -1.9800$
$v_3 = v_1 + v_2$	$= 3.0 + 0.1411$	$= 3.1411$
$\dot{v}_3 = \dot{v}_1 + \dot{v}_2$	$= 2.0 - 1.98$	$= 0.0200$
$y = v_3$	$= 3.1411$	
$\dot{y} = \dot{v_3}$	$= 0.0200$	

it is

$$\overline{v}_1 = \frac{\partial y}{\partial v_1} = \frac{\partial y}{\partial v_3}\frac{\partial v_3}{\partial v_1} \tag{5.29}$$

$$= \overline{v}_3 \frac{\partial(v_1 + v_2)}{\partial v_1}$$

$$= \overline{v}_3 \frac{\partial v_1}{\partial v_1} + \overline{v}_3 \frac{\partial v_2}{\partial v_1} \tag{5.30}$$

$$= \overline{v}_3 (1 + \frac{\partial(\sin v_1)}{\partial v_1})$$

$$= \overline{v}_3 + \overline{v}_3 \cos(v_1) \ . \tag{5.31}$$

This can also be evaluated by the iterative equations

$$\overline{v}_1 = \overline{v}_3$$
$$\overline{v}_2 = \overline{v}_3$$
$$\overline{v}_1 = \overline{v}_1 + \overline{v}_2 \cos(v_1) \ .$$

Thus applying the chain rule to every line in the evaluation trace of Table 5.1, we obtain the reverse differentiated code in Table 5.3. Note that the adjoint statements are lined up vertically underneath the original statements that spawned them.

Note also that line 7 of Table 5.3 belongs to the expansion of Eq. (5.29), lines 8 and 9 to the expansion of Eq. (5.30) and line 10 to the expansion of Eq. (5.31).

As with the forward propagation method, the floating-point operation count of the added lines is a small multiple of that for the underlying code to evaluate y. But this time the complete gradient has been computed.

The main advantage of the AD method over the above two mentioned methods is that gradients can be computed with the best possible accuracy. In forward mode the gradient computation speed is dependent on the number of design variables which can be expensive if the evaluation trace for the cost function is long. In reverse mode the gradient computation is independent on any input and therefore is very efficient when numerous design variables are needed.

There are mainly two possible implementations of AD methods. The first is the so-called source to source which means that the primal evaluation trace is transferred into a differentiated evaluation trace. The second is based on operator overloading which only yields an executable.

However, the problem in reverse mode for both implementation strategies is that primal computation informations have to be recomputed and/or stored to compute backwards again. A fair amount of memory is thus essential for the chosen approach.

For more information on multiple output and input variables including vector valued functions, please refer to [13].

5.5 Adjoint Flow Solvers

5.5.1 Continuous Adjoint Flow Solvers

Within the MEGAFLOW project [22], an adjoint solver following the continuous adjoint formulation has been developed and widely validated for the block-structured flow solver FLOWer [7, 8]. The adjoint solver, which was implemented by hand, can deal with the boundary conditions for drag, lift

Table 5.3 Reverse differentiated evaluation trace

1	$v_{-1} = x_1 = 1.5$
2	$v_0 = x_2 = 0.5$
3	$v_1 = v_{-1}/v_0 = 1.5/0.5 = 3.0$
4	$v_2 = \sin(v_1) = \sin(3.0) = 0.1411$
5	$v_3 = v_1 + v_2 = 3.0 + 0.1411 = 3.1411$
6	$y = v_3 = 3.1411$
7	$\overline{v}_3 = \overline{y} = 1.0$
8	$\overline{v}_1 = \overline{v}_3 = 1.0$
9	$\overline{v}_2 = \overline{v}_3 = 1.0$
10	$\overline{v}_1 = \overline{v}_1 + \overline{v}_2 \cos(v_1) = 1.0 + 1.0 \cos(3.0) = 0.01$
11	$\overline{v}_0 = -\overline{v}_1 v_1/v_0 = -0.01 \times 3.0/0.5 = -0.06$
12	$\overline{v}_{-1} = \overline{v}_1/v_0 = 0.01/0.5 = 0.02$
13	$\overline{x}_2 = \overline{v}_0 = -0.06$
14	$\overline{x}_1 = \overline{v}_{-1} = 0.02$

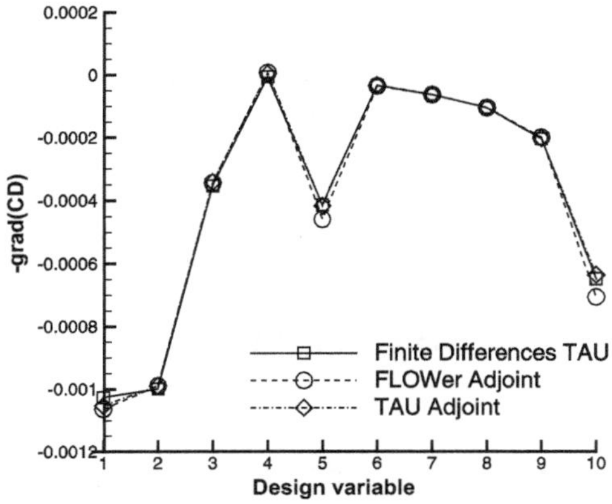

Fig. 5.3 Gradient of the drag computed for 20 B-spline variables by finite differences with TAU and by the continuous adjoint approach with FLOWer and TAU (RAE 2822, $M_\infty = 0.73$ and $\alpha = 2.0°$)

and pitching-moment sensitivities. The adjoint option of the FLOWer code has been validated for several 2D, as well as 3D optimization problems [2, 9] controlled by the (adjoint) Euler equations. Within MEGADESIGN the robustness and efficiency of the adjoint solver will be improved, especially for the Navier-Stokes equations. In case of Navier-Stokes applications, currently the turbulence model is frozen in the adjoint mode. It has been planned that AD be used to create adjoint turbulence models, which will then be linked to the hand coded adjoint solver.

Furthermore, it has been planned that the adjoint solver implemented in FLOWer, be transferred to the unstructured Navier-Stokes solver TAU. Here, the implementation work is already completed and validated for the inviscid adjoint solver [28] (see also Fig. 5.3).

5.5.2 Discrete Adjoint Flow Solvers

In addition to the continuous one, a discrete adjoint flow solver has been developed by hand within the unstructured Navier-Stokes solver TAU [1, 3]. The implementation consists of the explicit construction of the exact Jacobian of the spatial discretization with respect to the unknown variables allowing the adjoint equations to be formulated and solved. Different spatial discretizations available in TAU have been differentiated, including the

Spalart-Allmaras-Edwards one-equation, and the Wilcox k–ω two-equation turbulence models.

For both solvers, FLOWer as well as TAU, first activities are launched for the automated generation of discrete adjoint solvers by the use of AD tools. For the FLOWer code, the AD tool TAF [11] is used while ADOL-C [14] is used for the TAU code.

5.6 Automatic Differentiation Applied to an Entire Design Chain

For the optimizations presented in this section, we used the following tools covering the four steps of the design chain:

For the surface deformation, the tool defgeo has been used. This tool has been implemented in order to compute deformations based on Hicks-Henne as well as cosine functions.

The grid deformation within our optimization chain is done by a tool named meshdefo. This tool uses a public domain linear equations solver to compute the above mentioned coefficients of the interpolation.

DLR flow solver TAUij is used to compute the flow around the deformed airfoil. TAUij is a quasi 2D version of TAUijk [17], which again is based on a cell centered developer version of the DLR TAU code [29]. TAUij solves the quasi 2D Euler equations. For the spatial discretization, the MAPS+ [26] scheme is used. To achieve second order accuracy, gradients are used to reconstruct the values of variables at the cell faces. A slip wall and a far-field boundary condition are applied. For time integration, a Runge-Kutta scheme is used. To accelerate the convergence, local time stepping, explicit residual smoothing and a multigrid method are used. The code TAUij is written in C and comprises approximately 6,000 lines of code distributed over several files.

To compute the difference vectors of the original to the transformed shape geometry, another program named difgeo had to be implemented.

The chain to compute the cost function value is illustrated in Fig. 5.4. This entire optimization chain has been differentiated by the use of ADOL-C, which operates in reverse (adjoint) mode [10, 27]. This differentiated chain can be written as

$$\frac{\partial \mathrm{C_D}}{\partial \mathrm{X}} = \frac{\partial \mathrm{C_D}}{\partial \mathrm{m}} \cdot \frac{\partial \mathrm{m}}{\partial \mathrm{dx}} \cdot \frac{\partial \mathrm{dx}}{\partial \mathrm{x}} \cdot \frac{\partial \mathrm{x}}{\partial \mathrm{X}} \,.$$

Note that the first term on the right side corresponds to the differentiation of TAUij, the second term to the differentiation of meshdefo, the third term to the differentiation of difgeo and the last term to the differentiation of defgeo. Since difgeo computes only the differences $dx = x - x_s$ and x_s is a static initial surface, its corresponding factor becomes the unit matrix and therefore

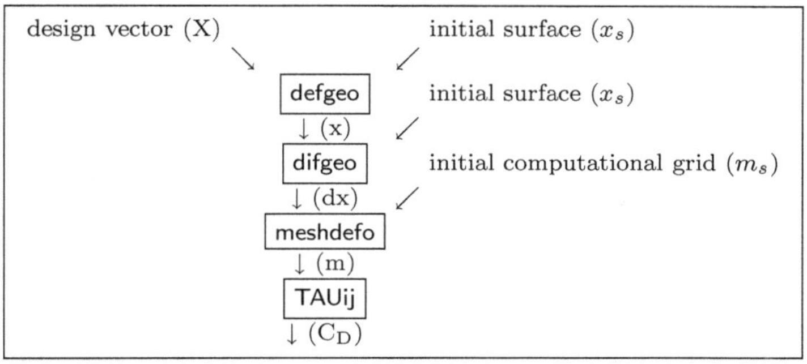

Fig. 5.4 Chain to compute the cost function value

$$\frac{\partial C_D}{\partial X} = \frac{\partial C_D}{\partial m} \cdot \frac{\partial m}{\partial dx} \cdot \frac{\partial x}{\partial X} \,. \tag{5.32}$$

The optimization strategy in the following computations is a steepest descent method which was implemented as an optimizer into the optimization framework Synaps Pointer Pro. This framework has the possibility to read in user-defined gradients. Therefore, the gradients are calculated by separate routines and are then submitted to the optimizer.

5.6.1 Test Case Definition

As test case for the validation and application of AD generated adjoint sensitivity calculations an RAE 2822 airfoil is chosen with a Mach number of 0.73 and an angle of attack of 2°. The drag coefficient for this test case has been optimized with both parameterizations, Hicks-Henne and cosine function parameterizations (see Sect. 5.2.1). In both optimizations, 20 design parameters have been used. The computational grid has 161 × 33 grid points.

5.6.2 Finite Differences

To compute the finite differences in order to have a validation framework for the AD sensitivities, the first task was to tune the stepsize h for the approximation

$$\frac{\partial I}{\partial x_i}(X) \approx \frac{I(X + he_i) - I(X)}{h} \quad (1 \le i \le n) \tag{5.33}$$

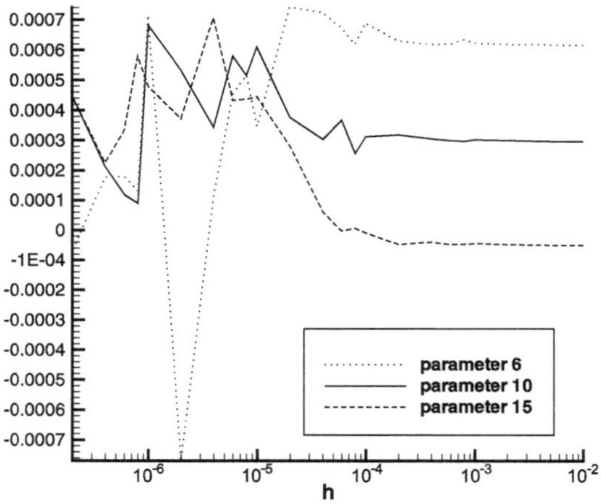

(a)

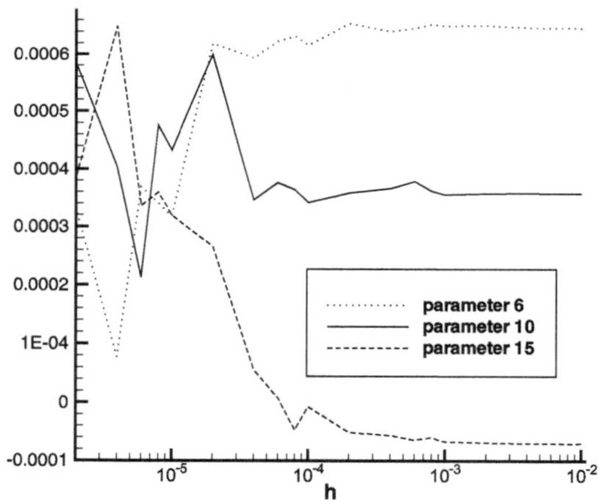

(b)

Fig. 5.5 Quotients of (**a**) Hicks-Henne and (**b**) cosine functions parameterization for parameters 6, 10 and 15 and varying stepsizes

as mentioned in Sect. 5.4. Therefore, the quotients of both parameterizations have been calculated for varying stepsizes with respect to all $n = 20$ parameters.

As can be seen in Fig. 5.5, a good choice for the stepsize h is 10^{-3} for both parameterizations. Another possibility which has not been used is to

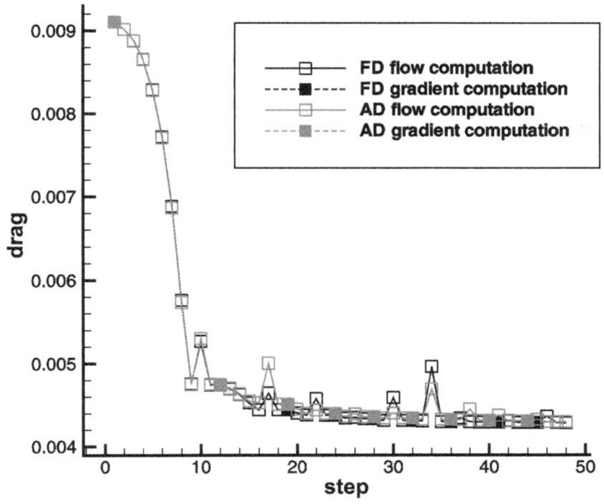

(a)

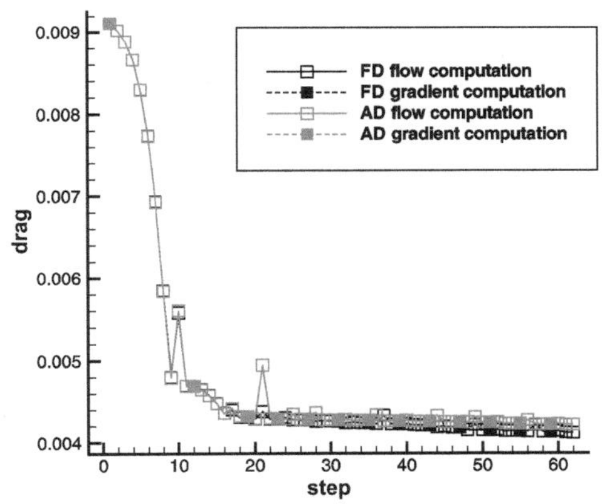

(b)

Fig. 5.6 Optimization history of FD and AD for (**a**) Hicks-Henne and (**b**) cosine functions parameterization

tune the stepsize for each parameter separately, which means selecting n stepsizes h_i for every approximation in (5.33). With this possibility, a more accurate result might be achieved for the original airfoil, but based on the fact that this tuning cannot be done for every optimization step due to the high computational effort, it might cause worse optimization results at the end. Therefore, a stepsize of 10^{-3} has been used for all gradient computations within the optimizations for both parameterizations.

In Fig. 5.6, the optimization history for both parameterizations can be seen. In case of the Hicks-Henne functions parameterization, the optimization

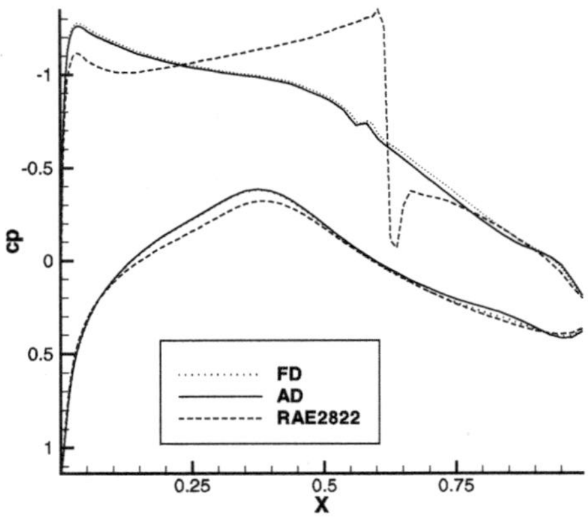

(a)

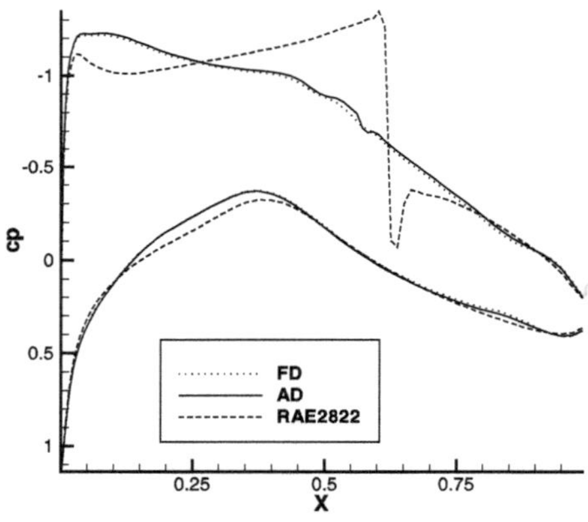

(b)

Fig. 5.7 Pressure distribution of the original test case and the optimum of FD and AD for (**a**) Hicks-Henne and (**b**) cosine functions parameterization

converges after nine gradient computations which are marked by a filled out square. The optimization with cosine functions converges after 13 gradient computations. The pressure distribution for both optimizations is drawn in Fig. 5.7 and the optimal geometries can be found in Fig. 5.8.

As one can see, the strong shock of the original baseline geometry nearly vanished in both cases and one ends up with more than 50% decrease in drag. In contrast to the optimum with the help of Hicks-Henne functions, the optimum with cosine functions parameterization shows slight oscillations in

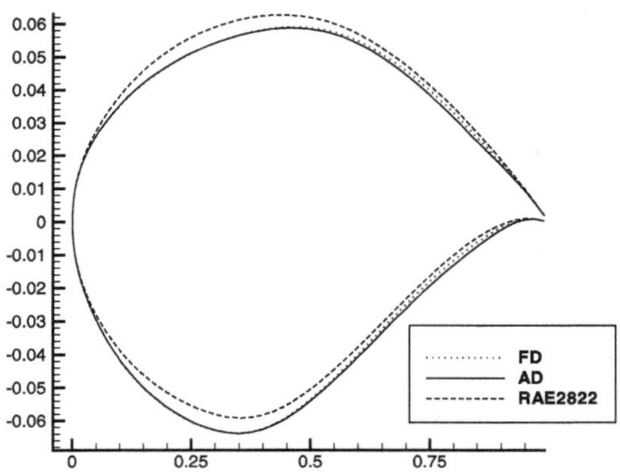

(a)

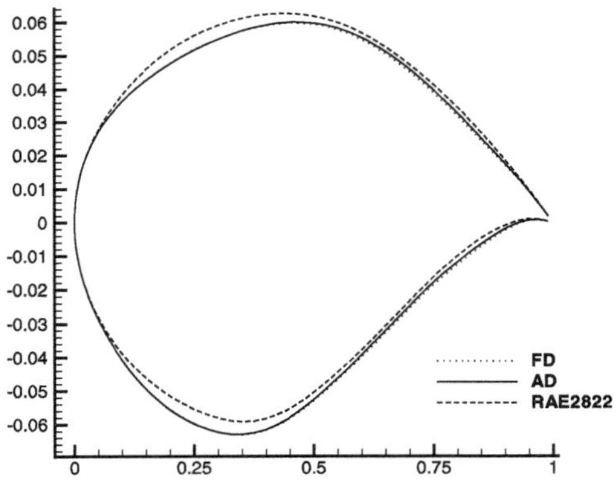

(b)

Fig. 5.8 Surface geometry of the original test case and the optimum of FD and AD for (**a**) Hicks-Henne and (**b**) cosine functions parameterization

the pressure distribution, which is due to the fact that cosine functions only have local impacts on the surface deformation whereas Hicks-Henne functions have global impacts.

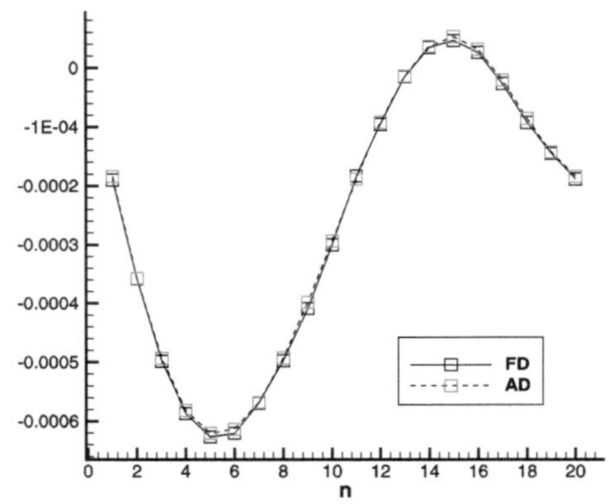

(a)

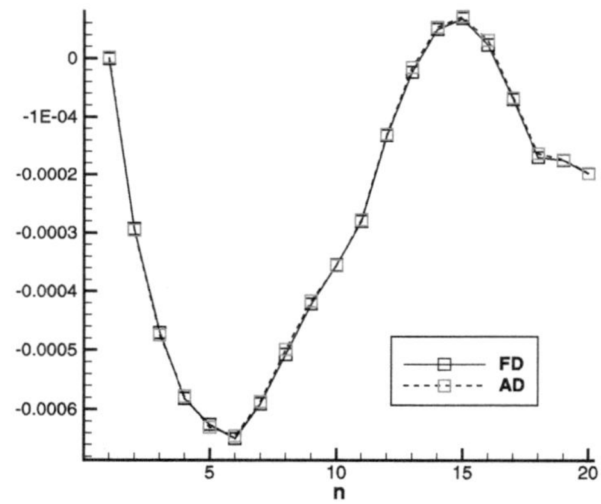

(b)

Fig. 5.9 Comparison between FD and AD gradients on RAE 2822 with (**a**) Hicks-Henne and (**b**) cosine functions parameterization

5.6.3 Automatic Differentiation

In Fig. 5.9, one can see the comparisons between the FD and AD gradients for the original RAE 2822 airfoil with Hicks-Henne and cosine functions parameterization. This validates the AD approach and shows also that the chosen stepsize for the FD gradient is accurate enough.

As with the finite difference method the optimizations have also been done with the AD approach. The optimization histories, the pressure distributions and the surface geometries of the optimum with AD are also shown in Figs. 5.6, 5.7 and 5.8, respectively. This clearly validates the AD approach.

5.7 Adjoint Approch for Aero-Structure Coupling

The use of successively performed single disciplinary optimizations in case of a multidisciplinary optimization problem is not only inefficient but in some cases has been shown to lead to faulty, non-optimal designs [24]. Although multidisciplinary optimization is possible with classical approaches for sensitivity evaluation by means of finite differences, this method is extremely expensive in terms of calculation time, requiring the reiterated solution of the coupled problem for every design variable.

A new approach that allows the evaluation of the gradient with low computational cost takes advantage of the adjoint formulation of the multidisciplinary optimization problem [23, 24]. Therefore, the FLOWer adjoint option has been coupled with the structure solver MSC Nastran for an efficient coupled aero-structure adjoint solver. The implementation and validation of this approach are described in detail in [4, 5, 6].

5.7.1 Adjoint Formulation for Aero-Structure Coupling

The derivation of the adjoint equations in case of a multidisciplinary problem is similar to what has been carried out for the pure aerodynamic case, with the difference that we will end up with a dual adjoint variable for each set of state variables of the problem. An adjoint formulation is possible for any problem involving the calculation of the gradient of a function of one or more sets of variables obeying one or more constraint equations. We will restrict ourselves to the case of two sets: one that represents the flow variables, the other representing the structure nodal displacement. As already seen $I(X, w, Z)$ denotes the cost function of the optimization problem, dependent now also on the displacement field Z, which is the solution of the structural problem. Then, the gradient takes the form

$$\frac{dI}{dX} = \frac{\partial I}{\partial X} + \frac{\partial I}{\partial w}\frac{\partial w}{\partial X} + \frac{\partial I}{\partial Z}\frac{\partial Z}{\partial X} \tag{5.34}$$

or, in terms of differentials

$$\delta I = \frac{\partial I}{\partial X}\delta X + \frac{\partial I}{\partial w}\delta w + \frac{\partial I}{\partial Z}\delta Z\ . \tag{5.35}$$

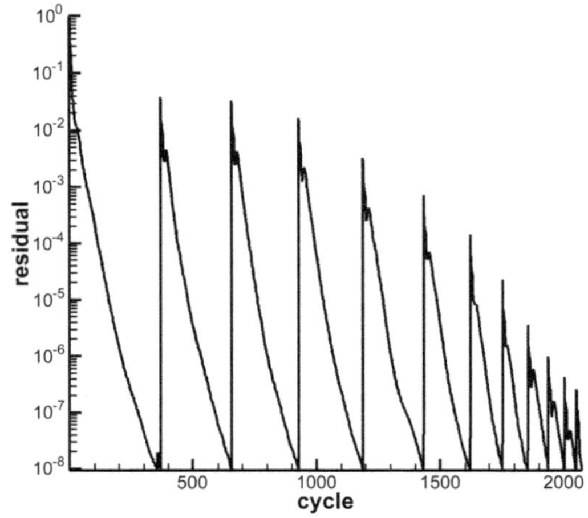

Fig. 5.10 Plot of residual (log scale) of flow equation during coupled computation (multigrid is used): AMP wing, $M_\infty = 0.78$, $\alpha = 2.83°$, 2-block structured grid of about 140,000 nodes each

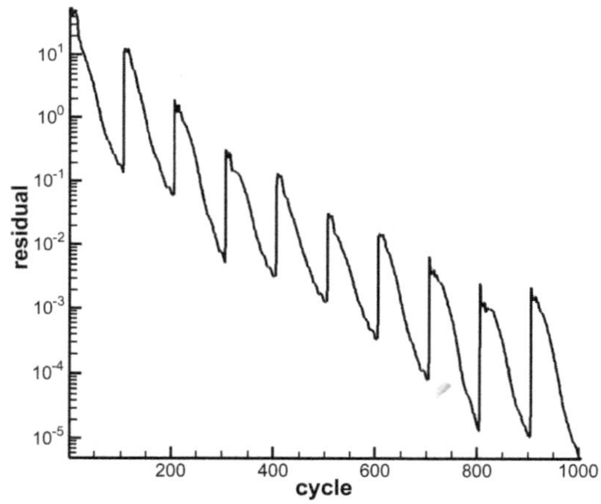

Fig. 5.11 Plot of residual (log scale) of adjoint flow equation during coupled computation: AMP wing, $M_\infty = 0.78$, $\alpha = 2.83°$, 2-block structured grid of about 140,000 nodes each. Each 100 iterations, the boundary conditions of the adjoint flow solver are updated

The fields (w, Z) are the solution of the system of partial differential equations

$$R(X, w, Z) = 0 \tag{5.36}$$

$$S(X, w, Z) = 0 \tag{5.37}$$

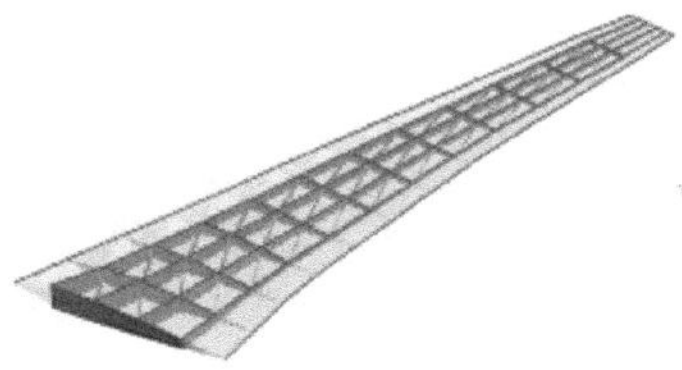

Fig. 5.12 Wing structure model

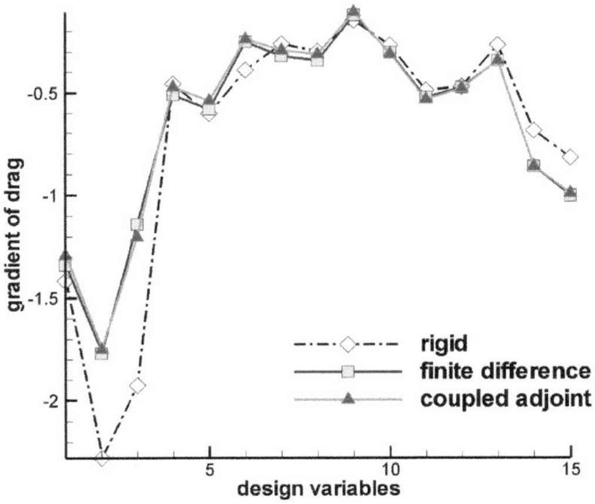

Fig. 5.13 Validation of the aero-structural coupled adjoint with finite differences (AMP wing, $M_\infty = 0.78$ and $\alpha = 2.83°$)

being (5.36) the flow and (5.37) the structural equations. We take the first variation of the PDEs. This yields

$$\delta R = \frac{\partial R}{\partial X}\delta X + \frac{\partial R}{\partial w}\delta w + \frac{\partial R}{\partial Z}\delta Z = 0 \, , \tag{5.38}$$

$$\delta S = \frac{\partial S}{\partial X}\delta X + \frac{\partial S}{\partial w}\delta w + \frac{\partial S}{\partial Z}\delta Z = 0 \, . \tag{5.39}$$

We multiply Eqs. (5.38) and (5.39) with the Lagrange multipliers ψ and ϕ respectively and add the result to the expression for the differential increment of I in terms of the differentials of the independent set (X, w, Z), obtaining

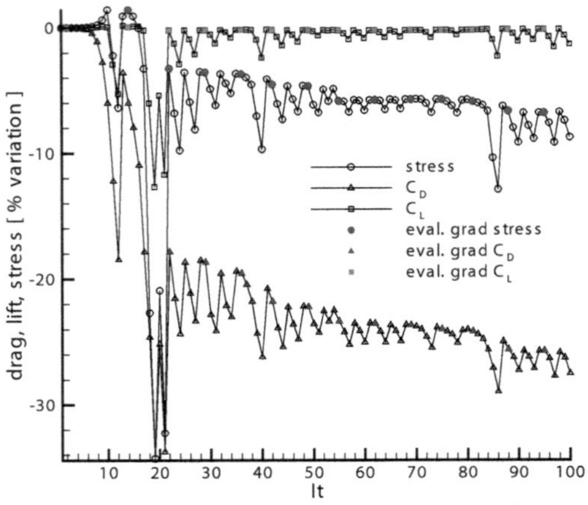

(a)

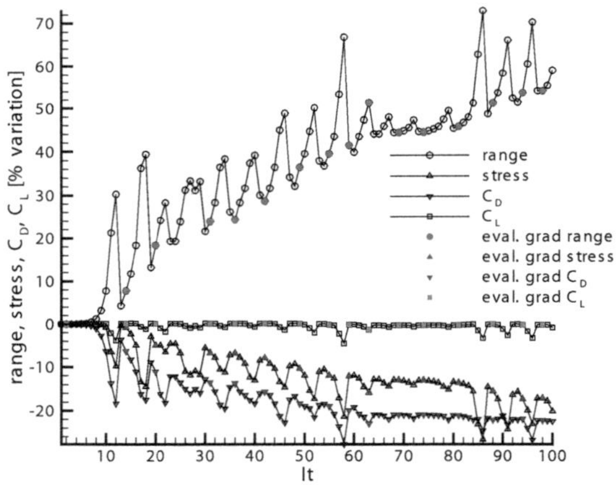

(b)

Fig. 5.14 Optimization history (**a**) for the drag reduction by constant lift while taking into account the static deformation and (**b**) for the range maximization (bottom picture) of the AMP wing ($M_\infty = 0.78$, $\alpha = 2.83°$). For both optimizations free-form deformation is considered, 240 design variables are used for parameterization and the optimization strategy relies on feasible directions

$$\delta I = \left(\frac{\partial I}{\partial X} + \psi^T \frac{\partial R}{\partial X} + \phi^T \frac{\partial S}{\partial X}\right) \delta X + \left(\frac{\partial I}{\partial w} + \psi^T \frac{\partial R}{\partial w} + \phi^T \frac{\partial S}{\partial w}\right) \delta w + \left(\frac{\partial I}{\partial Z} + \psi^T \frac{\partial R}{\partial Z} + \phi^T \frac{\partial S}{\partial Z}\right) \delta Z \quad (5.40)$$

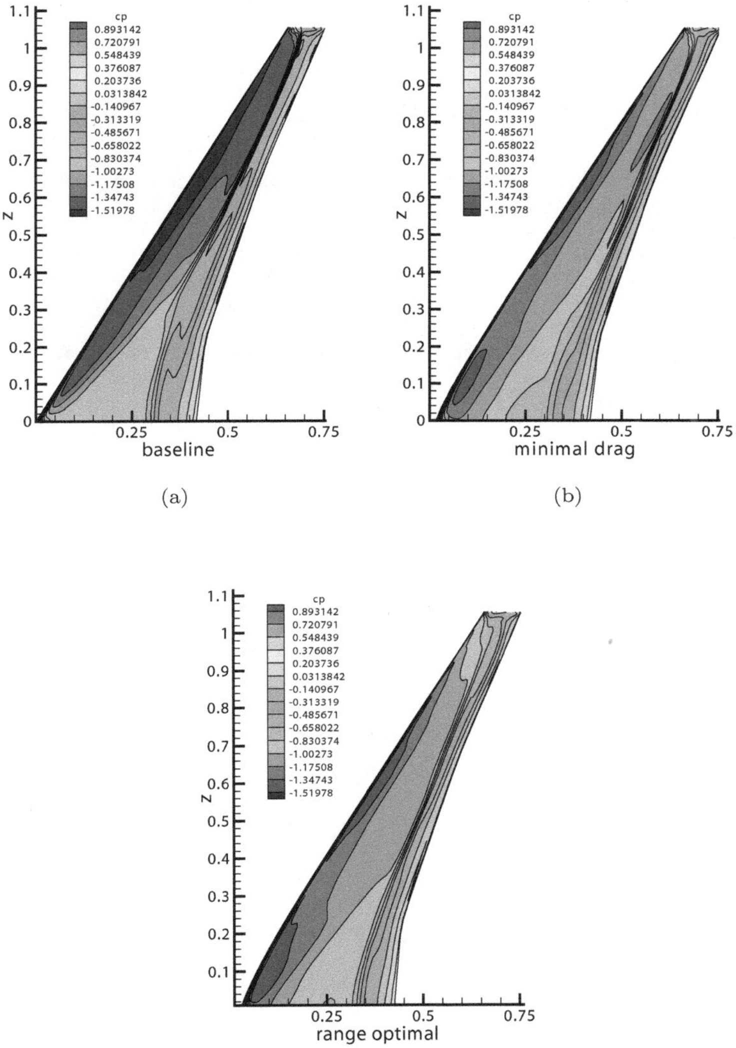

(c)

Fig. 5.15 Pressure distribution for the baseline AMP wing shape and for the optimal wing shapes for drag minimization and range maximization ($M_\infty = 0.78$, $\alpha = 2.83°$)

Since we want to avoid recalculation of the (w, Z) fields, we cancel the terms in δw and δZ from δI by imposing the fields ϕ and ψ, to be the solution of the equations

$$\left(\frac{\partial I}{\partial w} + \psi^T \frac{\partial R}{\partial w} + \phi^T \frac{\partial S}{\partial w}\right) = 0 \ , \tag{5.41}$$

$$\left(\frac{\partial I}{\partial Z} + \psi^T \frac{\partial R}{\partial Z} + \phi^T \frac{\partial S}{\partial Z}\right) = 0 \ . \tag{5.42}$$

These are the adjoint equations for the problem of coupled aeroelasticity. After their solution, the gradient can be recovered from the expression

$$\delta I = \left(\frac{\partial I}{\partial X} + \psi^T \frac{\partial R}{\partial X} + \phi^T \frac{\partial S}{\partial X}\right) \delta X \ . \tag{5.43}$$

We can assume the cost function to be a functional in the form

$$I(X, w, Z) = \int_V i(X, w, Z) dV \tag{5.44}$$

with

$$i(X, w, Z) = \frac{C_p}{C_{\text{ref}}}(n_x \cos\alpha + n_y \sin\alpha)\delta(\eta) \tag{5.45}$$

where $\delta(\eta)$ is the Dirac delta function. The equation $\eta = 0$ defines the airfoil shape in the body fitted coordinates (ξ, η). For the Dirac delta function under integration the following equation holds

$$\int \delta(\eta) f(\eta) d\eta = f(0) \ . \tag{5.46}$$

In the context of Eq. (5.44), it reduces the volume integral to a surface integral. We suppose that the fluid obeys the Euler equations, which in body fitted coordinates take the form

$$\frac{\partial F}{\partial \xi} + \frac{\partial G}{\partial \eta} = 0 \ , \tag{5.47}$$

where the transformed F, G are appropriate combinations of f and g, e.g.,

$$F = J\frac{\partial \xi}{\partial x} f + J\frac{\partial \xi}{\partial y} g = J \begin{bmatrix} \rho U \\ \rho u U + \frac{\partial \xi}{\partial x} p \\ \rho v U + \frac{\partial \xi}{\partial y} p \\ \rho H U \end{bmatrix} . \tag{5.48}$$

Since our cost function I is of the form shown in Eq. (5.44), as first step we have to formulate Eqs. (5.41) and (5.42) in an appropriate way, using the following property

$$\delta I(X,w,Z) = \int\limits_V \delta i(X,w,Z)dV$$
$$= \int\limits_V \left(\frac{\partial i(X,w,Z)}{\partial X}\delta X + \frac{\partial i(X,w,Z)}{\partial w}\delta w + \frac{\partial i(X,w,Z)}{\partial Z}\delta Z \right) dV \; . \quad (5.49)$$

The derivation is identical to what has already been seen, and gives the adjoint equations

$$\int\limits_V \left(\frac{\partial i}{\partial w} + \psi^T \frac{\partial R}{\partial w} + \phi^T \frac{\partial S}{\partial w} \right) dV = 0 \; , \quad (5.50)$$

$$\int\limits_V \left(\frac{\partial i}{\partial Z} + \psi^T \frac{\partial R}{\partial Z} + \phi^T \frac{\partial S}{\partial Z} \right) dV = 0 \; . \quad (5.51)$$

For the gradient we get

$$\delta I(X,w,Z) =$$
$$\int\limits_V \left(\frac{\partial i(X,w,Z)}{\partial X}\delta X + \psi^T \frac{\partial R(X,w,Z)}{\partial X}\delta X + \phi^T \frac{\partial S(X,w,Z)}{\partial X}\delta X \right) dV \quad (5.52)$$

It can be shown that Eq. (5.50) is equivalent to the equation

$$\int\limits_V \left(\left(\frac{\partial \psi}{\partial \xi}\right)^T \frac{\partial F}{\partial w} + \left(\frac{\partial \psi}{\partial \eta}\right)^T \frac{\partial G}{\partial w} \right) dV = 0 \quad (5.53)$$

and the boundary condition (in the case of the drag)

$$\psi_2 n_x + \psi_3 n_y + n_x \cos(\alpha) + n_y \sin(\alpha) - \mathbf{n}^T \phi = 0 \; . \quad (5.54)$$

Note that the structural adjoint variables appear only in the boundary condition (5.54), while the adjoint flow equation (5.53) is unchanged. This implies that in order to implement the coupling, only the boundary condition treatment in the FLOWer code has to be modified. Equation (5.42) represents the structural adjoint equation and its boundary conditions. The structural equation reads in the case of linear elasticity

$$S(X,w,Z) = K \cdot Z - a = 0 \quad (5.55)$$

where K is the symmetric stiffness matrix and a is the aerodynamic force. The derivative $\frac{\partial S}{\partial Z}$ in (5.42) can thus be replaced by K and the product $\phi^T K$ by $K\phi$. In this way, the same solver can be used for the structural direct and adjoint equation, with different boundary conditions, given by the first and second term in Eq. (5.42). The first term is reduced to a surface integral by the presence of the Dirac delta function, giving a vector defined by

$$V_i = \frac{\partial \int\limits_S I(X, w, Z) dS}{\partial Z_i} \tag{5.56}$$

that is the derivative of the cost function with respect to a structural degree of freedom. The second term, namely

$$\int\limits_V \left(\psi^T \frac{\partial R}{\partial Z} \right) dV \tag{5.57}$$

represents the integral of the scalar product of the adjoint field ψ and the partial derivative of the flow operator $R(X, w, Z)$ with respect to a structural degree of freedom, thus keeping the flow field and the design variables constant. It is evaluated by making use of the finite volume formulation implemented in FLOWer. A similar term appears in the expression for gradient (5.52), which explicitly becomes

$$\frac{dI}{dX} = \frac{\partial I}{\partial X} + \int\limits_V \left(\psi^T \frac{\partial R}{\partial X} \right) dV + \int\limits_V \left(\phi^T \frac{\partial S}{\partial X} \right) dV \,. \tag{5.58}$$

We already know how to evaluate the first two terms. The third term reduces to the surface integral of the adjoint field ϕ multiplied by the term

$$\frac{\partial S}{\partial X} = \frac{\partial K}{\partial X} Z - \frac{\partial a}{\partial X} \,. \tag{5.59}$$

Of the two terms on the right hand side, the first has been neglected, which is equivalent to assuming that shape deformations do not act on the structural mesh and thus on the stiffness matrix.

5.7.2 Implementation

In order to solve the coupled equations of the aero-structural system, a sequential staggered method has been implemented, where forces are transferred from the flow mesh to the structure mesh and give the nodal loads, and deflections are transferred back from the structure mesh to the flow mesh which is consequently deformed. The flow around the body described by the Euler equation is solved by the DLR solver FLOWer, while the structural problem is solved by MSC Nastran. The transfer of information between the two meshes is managed by a module developed in-house based on B-spline volume interpolation. Typically, 20 exchanges of information between the two codes are more than enough to reach a converged aeroelastic solution as shown in Fig. 5.10.

The same staggered scheme has been used to solve the systems of the coupled adjoint equations, with the difference that now only adjoint deflections are interpolated from the structural mesh to the flow mesh, in order to evaluate the boundary condition (5.54) for the new adjoint flow computation. Boundary conditions coming from the coupling are exchanged and updated for every 100 steps of the adjoint flow solver as shown in Fig. 5.11.

5.7.3 Validation and Application

The validation of both the theory and the implementation of the adjoint formulation for the aeroelastic system has been achieved by comparison with the FD method.

As test case for the validation, the AMP wing has been chosen (Fig. 5.12). The structure has been modeled with a simplified model of 126 nodes, all lying on the wing surface, connected by 422 tria/quad shell and 198 beam elements. Such a model, unlike its fluid counterpart, is not state of the art, but is sufficient to demonstrate the features of the method. In order to underline the effect of aeroelasticity, the thickness of the beam elements of the wing has been tuned to reach a deflection of about 10% of the wing span at the wing tip.

Making use of the FD method, the gradient of the drag with respect to the shape parameters has been calculated, this time including the effect of aeroelastic interaction. This means that after a deformation of the jig shape (undeflected shape), an aeroelastic coupling was called and a stationary state was reached as already shown in Fig. 5.10. This operation was repeated for every design parameter.
On the other hand, after the solution of the coupled adjoint equations, both the flow and structural adjoint fields have been used to reconstruct the gradient according to Eq. (5.58). The comparison of both methods is shown in Fig. 5.13, together with the gradient obtained when neglecting the aeroelastic coupling (rigid).

Finally, Figs. 5.14 and 5.15 illustrate the application of the coupled aero-structural adjoint approach to the drag reduction of the AMP wing by constant lift while taking into account the static deformation of this wing caused by the aerodynamic forces. Additionally, the Breguet formula of aircraft range is considered where in addition to the lift to drag ratio, the weight of the wing is taken into account as well.

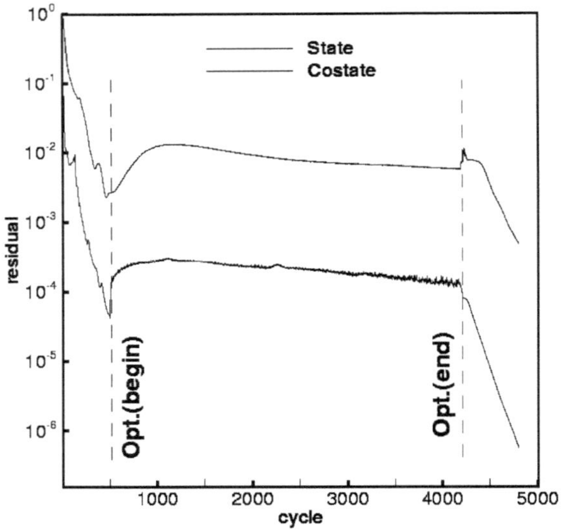

Fig. 5.16 Convergence history of state and costate (RAE 2822, $M_\infty = 0.73$ and $\alpha = 2.0°$)

5.8 One-shot Methods

The algorithmic approach of the so-called one-shot methods is based on an embedding of optimization strategies within the iterations of the respective flow solver. A continuous reduced SQP method is developed to solve the optimization problem in one joint pseudo-timestepping iteration for all variables (flow state, adjoint and wing design variables) [15, 16]. In this way, we look for the steady states of the pseudo-time embedded non-stationary system of state, costate (or adjoint state) and design equations. The preconditioner used corresponds to Karush-Kuhn-Tucker matrices, which are used in an approximate reduced SQP method.

A first demonstration of the capability of the one-shot method is given for the drag reduction of the RAE 2822 airfoil in inviscid flow with $M_\infty = 0.73$ and $\alpha = 2.0°$. Figure 5.16 presents the convergence history of the optimization iterations. The optimization is started with the initial solution of the state and costate equations obtained after 500 steps with Runge-Kutta time integration. The convergence of the optimization is achieved after 3,700 optimization iterations. After convergence is achieved for optimization, we perform another 600 time iterations for state and costate solvers to reduce the residual of these two variables further to get more accurate values of surface pressure and force coefficients. Figure 5.17 shows the inexact and exact drag reduction during the optimization iterations. Inexact here means that the

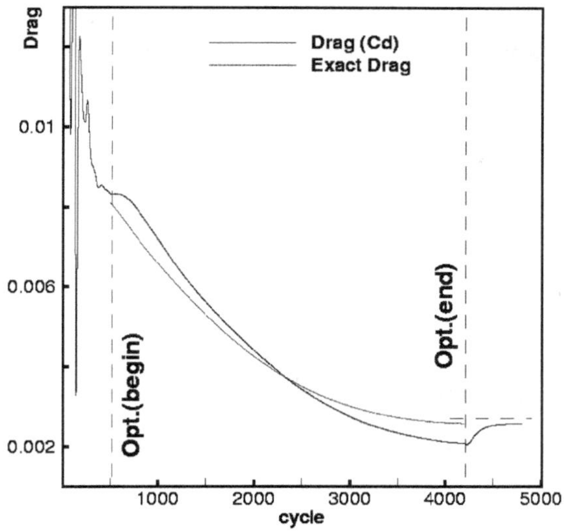

Fig. 5.17 Convergence history of design (inexact/exact drag), RAE 2822, $M_\infty = 0.73$ and $\alpha = 2.0°$

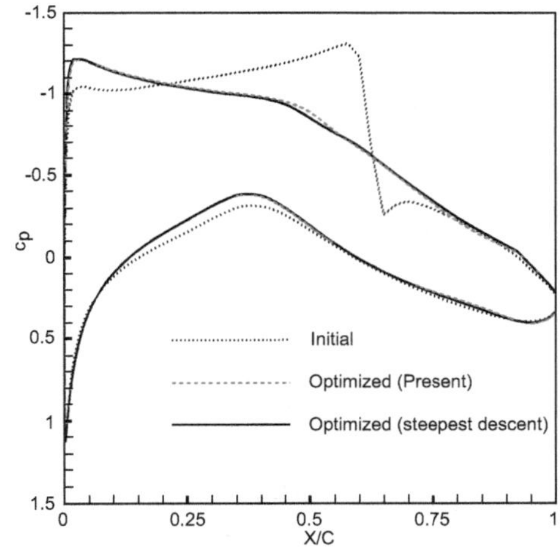

Fig. 5.18 Initial and optimized pressure distribution (RAE 2822, $M_\infty = 0.73$ and $\alpha = 2.0°$)

drag is evaluated for the less converged state and costate variables used in the design loop. Afterwards, on the trace of modified shapes generated during the one-shot approach, the drag was recomputed up to an accuracy of 7 digits and compared with the inexact one. The final drag reduction after the optimization is about 68% and the shock completely vanished (Fig. 5.18) as expected for inviscid cases. Figure 5.18 presents the comparison of the initial and final surface pressure distributions achieved with the one-shot approach (present) and with the conventional gradient based adjoint approach (steepest descent).

Altogether, the numerical cost of the one-shot optimization is of the magnitude of just 4 flow simulations, which is a dramatic reduction in computation time compared to the conventional approach.

Acknowledgements The author thanks his colleagues at DLR, in particular A. Fazzolari, J. Brezillon and M. Widhalm, as well as the MEGADESIGN partners V. Schulz and S. Hazra from University of Trier for their contributions to this chapter. Furthermore, the author thanks A. Walther and C. Moldenhauer from TU Dresden for their support and contributions w.r.t. algorithmic differentiation.

References

1. Brezillon, J., Dwight, R.: Discrete adjoint of the Navier-Stokes equations for aerodynamic shape optimization. In: Proceedings of EUROGEN05 (2005)
2. Brezillon, J., Gauger, N.R.: 2D and 3D aerodynamic shape optimization using the adjoint approach. Aerospace Science and Technology **8**(8), 715–727 (2004)
3. Dwight, R.: Efficiency improvments of RANS-based analysis and optimization using implicit and adjoint methods on unstructured grids. Ph.D. thesis, DLR-Report No. DLR-FB–2006-11 (ISSN 1434-8454) (2006)
4. Fazzolari, A.: An aero-structure adjoint formulation for efficient multidisciplinary wing optimization. Ph.D. thesis, TU Braunschweig, Germany (2006)
5. Fazzolari, A., Gauger, N.R., Brezillon, J.: An aero-structure adjoint formulation for efficient multidisciplinary wing optimization. In: Proceedings of EUROGEN05 (2005)
6. Fazzolari, A., Gauger, N.R., Brezillon, J.: Efficient aerodynamic shape optimization in mdo context. Journal of Computational and Applied Mathematics **203**, 548–560 (2007)
7. Gauger, N.R.: Aerodynamic shape optimization using the adjoint Euler equations. In: Proceedings of the GAMM Workshop on Discrete Modelling and Discrete Algorithms in Continuum Mechanics, pp. 87–96. Logos Verlag, Berlin (2001)
8. Gauger, N.R.: Das Adjungiertenverfahren in der aerodynamischen Formoptimierung. Ph.D. thesis, DLR-Report No. DLR-FB–2003-05 (ISSN 1434-8454) (2003)
9. Gauger, N.R., Brezillon, J.: Aerodynamic shape optimization using adjoint method. Journal of the Aeronautical Society of India **54**(3), 247–254 (2002)
10. Gauger, N.R., Walther, A., Moldenhauer, C., Widhalm, M.: Automatic differentiation of an entire design chain for aerodynamic shape optimization. In: Notes on Numerical Fluid Mechanics and Multidisciplinary Design (to appear), vol. 96. Springer Verlag (2007)
11. Giering, R., Kaminski, T., Slawig, T.: Applying TAF to a Navier-Stokes solver that simulates an Euler flow around an airfoil. Future Generation Computer Systems **21**(8) (2005)
12. Giles, M.B., Duta, M.C., Müller, J.D., Pierce, N.A.: Algorithm developments for discrete adjoint methods. AIAA Journal **41**(2), 198–205 (2003)

13. Griewank, A.: Evaluating Derivatives, Principles and Techniques of Algorithmic Differentiation. Society for Industrial and Applied Mathematics, Philadelphia (2000)
14. Griewank, A., Juedes, D., Mitev, H., Utke, J., Vogel, O., Walther, A.: ADOL-C: A package for the automatic differentiation of algorithms written in C/C++. Tech. rep., Technical University of Dresden, Institute of Scientific Computing and Institute of Geometry (1999)
15. Hazra, S.B., Schulz, V.: Simultaneous pseudo-timestepping for PDE-model based optimization problems. Bit Numerical Mathematics **44**(3), 457–472 (2004)
16. Hazra, S.B., Schulz, V., Brezillon, J., Gauger, N.R.: Aerodynamic shape optimization using simultaneous pseudo-timestepping. Journal of Computational Physics **204**(1), 46–64 (2005)
17. Heinrich, R.: Implementation and usage of structured algorithms within an unstructured CFD-code. In: Notes on Numerical Fluid Mechanics and Multidisciplinary Design, vol. 92. Springer Verlag (2006)
18. Hounjet, M.H.L., Prananta, B.B., Zwaan, R.: A thin layer Navier Stokes solver and its application for aeroelastic analysis of an airfoil in transonic flow. Netherlands, DLR-Publication (1995)
19. Jameson, A.: Aerodynamic design via control theory. Journal of Scientific Computing **3**, 233–260 (1988)
20. Jameson, A., Martinelli, L., Pierce, N.A.: Optimum aerodynamic design using the navier-stokes equations. Theoretical and Computational Fluid Dynamics **10**, 213–237 (1998)
21. Jameson, A., Reuther, J.: Control theory based on airfoil design using the Euler equations. AIAA Proceedings 94-4272-CP (1994)
22. Kroll, N., Rossow, C.C., Schwamborn, D., Becker, K., Heller, G.: MEGAFLOW - A numerical flow simulation tool for transport aircraft design (2002)
23. Martins, J.R., Alonso, J.J., Reuther, J.J.: Complete configuration aero-structural optimization using a coupled sensitivity analysis method. AIAA Paper 2002-5402 (2002)
24. Martins, J.R., Alonso, J.J., Reuther, J.J.: High-fidelity aero-structural design optimization of a supersonic business jet. AIAA Paper 2002-1483 (2002)
25. Nocedal, J., Wright, S.J.: Numerical Optimization. Springer Series in Operations Research. Springer (1999)
26. Rossow, C.C.: A flux splitting scheme for compressible and incompressible flows. Journal of Computational Physics **164**, 104–122 (2000)
27. Schlenkrich, S., Walther, A., Gauger, N.R., Heinrich, R.: Differentiating fixed point iterations with ADOL-C: Gradient calculation for fluid dynamics. In: Proceedings of the International Conference on High Performance Scientific Computing (2006)
28. Widhalm, M., Gauger, N.R., Brezillon, J.: Implementation of a continuous adjoint solver in TAU. DLR-Report (in press) (2007)
29. Widhalm, M., Rossow, C.C.: Improvement of upwind schemes with the least square method in the DLR TAU code. In: Notes on Numerical Fluid Mechanics, vol. 87, pp. 398–406. Springer Verlag (2004)

Chapter 6
Numerical Optimization for Advanced Turbomachinery Design

René A. Van den Braembussche

Abstract The multilevel-multidisciplinary-multipoint optimization system developed at the von Kármán Institute and its applications to turbomachinery design is presented. To speed up the convergence to the optimum geometry, the method combines an Artificial Neural Network, a Design Of Experiment technique and a Genetic Algorithm. The different components are described, the main requirements are outlined and the basic method is illustrated by the design of an axial turbine blade.

A procedure for multipoint optimization, aiming for optimal performance at more than one operating point, is outlined and applied to the optimization of a low solidity diffuser.

The extension to a multidisciplinary optimization, by combining a Navier-Stokes solver with a Finite Element Analysis, allows an efficient search for a compromise between the sometimes conflicting demands of high efficiency and respect of mechanical constraints. It is shown that a significant reduction of the stresses is possible with only a small penalty on the performance and that this approach may lead to geometries that would normally be excluded when using less sophisticated methods.

6.1 Introduction

Computational Fluid Dynamics (CFD) has seen a very important development over the last 30 years. Navier-Stokes (NS) solvers have reached a high level of reliability at affordable cost. They are now routinely used to ana-

René A. Van den Braembussche
von Kármán Institute for Fluid Dynamics,
Turbomachinery and Propulsion Department,
Waterloose steenweg, 72, 1640 Sint-Genesius-Rode, Belgium
(e-mail: vdb@vki.ac.be)

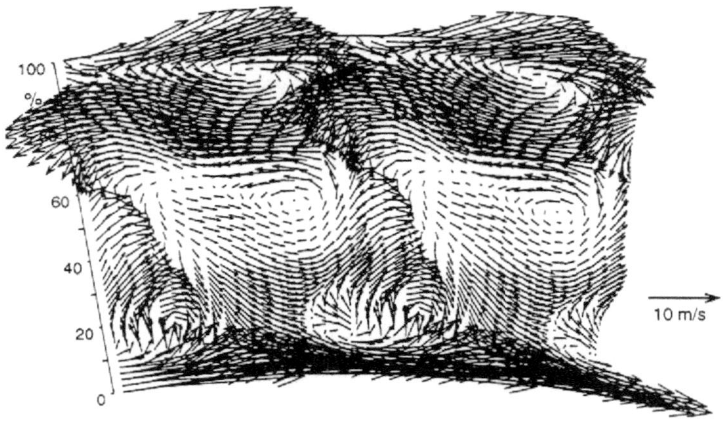

Fig. 6.1 2D view of the 3D flow at the exit of a turbine stage

lyze the fluid flows in the same way Finite Element Analysis (FEA) is used for stress predictions. They provide detailed information about the 3D flow around existing blade shapes and constitute an attractive alternative for detailed flow measurements. Complex flow phenomena can now be studied in what is called "Numerical Laboratories". Although this has resulted in a drastic decrease of the number of prototype testing, there are still two problems that prevent a more efficient use of CFD in the turbomachinery design process.

The first one results from the difficulty to analyze 3D flows on 2D screens or drawings. 2D vector plots are only a poor representation of the reality. They can be very misleading as they may suggest that the flow is penetrating the solid walls (Fig. 6.1). Synthetic environments, also called virtual reality, are very promising in this respect. These techniques are not only applicable to mere computer games but will become part of everyday reality for engineers in the next decade [16]. Designers will walk inside blade rows and diffusers to inspect the complex 3D flow structures by tracing the streamlines and to find out what geometrical changes may be needed to improve the performance.

The second problem relates to the abundance of information provided by the NS calculation. The output of an NS solver contains all the information needed to improve the performance. However, it does not provide any information on what modifications are needed to reach that goal. Three velocity components, the pressure and the temperature in typically 10,000 points (2D flows) or in more than 1,000,000 points (3D flows) are more than what the human brain is able to grasp and fully exploit in new designs. Most of the available information remains unused as the designer will often calculate a global parameter to find out if one geometry performs better than another.

The traditional design procedures in which standard 2D blade sections are selected and scaled up or down to adapt them to the different operating conditions are no longer acceptable. The designer is now faced with the development of new and better performing 2D and 3D blade shapes [10]. He needs new tools to use the available information in a more efficient way than with the traditional trial and error procedure in which the systematic testing of blade shapes has only been replaced by NS calculations. Those manual designs are very time consuming and the outcome depends on the expertise of the designer. This may become problematic since experienced designers are replaced by young engineers, who may have expertise in CFD but limited experience in turbomachinery design. Moreover, they can hardly be expert in all disciplines that interfere with a design (aerodynamics, mechanics, manufacturing etc.). Hence there is a need for automated and computerized design systems.

The main goal when designing turbines or compressors is to achieve light, compact and highly efficient systems while reducing the cost and the duration of the design cycle. Existing computerized design systems are often too expensive in terms of computational effort. Too many design processes have been concluded not because the target has been obtained, but because the deadline has come up. New design systems should therefore aim to be *fast* and affordable.

Turbomachines often operate outside the nominal or design conditions. Compressors for air-conditioning applications must be able to operate efficiently in all seasons, i.e., at different mass flows but constant pressure ratio. Low Solidity Diffusers (LSDs) are specially designed to increase the performance at a large variety of inlet flow conditions. Optimizing those geometries for one operating point is only part of the job. *Multipoint* design systems are needed to find a global optimum, i.e., maximizing the performance at all operating points to minimize the lifetime operating cost of the device.

Optimum performance is of no use if the mechanical integrity of the turbomachine cannot be guaranteed. This requires a stress and/or heat transfer analysis to verify that the stress constraints are not violated. Lower material and manufacturing costs are also important design criteria. Designing turbomachines is therefore a complex *multidisciplinary* exercise.

Inverse design methods define the geometry corresponding to a prescribed pressure or velocity distribution. However specifying the input of such a method, that satisfies mechanical and geometrical constraints and results in high performance in all operating points, is not an easy task. This is particularly difficult for 3D flows where secondary flow phenomena play a dominant role. A lot of insight is required to foresee the mechanical and geometrical consequences of a velocity variation. Adjustment of the target pressure distribution during the optimization process may be needed [7, 20].

Optimization systems searching for the geometry that best satisfies more global requirements in terms of performance, mechanical constraints or any other design criterion are a valuable alternative and have experienced a lot of attention in recent years. In what follows one will describe the fast, multipoint and multidisciplinary optimization method, developed at the von Kármán Institute, and its application to different turbomachinery designs.

6.2 Optimization Methods

Optimization methods attempt to determine the design variables $X_i (i = 1, n)$ that minimize an objective function $OF(U(X_i), X_i)$ where $U(X_i)$ is the solution of the flow equations $R(U(X_i), X_i) = 0$ and subject to n_A performance constraints $A_j(U(X_i), X_i) \leq 0$ $(j = 1, n_A)$ and n_G geometrical constraints $G_k(X_i) \leq 0$ $(k = 1, n_G)$.

Specifying as Objective Function (OF) the difference between a prescribed and calculated pressure distribution results in an inverse design method. Aiming for the improvement of the overall performance leads to a global optimization technique.

Numerical optimization procedures consist of the following components, described in more detail in the following paragraphs:

- a choice of independent design parameters and the definition of the addressable part of the design space.
- definition of an OF quantifying the performance. Any standard analysis tool can be used to calculate the components such as lift, drag, efficiency, mass flow, manufacturing cost or a combination of all of them.
- a search mechanism to find the optimum combination of the design parameters, i.e., the one corresponding to the minimum of the OF.

6.2.1 Search Mechanisms

There are two main groups of search mechanisms:

- Analytical ones calculate the required geometry changes in a deterministic way from the output of the performance evaluation. A common one is the steepest descent method approaching the area of minimum OF by following the path with the largest negative gradient on the OF surface (Fig. 6.2). This approach requires the calculation of the direction of the largest gradient of the OF and the step length. A comprehensive overview of gradient based optimization techniques is given by Vanderplaats [17].
- Zero-order or stochastic procedures require only function evaluations. They make a random or systematic sweep of the design space or use evolutionary theories such as Genetic Algorithms (GA) or Simulated Annealing (SA).

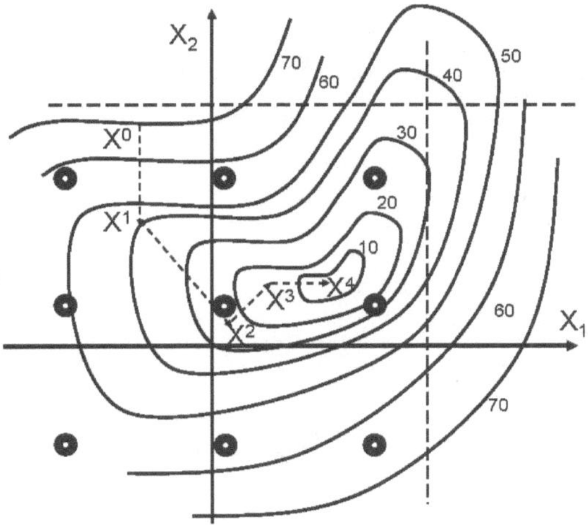

Fig. 6.2 Gradient method (- - -) and zero-order sweep of the design space (o)

To minimize the OF, most numerical algorithms require a large number of performance evaluations and are often very expensive in terms of computer resources. Zero order methods may require even more evaluations than gradient methods but the latter may get stuck in a local minimum. The method presented here uses a zero order search mechanism based on an evolutionary theory.

6.2.1.1 Zero-order Search

A systematic sweep of the design space, defining v values between the minimum and maximum limits of each of the n design parameters, requires v^n function evaluations. Figure 6.2 illustrates how such a sweep, calculating the OF for 3 different values of X_1 and X_2, provides a very good estimation of where the optimum is located with only 9 function evaluations. The risk of converging to a local minimum is low and such a systematic sweep is a valid alternative for analytical search methods for small values of n. However it requires more than 14×10^6 evaluations for $n = 15$.

Evolutionary strategies such as GA and SA can accelerate the procedure by replacing the systematic sweep with a more intelligent selection of new geometries using the information obtained during previous calculations in a stochastic way.

SA is derived from the annealing of solids [1]. At a given temperature, the state of the system varies randomly. It is immediately accepted if the new state has a lower energy level. If however, the variation results in a higher state, it is only accepted with a probability Pr that is a function of the temperature.

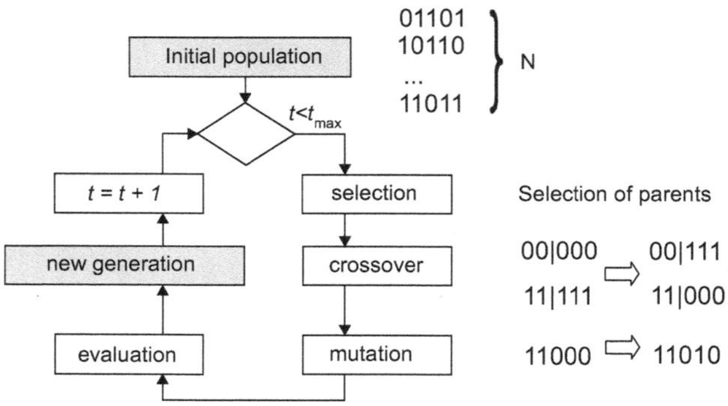

Fig. 6.3 GA operating principle

$$\mathrm{Pr} = \exp \frac{(E_{opt} - E_{act})}{T}$$

As the temperature decreases, the probability of accepting a higher state becomes lower. In a SA algorithm, the design parameters characterize the state of the system whereas the OF characterizes the energy level.

6.2.1.2 Genetic Algorithm (GA)

The method presented here uses a GA to find the optimum. This is a numerical technique that simulates Darwin's evolutionary theory stating that the fittest survives [4]. According to this theory, an individual (a geometry) with favorable genetic characteristics (design variables) is most likely to produce better offsprings. Selecting them as parent, increases the probability that the individuals of next generation will perform better than the previous one.

The operational principle of a standard GA is shown in Fig. 6.3. Pairs of individuals (parents) are selected from an initially random population of N geometries, each represented by a binary coded string of length l. Genetic material is subsequently exchanged between them (crossover) and altered within the offspring (mutation). It is followed by an evaluation of each new individual. This process is used to create the N individuals of the next generation. Such a procedure is repeated for t generations and it is assumed that the best individual of the last generation is the optimum.

Quality of the GA optimizer is measured by:

- the required computational effort, i.e., the number of performance evaluations that are needed to find that optimum (GA efficiency).
- the value of the optimum (GA effectiveness).

The GA software used in the VKI design system is the one developed by David L. Carroll [5]. The optimum parameter setting has been defined by means of a systematic study on two typical design cases: one geometry defined by 7 parameters and one defined by 27 parameters [8]. Conclusions are based on the solution quality q, i.e., the degree to where the GA optimum approaches the real one within a given effort (5,000 function evaluations). It is defined by:

$$q = \frac{OF_{\mathrm{AV}} - OF_{\mathrm{GA}}}{OF_{\mathrm{AV}} - OF_{\mathrm{min}}} \cdot 100\%$$

where OF_{AV} is the average of the OF over the complete design space.

OF_{min} is the global minimum value of the OF obtained from a systematic (numerically very expensive) scanning of the whole design space.

OF_{GA} is the minimum value of the OF obtained from the GA optimization.

A q value of 100% indicates that the global minimum has been found. The function evaluations for the numerical experiments are made by means of an approximation of the NS solver based on Artificial Neural Networks (ANN, explained in Sect. 6.3.1).

Optimum substring length

In a standard binary-coded GA, the n real-valued design parameters X_i, defining a geometry, are jointly represented by one binary string:

$$\underbrace{1101...0}_{X_1}\ \underbrace{1001...1}_{X_2}\ \underbrace{0011...0}_{X_3}\ \dots\dots\dots\ \underbrace{0101...1}_{X_n}$$

The substring length, denoted by l (number of bits per variable), determines the total number of values (2^l) that each design parameter can take.

The minimum substring length l_i for the ith design variable depends on the lower and upper bound respectively X_i^{min} and X_i^{max}, as well as on the desired resolution (ε_i) for this variable:

$$l_i = \log_2 \frac{X_i^{\min} - X_i^{\max}}{\varepsilon_i}$$

Very short substrings ($l < 3$) result in a too low resolution and the GA may not be able to accurately locate the minimum. Longer substrings ($3 < l < 10$) enable a higher resolution but cause a larger search space, making it difficult to find the complete optimal binary string. Systematic testing shows that $l = 8$ is the optimum substring length, independent of the number of unknown parameters.

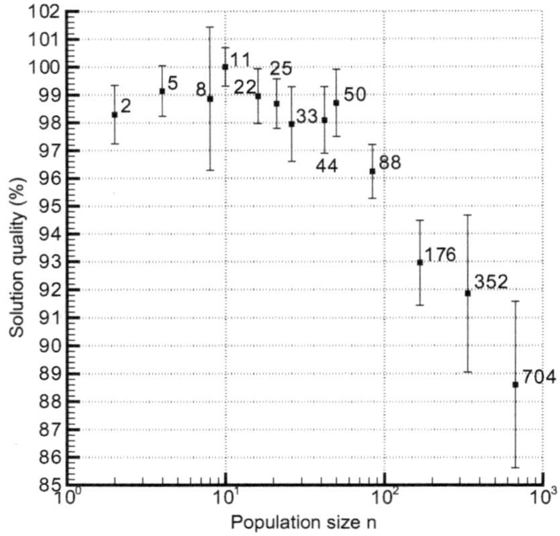

Fig. 6.4 Dependence of GA solution quality on population size for the 27 parameter test case

Selection scheme

Different selection schemes have been proposed. One is the roulette: a system in which the chance that an individual is selected increases proportional with $1/OF$. This scheme favors the best individuals as parents. It is elitist and has larger chances to get stuck in a local optimum.

In the tournament selection, "s" individuals are chosen randomly from the population and the individual with the highest fitness (lowest OF) is selected as parent. The same process is repeated to find the second parent. The parameter s is called the tournament size and can take values between 1 and N (population size). Larger values of s give more chances to the best samples to be selected and to create offsprings. It favors a rapid, although perhaps premature, convergence to a local optimum. Very small values of s result in a more random selection of parents. Tests have shown that a standard value of $s = 2$ gives the best results.

Population size

Fixing the total number of function evaluations at 5,000, the number of generations t is a consequence of the population size N ($N \times t = 5,000$). Figure 6.4 shows the solution quality at the end of the GA run in function of the population size. The solution quality is maximum for $N = 11$ to 20. Small populations ($N < 10$) converge prematurely to suboptimal solutions, due to a lack of high performing samples in the initial population. Larger populations ($N > 25$) have a sluggish convergence to the optimal geometry because less generations are allowed.

Crossover probability

In a single-point crossover operator, both parent strings are cut at a random place and the right-side portions of both strings are swapped with the probability p_{c} (Fig. 6.3). In case of uniform crossover, the value of p_{c} defines the probability that crossover is applied per bit of the complete parent string. High values of p_{c} increase mixing of string parts but at the same time, increase the disruption of good string parts. Low values limit the search to combinations of samples in the existing design space. Experiments confirm that a single point crossover with probability $p_{\mathrm{c}} = 0.5$ is optimal.

Mutation probability

The mutation operator creates new individuals by changing in the offspring strings a "1" to a "0" or vice versa. The mutation probability p_{m} is defined as the probability that a bit of a string is flipped. Systematic numerical experiments confirm that the optimum setting for the mutation probability is $p_{\mathrm{m}} = 1/(N \times l)$ for all optimizations. This corresponds to changing on average one bit at every generation.

Figure 6.5 shows how an optimization of the GA parameter settings can lead to an improved and smoother GA convergence.

Creep mutation and Gray coding

Changing one digit in a binary code may result in a large variation of the corresponding digital value: i.e., the small difference between 0111 and 1111 corresponds to a doubling of the digital value. Small variations of the digital value may require a large number of binary digits to be changed: i.e., 0111 and 1000 are adjacent digital values but all four digits are different. This discontinuous relation between the digital value and binary string may confuse the GA optimizer. Creep mutation tries to avoid this by limiting the change of the real value to a binary step length [5]. Gray coding uses an algorithm in which similar binary strings correspond to adjacent digital values. In contrast to what could be expected, no acceleration of convergence was obtained with either one of these approaches.

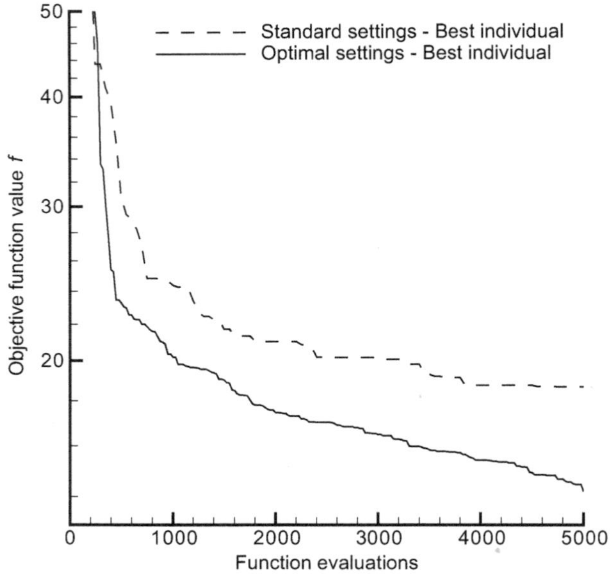

Fig. 6.5 GA convergence for a 27 parameter test case (standard versus optimal parameter setting)

6.2.2 Objective Function

The *OF* measures how far a geometry satisfies the aero-requirements and if the performance goals that have been set forward are reached.

High aero-performance is not the only objective of an optimization. A good design must also provide good off-design performance (multipoint optimization) and respect the mechanical and manufacturing constraints (multidisciplinary optimization). Some constraints must be satisfied without any compromise (i.e., maximum stress level). They result in an inequality and a more detailed discussion is given in Sect. 6.6.2. Others tolerate some margin (i.e., cost or weight) that can be corrected for after the design is finished (i.e., by adjusting the blade length to achieve the required mass flow). A possible alternative for these inequalities is to add penalty terms to the *OF* that increase when the constraints are violated [13].

The following lists some contributions to the global *OF* that are common for the different applications. Each term is multiplied by a weight factor to adjust its relative importance in the optimization procedure.

$$OF_{2D} = w_\eta \cdot P_{\mathrm{perf}} + w_{\mathrm{a}} \cdot P_{\mathrm{aeroBC}} + w_{\mathrm{m}} \cdot P_{\mathrm{mech}} + w_{\mathrm{M}} \cdot P_{\mathrm{Mach}} \\ + w_{\mathrm{d}} \cdot P_{\mathrm{discharge}} + w_{\mathrm{G}} \cdot P_{\mathrm{Geom}} + w_{\mathrm{S}} \cdot P_{\mathrm{Side}}$$

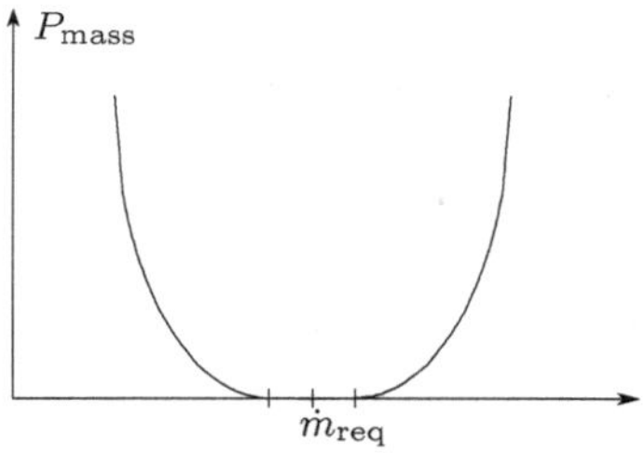

Fig. 6.6 Variation of penalty function for incorrect mass flow

P_{perf} is the penalty for non-optimum performance and increases with decreasing efficiency (η)

$$P_{\text{perf}} = \max\left[\left|\eta_{\text{req}} - \eta\right|, 0.0\right]$$

Minimizing this term corresponds to maximizing the efficiency. The required efficiency η_{req} is set to an unachievable value (for instance 1.0) so that this penalty never goes to zero. The argument is that, after all other requirements are met, the efficiency should still be maximized.

P_{AeroBC} is the penalty for violating the aerodynamic boundary conditions. The purpose of this penalty is to enforce the boundary conditions and requirements at the inlet and outlet of the computational domain that cannot be imposed such as: the outlet flow angle (β_2), the mass flow or pressure ratio, etc. The penalties for not respecting the boundary conditions start increasing when the actual values differ from the target values by more than a predefined tolerance. Following penalty for incorrect mass flow increases when the mass flow differs more than 2% from the required value (Fig. 6.6):

$$P_{\text{mass}} = \left(\max\left[\frac{\left|\dot{m}_{\text{act}} - \dot{m}_{\text{req}}\right|}{\dot{m}_{\text{req}}} - 0.02\ ,\ 0.\right]\right)^2$$

P_{mech} is the penalty for not respecting the mechanical constraints. The latter must be satisfied without compromise because exceeding the maximum stress level cannot be tolerated as it may destroy the device. A rigorous respect of the minimum stress limits requires a Finite Element stress Analysis (FEA) and will be discussed in detail in Sect. 6.6. The computational effort can be drastically reduced if one can replace the mechanical constraints by simpler geometrical ones that are much easier to verify.

The large stresses in the blade root section of radial impellers are a complex function of the blade curvature and lean. The subsequent deformations can reduce the tip clearance to zero which may lead to the destruction of the optimized geometry. Traditional design systems limit the lean (Fig. 6.7) to a maximum value based on experience and simple stress models.

Prescribing the radial variation of the cross section area of a fan blade or low-pressure (LP) turbine blade is a common way to control the centrifu-

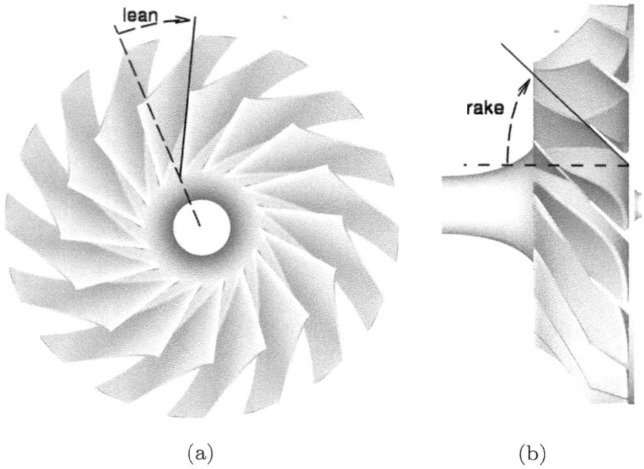

Fig. 6.7 Definition of (**a**) lean and (**b**) rake in radial impellers

gal stresses. The parameters of primary importance in controlling the blade bending due to the static and dynamic load in an axial compressor or turbine blade are: minimum and maximum moment of inertia ($I_{\min}$) and ($I_{\max}$) of the cross sections and the direction κ of its maximum value.

P_{Mach} is the penalty for a non-optimum Mach number distribution. Analyzing the Mach number distribution may help to rank blades that have nearly the same loss coefficient. NS solvers are not always reliable in terms of transition modeling and erroneous penalty function may occur when the transition point is incorrectly located. Transition criteria based on the Mach number distribution may help to relieve this uncertainty.

One is also not interested in designing blades that have very good performance at design point but for which the flow is likely to separate (with large increase in losses) at slightly off-design conditions. A rigorous way of verifying the operating range is discussed in Sect. 6.5. A simpler approach accounts for the changes in the Mach number distribution that can be expected at off-design. It increases the chances for good performance of the blade over a wide range of operating conditions without the cost of extra NS computations.

The Mach number penalties that have been formulated for turbine blade optimizations are presented in Sect. 6.4. They tend to achieve a continuous flow acceleration with minimum deceleration. Mach number penalties for radial compressor impellers are presented in Sect. 6.6.

$P_{\mathrm{discharge}}$ can be used to penalize the spanwise distortion of the flow at the exit. It results from the idea that a more uniform exit flow has a favorable effect on the downstream diffuser or blade row and hence, on the stage efficiency. The distortion penalty quantifies the difference between the average flow angle at 20% (hub) and 80% (shroud) span with the one at midspan

$$P_{\text{dist}} = \left| 1 - \frac{2.\alpha_{\text{midspan}}}{\alpha_{\text{hub}} + \alpha_{\text{shroud}}} \right|$$

The penalty for flow skewness is proportional to the difference between the flow angle, or any other flow quantity, at 20% and 80% span, non dimensionalized by the average value

$$P_{\text{skew}} = \frac{2.(\alpha_{\text{hub}} - \alpha_{\text{shroud}})}{\alpha_{\text{hub}} + \alpha_{\text{shroud}}}$$

P_{Geom} is the penalty for violating the geometrical constraints. They either are related to the mechanical integrity or assure dimensional agreement with other components.

P_{Side} is the penalty for violating the side constraints. Depending on the application, the weight, manufacturing and maintenance cost may be important issues and some geometries can be favored by formulating an appropriate penalty.

The penalties and weight factors are specific for each design. An appropriate choice of the weights can be based on overall design criteria such as energy savings, total lifetime cost, etc. Plotting the different contributions to the OF as a Pareto front in the fitness space (Fig. 6.8) facilitates a trade-off between different counteracting goals. This is particularly useful for problems with only 2 or 3 groups of OF where the Pareto front can easily be visualized. However it is much more cumbersome in higher order problems or when the Pareto front is not convex.

6.2.3 Parameterization

The number of coordinates needed for the complete definition of an arbitrary geometry is infinite and a direct calculation of all of them by a numerical optimization procedure is not feasible. A reduction of the number of variables by an adequate parameterization of the geometry is required. It is important that the parameterization does not exclude any physically-acceptable geometry. A blade shape that can not be generated can of course not be found by the optimizer, even if it would be the optimum one. The parameterization should be sufficiently simple to limit the number of variables that need to be defined. The use of Bézier curves or B-splines to describe the geometry is recommended. This assures smoothness of the surface and facilitates the transfer of the data to Computer-Aided Design and Manufacturing (CAD-CAM) systems. Different parameterizations will be explained in Sects. 6.4 to 6.6.

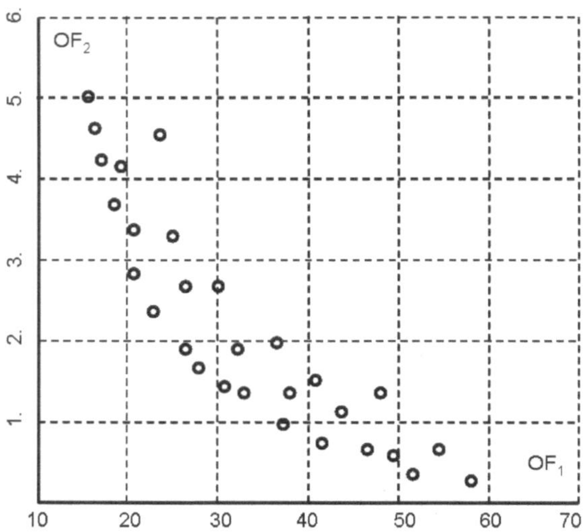

Fig. 6.8 Convex Pareto front

6.3 Two-level Optimization

The system presented here (Fig. 6.9) is developed at the von Kármán Institute [13] and makes use of a GA to minimize the *OF*. A GA requires a large number of function evaluations. Using an expensive NS solver for all function evaluations is in most cases, prohibitive in terms of computer effort.

One way to reduce the computational effort is by working on different levels of sophistication. Fast but approximate prediction methods can be used to find a near optimum geometry, which is then further verified and refined by a more accurate but also more expensive analysis. Approximations of the NS solver and FEA, called meta-functions, are used for the first level optimization. The more accurate but expensive NS and FEA are used only to verify the accuracy of the meta-function predictions.

Meta-functions not only need to be fast but must also be accurate. The GA can only converge to the real optimum if it receives accurate information about the impact of a geometry change on the performance. Different type of meta-functions have been proposed. The main problem is the risk that the discrepancies between the predictions by the meta-function and the NS results drive the optimizer to a false optimum. Euler and NS solutions on coarse grids are sometimes proposed as meta-functions. They are fast but inaccurate and using them for performance predictions may drive the GA to a non-optimum combination of design parameters. Any further control by an accurate NS solver will reveal the inherent inaccuracy of the fast calculation methods but there is no mechanism to correct for it.

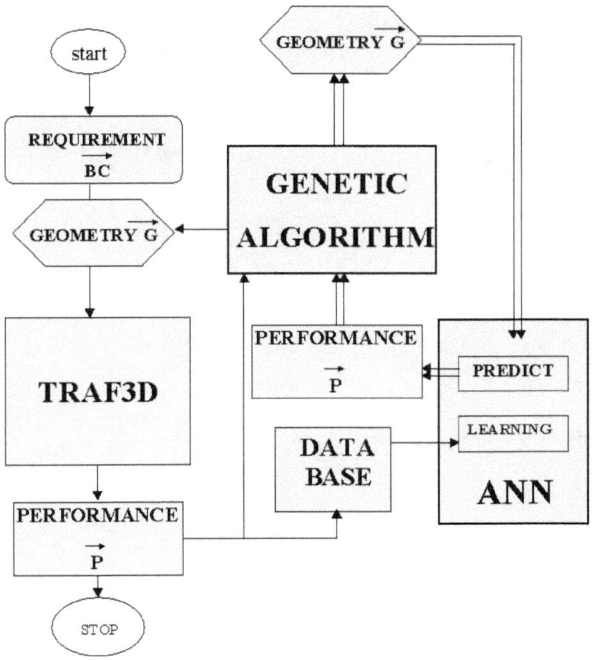

Fig. 6.9 Flowchart of optimization system

The meta-function used in the present method is an Artificial Neural Network (ANN). This interpolator uses the information contained in the database to correlate the performance to the geometry, similar to what is done by an NS solver. However, an ANN is a very fast predictor and allows the evaluation of the numerous geometries generated by the GA with much less effort than an NS solver. Unfortunately, a verification by means of a more accurate but time consuming NS solver indicates that such a fast prediction is not always very accurate. The results (geometry and performance) of this verification are added to the database and a new optimization cycle is started. It is expected that the new learning on an extended database will result in a more accurate ANN. This procedure is repeated until the ANN predictions are in agreement with the NS calculations, i.e., once the GA optimization has been made with an accurate performance predictor. In this way, there will be no discrepancy between the optimum found by a GA, driven by the meta-function or by the results of NS analyses. However, the number of time-consuming NS analyses is much smaller than what would have been required by a GA and NS combination.

The TRAF3D NS solver [3] is used to predict the aerodynamic performance. Similar grids with the same number of cells are used for all computations to guarantee a comparable accuracy for all the predictions.

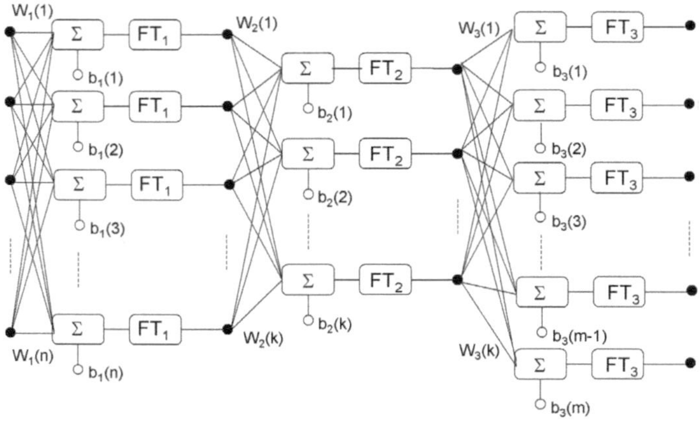

Fig. 6.10 Architecture of a three-layer ANN

6.3.1 Artificial Neural Networks

Artificial Neural Networks (ANN) are used to predict the performance of a new geometry by means of the information contained in the database. This requires the learning of the relation between the n input data (geometry parameters) of a process (NS solver) and the m outputs of the process (mass flow, efficiency, local pressures and temperatures, velocities, etc.). The use of an exact ANN predictor could reduce the effort to one design cycle by the GA. Hence, improving the accuracy of the ANN will shorten the design process.

An ANN (Fig. 6.10) is composed of n, k, m elementary processing units called neurons or nodes. These nodes are organized in layers and joined with connections (synapses) of different intensity, called the connection weight (W) to form a parallel architecture. Each node performs two operations: the first one is the summation of all the incoming signals and a bias b_i, the second one is the transformation of the signal by using a transfer function (FT). For the first layer this corresponds to:

$$a_1(i) = FT_1\left(\sum_{j=1}^{n} W_1(i,j).n(j) + b_1(i)\right)$$

A network is generally composed of several layers: an input layer, zero, one or more hidden layers and one output layer. The coefficients are defined by a learning procedure relating the output to the input data.

The main purpose of ANN is not to reproduce the existing database with maximum accuracy but to predict the performance of new geometries it has not seen before, i.e., to generalize. A well-trained ANN may show a less

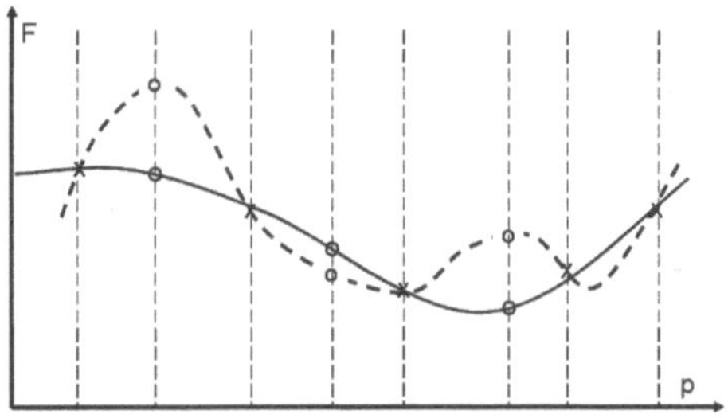

Fig. 6.11 Comparison between well trained —— and over-trained - - - - ANN (x database samples, o new geometries)

accurate reproduction of the database samples but predicts more realistic values for new geometries. This is illustrated in Fig. 6.11 comparing an over-trained ANN function with a well trained one. The first one reproduces the database samples exactly but the large oscillations of the function between the database values result in unrealistic predictions of the function at the intermediate locations.

Three conditions are necessary, although not sufficient, for a good generalization.

The first one is that the inputs to the ANN contain sufficient information pertaining to the target, so that one can define a mathematical function relating correctly outputs to inputs with the desired degree of accuracy. Hence the designer should select design parameters that are relevant, i.e., that have an influence on performance.

The second one is that the function to be learned (relating inputs to the outputs) is smooth. Small changes in the input should produce a small change in the outputs. Most physical problems are well defined in this respect. However, the appearance of large separation zones or large changes in shock position and strength for small geometrical changes may result in discontinuous changes of the output and complicate the problem.

The third one requires that the training set is sufficiently large and contains representative samples of all cases that one wants to generalize (the "population" in statistical terminology). It is difficult to define the minimum size of the training set that is required. Experience has shown that there is no advantage in creating a very large database because the design system itself will generate new geometries until the ANN achieves the required accuracy.

The standard back-propagation technique is the most widely used algorithm for ANN training. The available samples are normally subdivided into "training", "test" and "validation" sets. Each of them has its own purpose.

- the training set contains the samples used for the training; that is to define the parameters (weights and bias).
- the test set contains the samples used to assess the generalization capacity of a fully-specified ANN with given weights and architecture.

- the validation set, if used, contains the samples used to tune the ANN architecture (not the weights), for example to choose the number of hidden unit layers and nodes.

6.3.2 Database

The main purpose of the database is to provide input to the ANN, i.e., information about the relation between the geometry and performance. The more general and complete this information, the more accurate the ANN can be and the closer the results of the GA optimization will be to the real optimum. Hence, a good database may considerably speed up the convergence to the optimum.

Making a database is an expensive operation because it requires a large number of 3D NS calculations. One therefore aims for the smallest possible database containing the maximum amount of information about the available design space. This means a maximum of relevant information with minimum redundancy so that the impact of every design parameter is included but only once.

Any information missing in the database may result in an erroneous ANN that could drive the GA towards a non optimum geometry. This will slow down the convergence but will not lead to an incorrect final result since the NS analysis of that geometry will provide the missing information when added to the database.

A more risky situation is the one where an incomplete database results in an erroneous extrapolation by the ANN, predicting a low performance (large OF) in that part of the design space where in reality the OF is low. As a consequence, the corresponding geometry will never be selected by the GA and no information will be generated to correct this error. This is an additional argument to assure that the initial database covers the whole design space.

Design Of Experiment (DOE) refers to the process of planning an experiment so that the appropriate data, when analyzed by statistical methods, result in valid and objective conclusions. It is used in the optimization process to select the most significant geometries to be stored in the database. The theory of DOE is explained in many excellent textbooks [11]. The advantages of using DOE, to construct the database for the optimization program, have been evaluated in detail by Kostrewa et al. [9].

Factorial designs make a systematic scan of the design space. The common one is where each of the n design parameters has only two values corresponding to the "high" or "low" level of the design variable. A complete coverage of such a design requires 2^n observations and is called full factorial DOE. Fractional factorial DOE requires 2^{n-p} analyses where p defines the fraction of lower order combinations that are not analyzed. A typical initial database

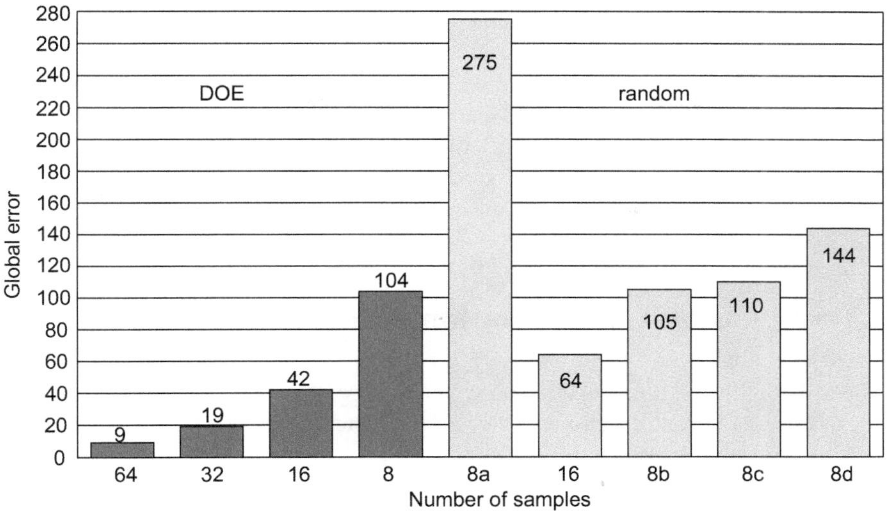

Fig. 6.12 Global error of ANN as a function of Database samples

contains a total of $2^6 = 64$ samples. The following evaluates the loss of information for a test function with 6 variables.

$$R = 1 - 0.001(A - D)^3 + 0.002(C + E)(F - B) - 0.06(A - F)^2(F + C)(E + A)$$

The results of the ANN predictions, trained on the databases defined by DOE, are compared to those of databases in which the variables are randomly generated between the prescribed boundaries.

The full factorial design requires $2^6 = 64$ runs to estimate all possible parameter combinations. The loss of information with fractional designs is measured by:

$$\text{Global error} = \sum_{i=1}^{2^6} \frac{\text{exact value} - \text{predicted value}}{\text{exact value}} \cdot 100$$

Comparing the error obtained by means of the DOE technique and by means of randomly selected samples (Fig. 6.12), clearly shows that for the same number of NS results in the database, the DOE based predictions are consistently more accurate than the ones based on the randomly generated samples.

Randomly generated databases are all different and so is the accuracy of the ANN predictions. The four randomly generated cases with 8 samples in the database, show an error that varies between 105 (8b), equal to the one obtained with the DOE defined database, up to an error that is almost 3 times larger (8a).

Only 2-level designs (every variable can take two values) and one central-point run (every value is at the center of the allowed range) are considered for the database used in following examples. The high and low value of each design variable are located at 75% and 25% of the parameter range, respectively. The central point is the mid value (50%). The range is defined by the designer based on his experience about feasible geometries and mechanical constraints.

6.4 Single Point Optimization of Turbine Blade

The optimizing design procedure has been successfully tested on a large number of designs and will be illustrated here by the design of a 2D highly loaded axial turbine blade section.

6.4.1 2D Blade Geometry Definition

The geometry definition, described here, makes use of Bézier curves to define the camber line and the blade suction and pressure side relative to the camber line.

The camber line is defined by 3 control points (Fig. 6.13.a). The first one coincides with the leading edge. The second one is at the trailing edge. Its position relative to the leading edge is defined by the stagger angle (γ) and axial chord length Lref. A third control point is located at the intersection between the tangent to the camber at leading and trailing edge. It is defined by the angles β_{1m} and β_{2m} with the axial direction.

The suction side is defined by a Bézier curve starting at the leading edge and ending at the tangent to a circle of radius Rte, defining the trailing edge thickness. The camber line is divided in a number of intervals which can be of equal or variable length using a stretching factor (Fig. 6.13.b). The lengths Tss(1), Tss(2), Tss(3), Tss(4), Tss(5), Tss(6) and Tss(7) measured in the direction perpendicular to the camber line determine the position of the 7 Bézier control points of the suction side.

The first Bézier control point coincides with the leading edge (start of the camber line) (Fig. 6.14.a) and the third point is the one defined by the parameter Tss(1). The second Bézier control point is defined by imposing that the suction side starts perpendicularly to the camber line at the leading edge and that the suction side curvature radius at the leading edge equals Rle. This condition defines the length Tss(0).

δ_{te} is the wedge angle between suction and pressure side at the trailing edge (Fig. 6.14.b). A Bézier control point is defined as the intersection of the

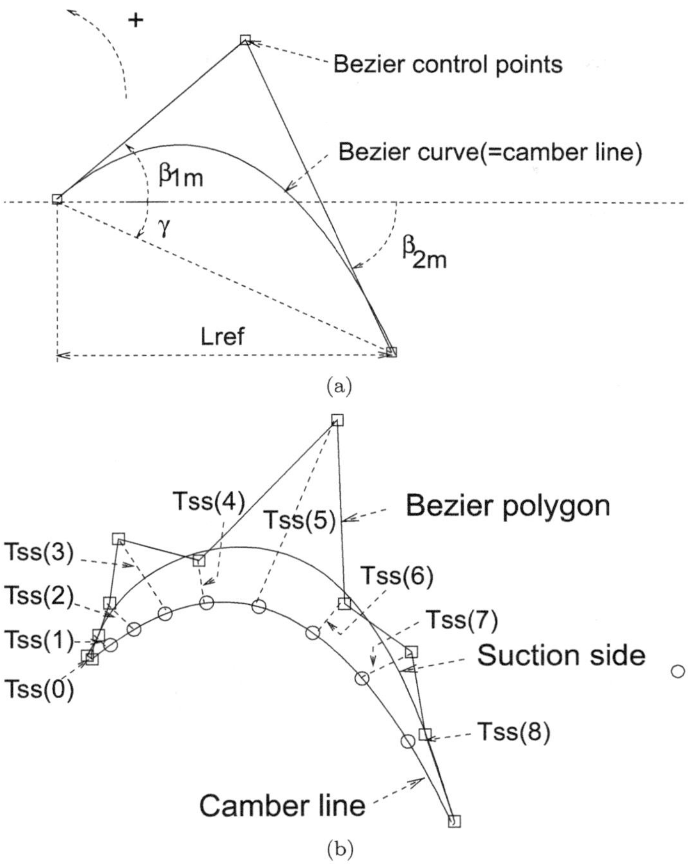

Fig. 6.13 Geometry model: (**a**) definition of the camber line and (**b**) the suction side

tangent at the trailing edge circle and the line perpendicular to the camber line at point 8.

The pressure side is defined in a way similar to the suction side. Imposing the same Rle as on the suction side assures a continuous curvature at the leading edge. Figure 6.15 demonstrates the capabilities of the method to represent the large variety of blades encountered in axial turbines.

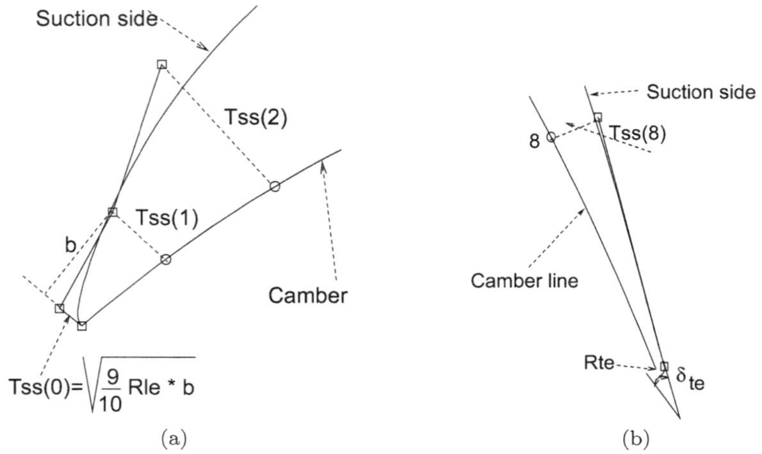

Fig. 6.14 Geometry model: (**a**) detailed view of the leading edge (**b**) and trailing edge

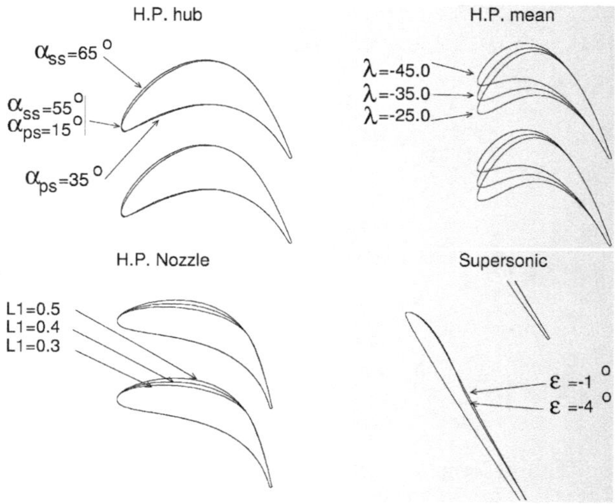

Fig. 6.15 Typical turbine blade sections generated by a 17-parameter model

6.4.2 Penalty for Non-optimum Mach Number Distribution P_{Mach}

The characteristics of the optimum Mach number M distribution for 2D turbine blades are well known and allow the definition of penalties for a non optimum Mach number distribution. These penalties can also be derived from experimental correlations.

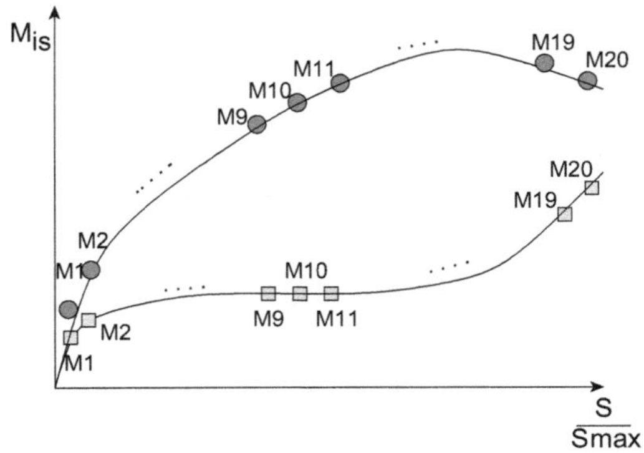

Fig. 6.16 Parametric representation of the Mach number distribution

The *OF* used in a turbine blade optimization [15] increases when there is a high probability of early transition, laminar or turbulent separation or poor off-design performances. It is a sum of penalties based on the Mach number in 20 points on each side of the blade (Fig. 6.16). The values may be defined from an NS calculation or predicted by the ANN.

- Penalty on the local slope of the Mach number distribution: a minimum positive slope of the Mach number distribution is required on the front part of the suction side in order to avoid deterioration of the performance at off-design incidence angle.
- Penalty on the second derivative of the Mach number distribution: the optimization process may result in a wavy Mach number distribution because of local changes in curvature radius of the blade surface. This may have a small impact on blade losses if the boundary layer is already turbulent but can deteriorate the off-design performances of the blade. One therefore penalizes the Mach number distribution for which the second derivative changes sign, i.e., with an inflection point in the suction side Mach number distribution.
- Penalty on the deceleration: it is important to limit the deceleration on the front part of the blade pressure side in order to avoid separation at negative incidence angles. This penalty is proportional to the difference between the first maximum Mach number found on the pressure side (starting from the stagnation point) and the minimum Mach number along the pressure side. It is also well known that the deceleration on the second half of the suction side has an important influence on the losses and in case of low Reynolds number may lead to flow separation.

Table 6.1 Imposed parameters

β_1^{flow} (deg.)	18.0
M_2^{is}	0.90
Re	5.8×10^5
$\gamma = Cp/Cv$	1.4
Tu (%)	4
Lref (m)	0.0520
Pitch/Lref	1.0393
TE thick. (m)	1.2×10^{-3}

Table 6.2 Mechanical and aerodynamical requirements

	Imposed		After
	Min.	Max.	18 modif.
Surface (m^2)	5.20×10^{-4}	6.80×10^{-4}	5.36×10^{-4}
$I_{\min}$(m^4)	7.50×10^{-9}	1.20×10^{-8}	7.45×10^{-9}
$I_{\max}$(m^4)	1.25×10^{-7}	2.20×10^{-7}	1.28×10^{-7}
$\kappa_{I_{\max}}$ (deg.)	-50.0	-30.0	-37.5
β_2^{flow} (deg.)	-57.8	-57.8	-57.62
Loss coef. (%)	0.0	0.0	1.90

6.4.3 Design of a Transonic Turbine Blade

The method is illustrated by the redesign of a transonic turbine blade with an outlet isentropic Mach number of 0.9. The design requirements imposed for this example are displayed in Tables 6.1 and 6.2.

The best blade of the initial database is used as starting geometry. The Mach number distribution has a shock at mid-chord (Fig. 6.17.a). The small constant velocity region on the suction side close to the leading edge and the low velocity on the pressure side close to the leading edge indicate that the incidence angle on the initial blade is too large. After the first modification (one GA and NS verification), this incidence angle has been partially reduced by decreasing the stagger angle (Fig. 6.17.b). The shock intensity is also smaller but the suction side Mach number distribution is still wavy. The shock completely disappeared after 13 design iterations. The stagger angle has decreased in order to adapt the blade geometry to the prescribed inlet flow angle. The smooth shock-free Mach number distribution is reflected in the low loss coefficient.

Figure 6.18 compares the value of the OF predicted by the ANN with the one predicted by the NS solver during the design process. The value of the OF computed by the approximate model decreases until iteration 13 after which only very small improvements are found. The value predicted by the NS solver shows large discrepancies between both predictions at iteration 2, 5 and

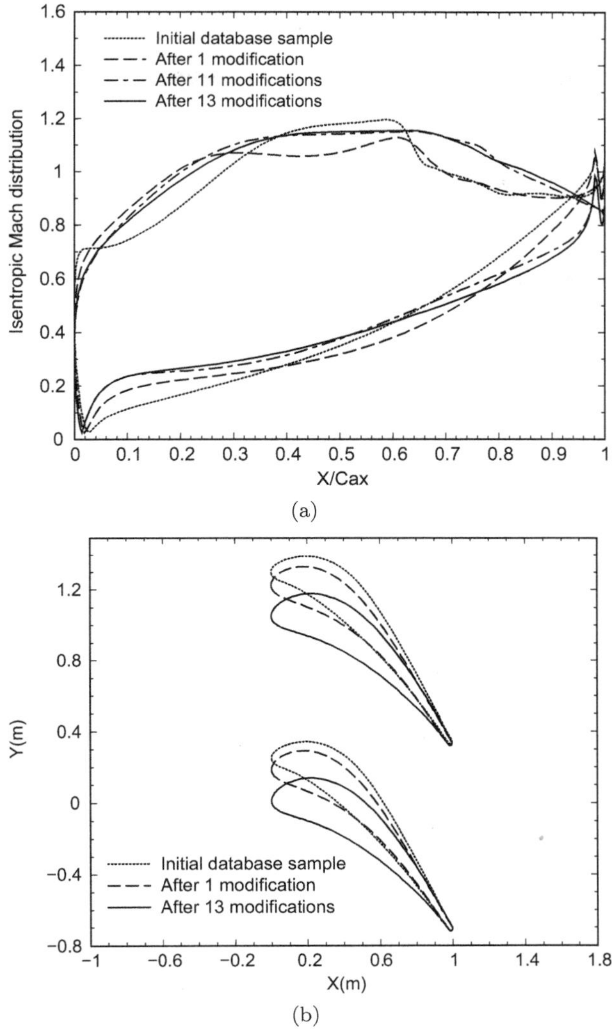

Fig. 6.17 Variation of (**a**) Mach number and (**b**) blade geometry during convergence

9. It indicates that during the first design iterations, the ANN predictions are not very accurate because the database does not sufficiently cover the relevant design space. However this shortcoming is remediated by adding new geometries to the database. Since these blades are close to the desired operating point they provide very valuable information and the ANN becomes more and more accurate. Starting from iteration 13, the ANN predictions are very reliable. This illustrates the self-learning capacity of the proposed

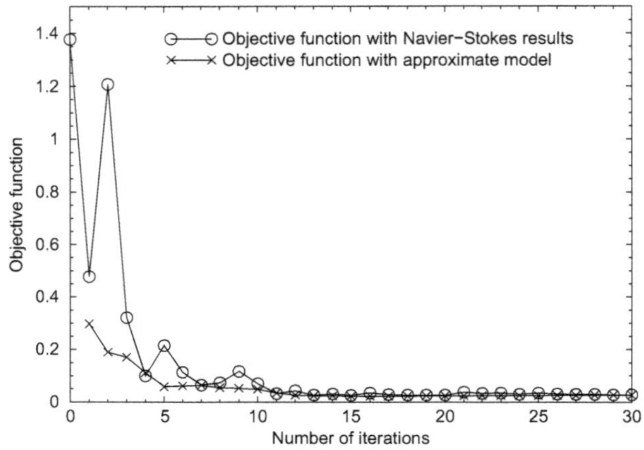

Fig. 6.18 Convergence history

Table 6.3 Inlet conditions at the 3 operating points

	Low $\dot{m}$	Medium $\dot{m}$	Large $\dot{m}$
α(deg)	68.	60.	54.
P_2^o(Pa)	218395.	213447.	187666.
P_3/P_2^o	0.9157	0.9276	0.9430
T_2^o (K)	365.4	360.3	350.4

procedure. The whole procedure could have been stopped after 15 iterations but has been continued to verify good convergence.

6.5 Multipoint Optimization of a Low Solidity Diffuser

The radial compressor vaned diffusers provide a higher pressure recovery and efficiency than vaneless ones but the operating range is limited by stall, at positive incidence, and diffuser throat choking, at negative incidence. Low Solidity Diffusers (LSD) are characterized by a small number of short vanes and do not show a well defined throat section. They intend to stabilize the flow at low mass flow (avoiding diffuser stall) without limiting the maximum mass flow by choking. The solidity (chord/pitch) is typically on the order of 1 or less (Fig. 6.19). A multipoint optimization is mandatory for the LSD design because a wide operating range is the major purpose of these devices.

The optimization of the LSD [12] is done for the 3 operating points listed in Table 6.3. Inlet conditions are different for each operating point and defined by the impeller exit flow at the corresponding mass flows.

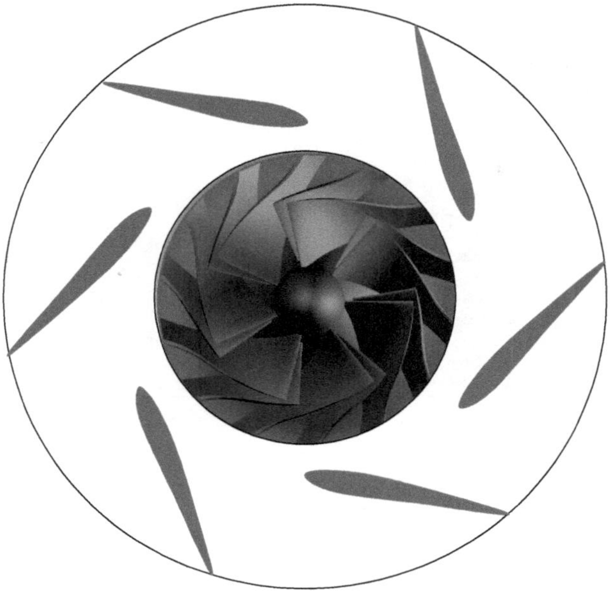

Fig. 6.19 Low Solidity Diffusor

The blade geometry is defined by a NACA 65 thickness distribution superposed on a camber line defined by a 4-parameter Bézier curve. A 5th design parameter is the scale factor for the thickness distribution (between 0.7 and 1.3). The 6th parameter is the number of blades (between 6 and 12).

The performance criteria are the static pressure rise and total pressure loss, non-dimensionalized by the diffuser inlet dynamic pressure

$$Cp = \frac{\bar{P}_3 - \bar{P}_2}{\bar{P}_2^{\circ} - \bar{P}_2} \qquad , \qquad \omega = \frac{\bar{P}_2^{\circ} - \bar{P}_3^{\circ}}{\bar{P}_2^{\circ} - \bar{P}_2}$$

Making the database is quite costly because it requires an NS analysis of each geometry at three operating points. The initial database is therefore limited to only 10 geometries requiring 30 NS calculations on a grid with 400,000 cells.

One wants to maximize what the diffuser is supposed to do: i.e., to increase the static pressure with minimum total pressure losses. The latter is the difference between the real pressure rise and the isentropic one.

$$Cp + \omega = Cp_{\text{isentropic}}$$

The maximum value $Cp_{\text{isentropic}}$ depends on the inlet conditions and diffuser geometry.

The best results have been obtained with the following OF

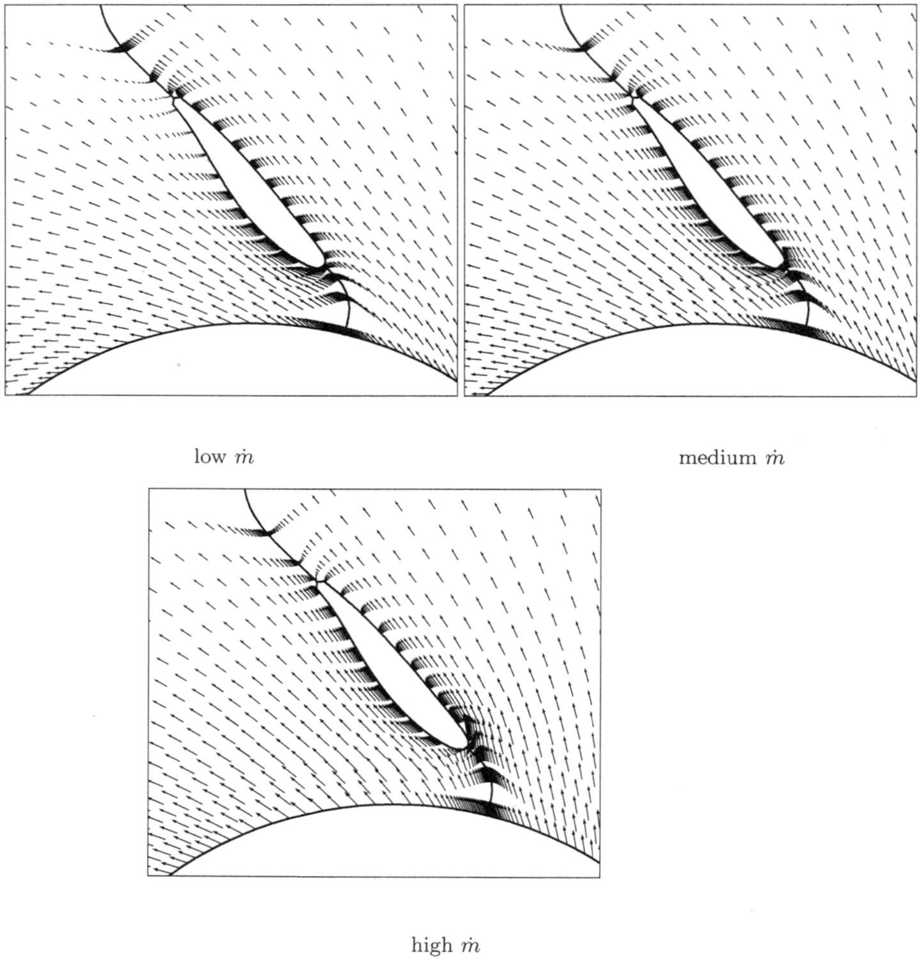

Fig. 6.20 Midspan velocity vector of optimized diffuser

$$OF = (1 - (w_{\text{low}} \cdot Cp_{\text{low}} + w_{\text{med}} \cdot Cp_{\text{med}} + w_{\text{high}} \cdot Cp_{\text{high}}))$$

$$+ w_{\text{low}} \cdot \omega_{\text{low}} + w_{\text{med}} \cdot \omega_{\text{med}} + w_{\text{high}} \cdot \omega_{\text{high}}$$

$$w_{\text{low}} = 0.25 \quad , \quad w_{\text{med}} = 0.5 \quad , \quad w_{\text{high}} = 0.25$$

Using three different ANN, dedicated to the performance prediction at the three operating points, improves the convergence.

The velocity vectors at midspan, shown in Fig. 6.20, indicate attached flow at all three operating points and hence stable operation at low mass flow as well as a large pressure rise at high mass flow.

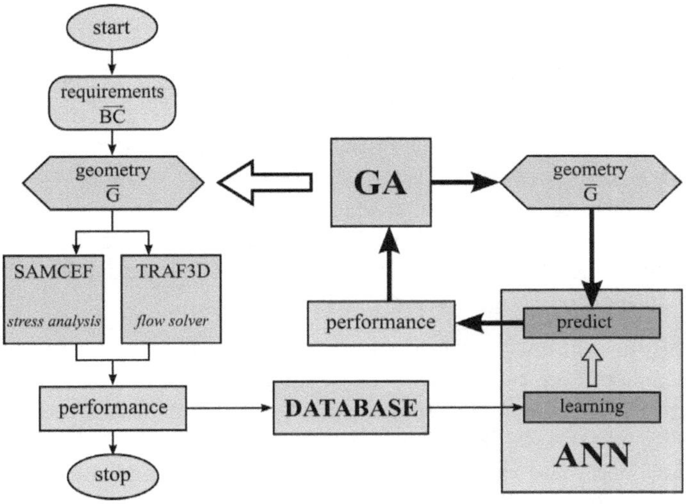

Fig. 6.21 Multidisciplinary optimization flow chart

6.6 Multidisciplinary Optimization

Mechanical constraints such as maximum stress and deformation have a direct impact on the turbomachinery integrity and must therefore be rigorously respected. Some of them can be guaranteed by a simple limitation of a design parameter and do not require any further analysis. Bird ingestion resistance may be accounted for by specifying a minimum leading edge radius (Rle) for fan blades. Corrosion may define the minimum trailing edge radius (Rte) and blade thickness.

However most of the mechanically unacceptable geometries result from a combination of different design parameters and cannot be avoided by restricting the individual parameters. A rigorous approach is the verification of the stress level by an FEA of the $N \times t$ geometries generated by the GA. This is possible by extending the two-level design method to more than one discipline [18]. The GA, searching for the optimum geometry, gets its input from the FEA as well as from the NS flow analysis (Fig. 6.21). The same type of extension can also be made to verify the constraints related to aero-acoustics or weight limitations.

The multidisciplinary optimization method requires the following extensions: stress predictions by the ANN, FEA stress analysis in parallel with the NS calculations, an extension of the OF to account for mechanical targets and the specification of additional design parameters that allow stress reductions. The main advantages of this approach (Fig. 6.21) are:

- The existence of only one "master" geometry, i.e., the one defined by the geometrical parameters used in the GA optimizer. This eliminates

possible approximations and errors when transmitting the geometry from one discipline to another.
- The existence of a global OF accounting for all design criteria. This allows a more direct convergence to the optimum geometry without iterations between the aerodynamically optimum geometry and the mechanically acceptable one.
- The possibility to do parallel calculations. The different analyses can be made in parallel if each discipline is independent, i.e., if stress calculations do not need the pressure distribution on the vanes and flow calculations are not influenced by geometrical deformations.

The multidisciplinary optimization method is illustrated by the design of a 20 mm diameter radial compressor for a micro-gas turbine rotating at 500,000 rpm [18]. The corresponding tip speed of 523.6 m/s results in very high centrifugal stresses. Titanium TI-6AL-4V has been selected for its high yield stress over mass density ratio (σ_{yield}/ρ). The characteristics used in the calculation are: Elasticity modulus = 113.8×10^9 Pa, Poisson modulus = 0.342 and mass density = 4.42×10^3 kg/m^3.

6.6.1 3D Geometry Definition

The hub and shroud meridional contour of radial impellers are defined by third-order Bézier curves, between the leading edge and trailing edge section (Fig. 6.22) [2, 6]. They are fully defined by the control points (X0,R0) to (X3,R3) at hub and shroud. The axial length (X3-X0) and outlet radius R3 are prescribed. They are the result of a preliminary 1D design where one can also account for the off-design operation.

The radius R1 should be larger than R0 at the shroud because otherwise, the unshrouded impeller cannot be mounted. One imposes that X2 is smaller than or equal to X3 at hub and shroud in order to avoid that the impeller exit bends forward. Restricting the possible variations of the design parameters to realistic values also accelerates the convergence.

Second-order curves are used for the upstream and downstream extensions. The points A and B at hub and shroud are automatically adjusted to obtain a smooth transition between the impeller and the radial inlet section. The 6 unknowns that need to be defined during the optimization process are indicated by arrows in Fig. 6.22.

The blade camber line β distribution at hub and shroud are defined by cubic Bézier curves in Bernstein polynomial form

$$\beta = \beta_0(1-u)^3 + \beta_1(1-u)^2u + \beta_2(1-u)u^2 + \beta_3u^3$$

β_0 is the blade camberline angle at the leading edge hub or shroud ($u = 0$) and β_3 is the blade trailing edge camber angle ($u = 1$). Similar curves are defined for the splitter vane β distributions.

The coordinates θ of the blade camber line are computed by integrating the β distribution along hub and shroud (Fig. 6.23):

$$d\theta = \frac{dm \tan\beta}{R}$$

The leading and trailing edge blade angles β_0 and β_3 can vary over $\pm 5°$ around a first estimate of the optimum value. Values of β_1 and β_2 are not constrained but the values of θ_3, obtained by integrating β along hub and shroud, should not be too different at the training edge, i.e., the trailing edge rake angle (Fig. 6.7) should be less than 45°.

The splitter trailing edge blade angles are equal to the full blade values at hub and shroud. This results in 14 design variables for the full and splitter blade camber line definition.

The impeller blade definition is completed by a parameterized thickness distribution (Fig. 6.24). Blade thickness distributions at hub and shroud are function of one parameter: the thickness LE of the ellipse defining the leading edge. Trailing edge thickness TE is related to LE. The blade thickness is fixed at the shroud (LE=TE=0.3 mm). The parameter defining the blade thickness at the hub can vary between 0.3 and 0.6 mm. The same value is used for the main and splitter blades. This increases the number of design parameters by 1.

The streamwise position of the splitter blade leading edge is also a design parameter. It is defined as a percentage of the main blade camber length and

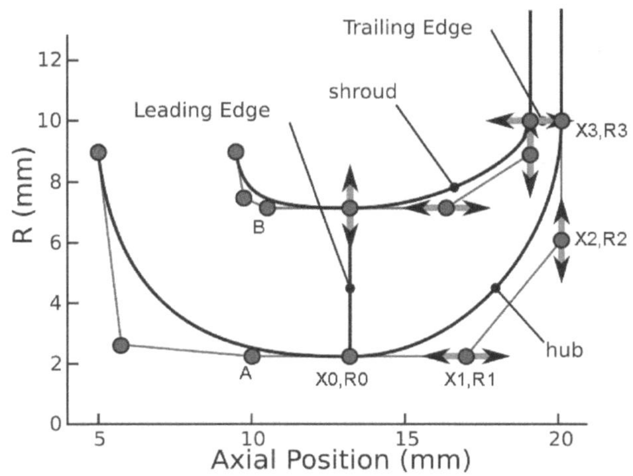

Fig. 6.22 Parameterization of meridional contour. Meridional contour defined by Bézier control points

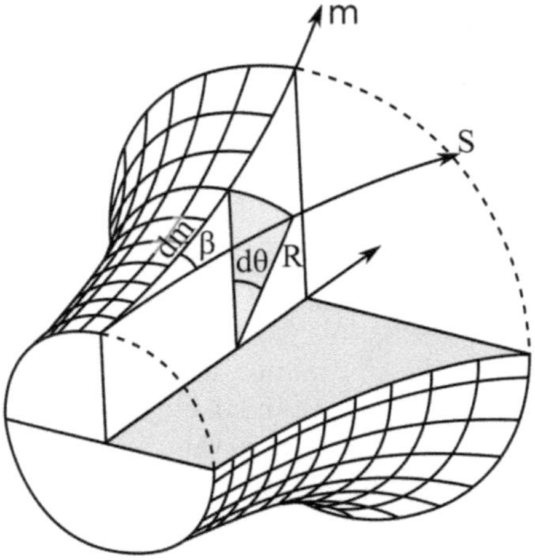

Fig. 6.23 Definition of θ distribution

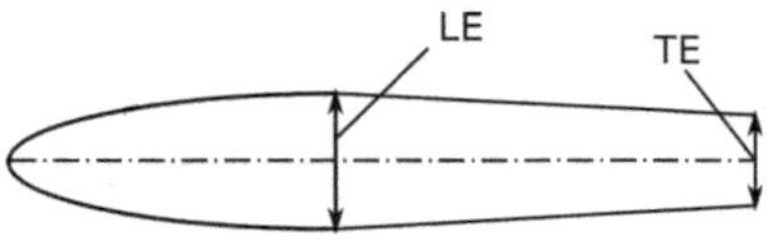

Fig. 6.24 Parameterized thickness distribution perpendicular to the blade camber line (not to scale)

can vary between 20% and 35% at hub and shroud (2 extra design parameters).

The number of blades could also be a design parameter to be optimized, but has been fixed to 7 for manufacturing reasons. This brings the total number of design parameters that need to be defined by the optimizer to 23.

6.6.2 Multidisciplinary Objective Function

The OF driving the multidisciplinary optimization increases with decreasing aero performance and when the aero and mechanical requirements are not met

$$OF = w_m \cdot P_{\text{mech}} + w_\eta \cdot P_{\text{perf}} + w_m \cdot P_{\text{mass}} + w_M \cdot P_{\text{Mach}}$$

The first penalty concerns the mechanical stresses. Limiting the maximum stress results in an inequality and geometries that do not satisfy this condition should be eliminated. However, this information about undesired geometries can also be used to guide the optimization algorithm towards acceptable ones. This is achieved by adding to the OF an extra term that increases

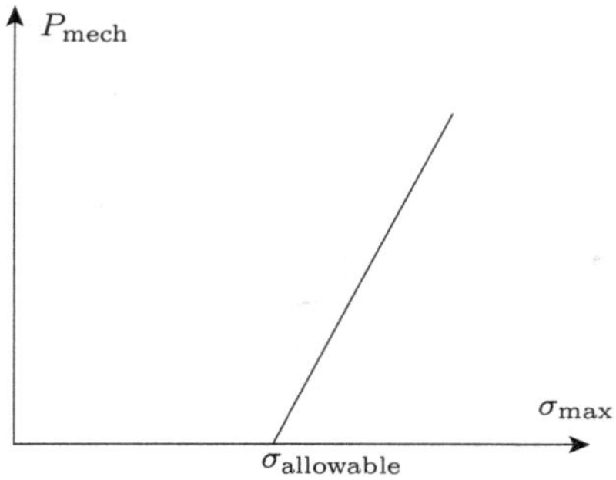

Fig. 6.25 Penalty function for not respecting stress limits

when the stress exceeds the prescribed maximum allowable value. Adding those geometries to the database makes this information available to the ANN which in turn informs the GA about what part of the design space is unacceptable in terms of stress level. Those geometries will therefore not be proposed for further investigation by the NS and FEA.

The penalty increases linearly when a prescribed value $\sigma_{allowable}$ is exceeded and is zero when the inequality $\sigma_{max} < \sigma_{allowable}$ is true (Fig. 6.25)

$$P_{mech} = \max\left[\frac{\sigma_{max} - \sigma_{allowable}}{\sigma_{allowable}}, 0\right]$$

This weak formulation of the stress constraint does not guarantee that the proposed geometry satisfies the stress limit in a strict way. It can not be excluded that an increase of the stress penalty is compensated by an equivalent decrease of another penalty term. However this risk can be minimized by increasing the weight factor of the stress penalty. The final selection of the optimum geometry is anyway made by the designer, on the basis of the NS and FEA results of all geometries produced by the optimization algorithm.

The second penalty term concerns the efficiency and has been already explained in Sect. 6.2.2

The third penalty verifies the mass flow and has now two contributions. The first one, explained in Sect. 6.2.2, increases when the total mass flow differs from the required one by more than a prescribed value. The second one penalizes the difference in mass flow on both sides of the splitter blade

$$P_{massdiff} = \left(\frac{\dot{m}_{upper} - \dot{m}_{lower}}{\dot{m}_{upper} + \dot{m}_{lower}}\right)^2$$

Having the same mass flow in every flow channel will result in a more uniform impeller exit flow and favors a more uniform distribution of blade loading.

The penalty on the Mach number is different from the one used for turbine blades (Sect. 6.4.2) and has two contributions. The first one penalizes negative

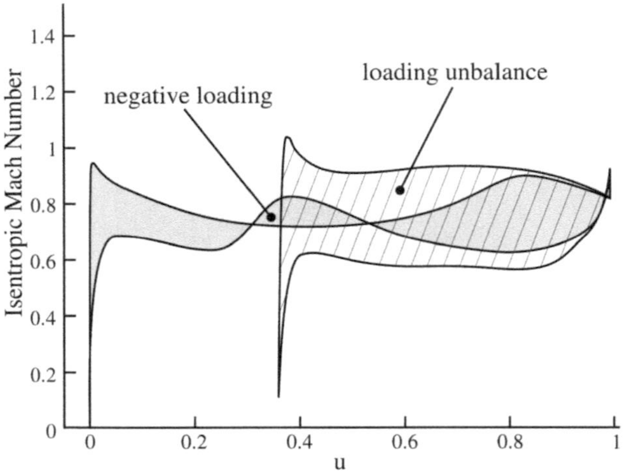

Fig. 6.26 Penalty function for negative loading and loading unbalance in a compressor with splitter vanes

loading and is proportional to the area between the suction and pressure side when the pressure side Mach number is higher than the suction side one (Fig. 6.26). Areas of negative loading are penalized because they result in extra friction losses without contribution to the pressure rise

$$P_{\text{negativeloading}} = \int_0^1 \max(M_{ps}(s) - M_{ss}(s), 0.0) \cdot ds$$

The second contribution to the Mach penalty increases with the load unbalance between main blade and splitter blade. This penalty compares the area between the suction and pressure side Mach number distribution of main blade A_{bl} and splitter blade A_{sp}, corrected for the difference in blade length (Fig. 6.26):

$$P_{\text{load unbalance}} = \left(\frac{A_{bl} - A_{sp}}{A_{bl} + A_{sp}} \right)^2$$

6.6.3 Design Conditions and Results

The computational domain starts at constant radius in the radial inlet and ends in the parallel vaneless diffuser at $R/\mathrm{R2} = 1.5$ (Fig. 6.22). Part of the hub surface at the inlet (for $R < 4$ mm) rotates because it connects the compressor shaft to the electric generator. The total inlet temperature is 293 K and the total inlet pressure is 1.013×10^5 Pa.

Table 6.4 List of penalty parameters and weight factors

stress penalty weight	24.0	mass flow difference weight	20.0
efficiency required	82.5	Mach negative loading weight hub	15.0
efficiency penalty weight	40.	Mach negative loading weight shroud	24.0
mass flow required	20.0 g/s	Mach loading unbalance weight	1.0
mass flow penalty weight	18.0	Mach loading unbalance weight	1.5

The disk thickness at the compressor rim is 1 mm. It is connected to a 8 mm diameter shaft with a fillet of 2 mm radius, all made of one piece. A fillet radius of 0.25 mm is applied at the blade hub to limit the local stress concentrations. The unshrouded impeller has a tip clearance of 0.1 mm, which is 10% of the exit blade height. This is typical for these small impellers and one of the reasons for the moderate efficiencies.

The weights in the OF depend on the application and allow emphasis on performance or on mechanical integrity. They have been determined from the knowledge gained in previous optimizations and are listed in Table 6.4. Taking into account the difference in weight factors, an efficiency drop of 1% is as penalizing as an excess in stress limit of $40/24 = 1.66\%$ (or 6.66 MPa).

The optimization starts from a "baseline" impeller which is the result of a simple aerodynamic optimization without stress analysis. Although this geometry has a good efficiency, it cannot be used because an FEA predicts von Mises stresses in excess of 750 MPa. It serves as a reference for further optimizations.

The TRAF3D solver [3] with an extension to calculate impellers with splitters is used to predict the aerodynamic performance of the radial impellers. Structured H-grids with $2 \times 216 \times 48 \times 52$ (1,090,000 cells) are used for all computations to guarantee a comparable accuracy for all the samples stored in the database. All computations are non-adiabatic with wall temperature fixed at 400 K, as found in a previous study on the heat transfer from the turbine to the compressor [19].

The commercial code SAMCEF [14] is used for the stress calculation. Quadratic tetrahedral elements are used as a compromise between element quality and automatic meshing. Similar grids with 250,000 nodes and 160,000 elements are used for all samples. The grid is refined in areas of stress concentrations. Periodic boundary conditions are applied, such that only a $1/7^{\text{th}}$ part of the geometry needs to be analyzed. The maximum allowable stress is a function of the material temperature. A large safety margin, to account for vibrations and possible local temperature peaks, results in a maximum allowable value of 400 MPa.

Eight individual ANNs are used: one to predict the efficiency, two to predict the mass flow in the channels on each side of the splitter, 4 ANNs predict the Mach number distribution respectively at hub and shroud of the full and splitter blades and one predicts the maximum stress in the geometry. This

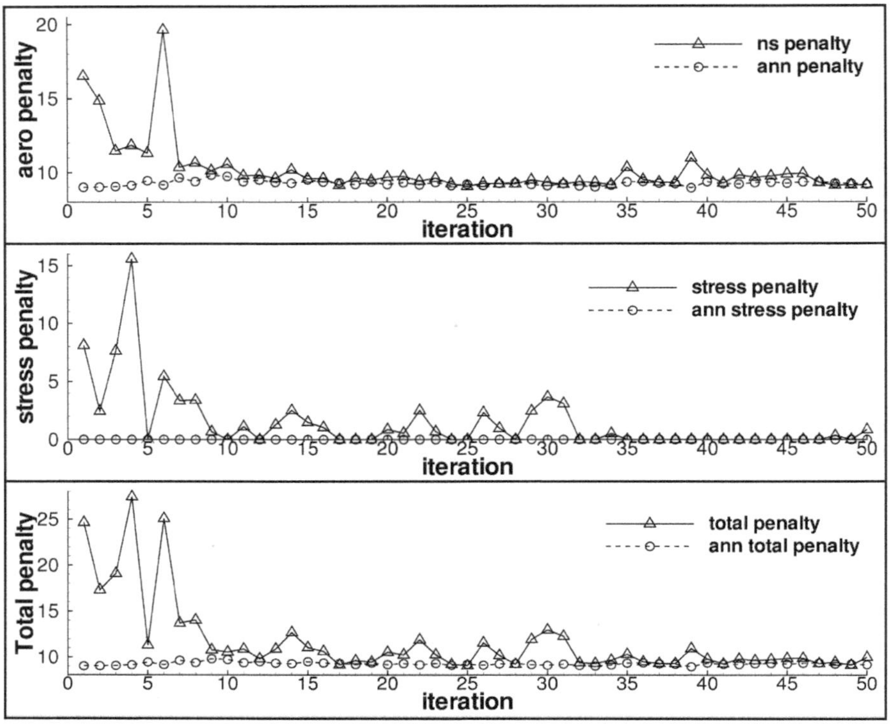

Fig. 6.27 Convergence history of the optimization

split into dedicated ANNs enhances the accuracy by which important values such as efficiency and maximum stress can be predicted.

A 2^{n-p} factorial design is used where n is 23 while p is fixed at 17. This results in a total of $2^6 = 64$ samples in the initial database. An initial database contains only 53 geometries since 13 geometries defined by the DOE technique could not be analyzed due to geometrical constraints (intersection between the main blade and the splitter blade). Two additional geometries have been added, namely the baseline geometry and the central case. The latter is a geometry with all parameters at 50% of their range.

Figure 6.27 shows the convergence history of the optimization. The "aero penalty" (based only on NS predictions of the efficiency, the Mach number distribution and mass flow), the "stress penalty" (based on the result of the FEA analyses) and the "total penalty" are all compared to the ones predicted by the ANNs. One observes a decrease in the discrepancy between both prediction methods with iteration number. This is the consequence of an increasing number of samples in the database, resulting in more accurate ANNs.

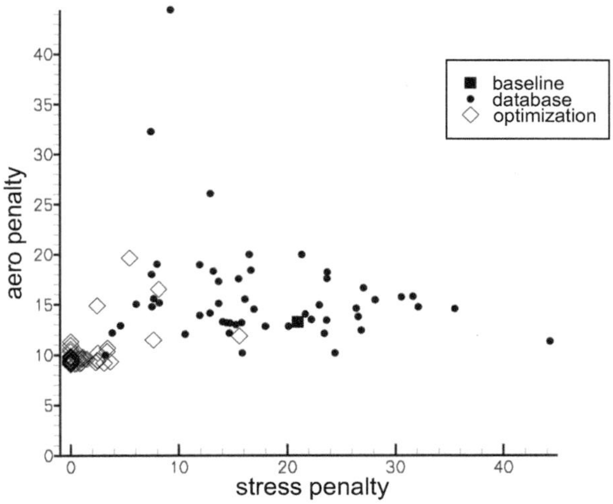

Fig. 6.28 Aero penalty versus stress penalty for baseline, database and optimized geometries

Only 10 iterations are needed to obtain a very good agreement for the aero penalty. The ANN predicted stress penalty is zero for every geometry proposed by the GA. However it takes more than 15 iterations before the FEA confirms that the proposed geometries do not violate the mechanical constraints.

The good agreement in both stress and aero penalties over the last 18 iterations indicates that the ANN predictions are reliable. It means that the same optimum geometry would have been obtained if the GA optimization had been based on the more sophisticated NS and FEA analyses. Hence no further improvement can be expected and the optimization procedure can be stopped after 35 iterations.

The aero penalty is plotted versus the stress penalty in Fig. 6.28. The database geometries show a good spread. Geometries created during the optimization process are all in the region of low penalties. Most of them outperform the geometries of the database. Only a few geometries have penalties of the same order as the database samples. Those geometries are the ones created during the first 10 iterations when the ANN is still inaccurate.

Figure 6.29 is a zoom on the low penalty region of Fig. 6.28. A large number of geometries have zero stress penalties but with a different aero penalty. The geometries corresponding to iteration 17, 49 and 25 have the lowest aero penalty and satisfy the stress constraints. Details are listed in Table 6.5. Iteration 2 has the highest efficiency (60.4%) but has a high aero penalty due to negative loading and loading unbalance. This geometry is also added to Table 6.5.

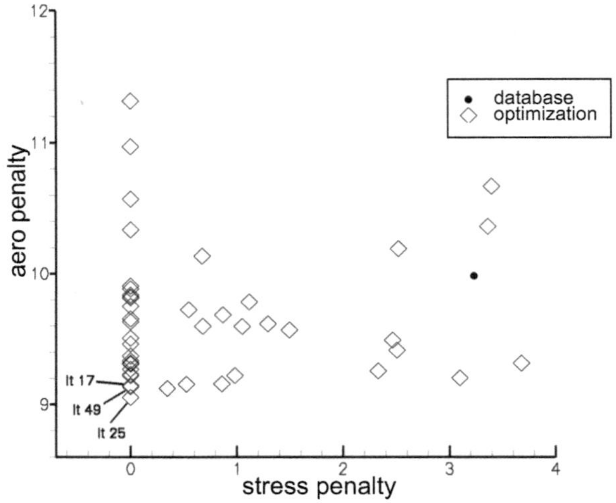

Fig. 6.29 Zoom on the low penalty region of Fig. 6.28

Table 6.5 Comparison between baseline and optimized impellers

	Baseline	Iter. 2	Iter. 17	Iter. 25	Iter. 49
$\eta_{TS}(\%)$	62.34	60.42	60.31	60.06	59.68
P_{aero}	13.24	14.86	9.14	9.05	9.14
$P_{\text{loadunbalance}}$	0.06	3.84	0.26	0.04	0.00
σ_{vonMises} (MPa)	749.	440	389	367	396
Blade lean (°)	-7.8	-11.8	-8.6	-7.3	-15.0
$\dot{m}$ (g/s)	25.9	19.6	20.2	20.2	20.1
Power (kW)	3.19	2.52	2.61	2.62	2.62
Spec.Pow. (W.s/kg)	1123.2	128.6	129.2	129.7	130.3

From all the geometries created during the optimization, iteration 25 performs best. It has an efficiency a little lower than iteration 17 but less loading unbalance and the stresses are 33 MPa below the limit. In spite of its high efficiency, the baseline impeller shows a high aero penalty because of a very high mass flow.

The influence of the stress penalty on the optimization is clear by comparing the values of the baseline impeller with the ones of iteration 25. The reduction of the maximum stress level with 370 MPa is at the cost of a 2.3% decrease of efficiency.

Figures 6.30 and 6.31 show the von Mises stresses in the baseline geometry and the one of iteration 25, respectively. The drastic reduction in stress of the optimized impeller is due to:

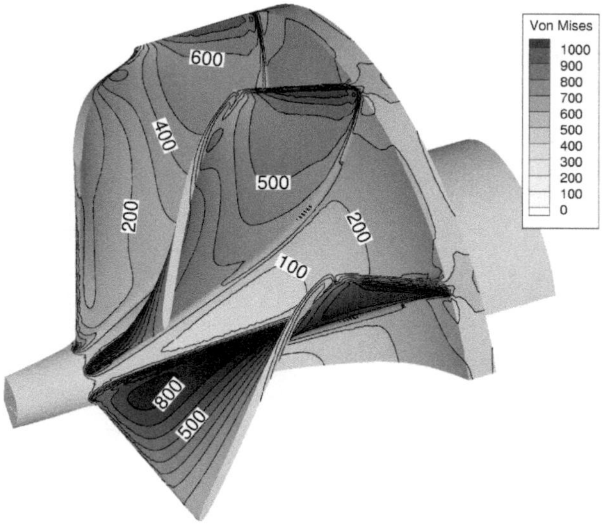

Fig. 6.30 von Mises stresses due to centrifugal loading in the baseline

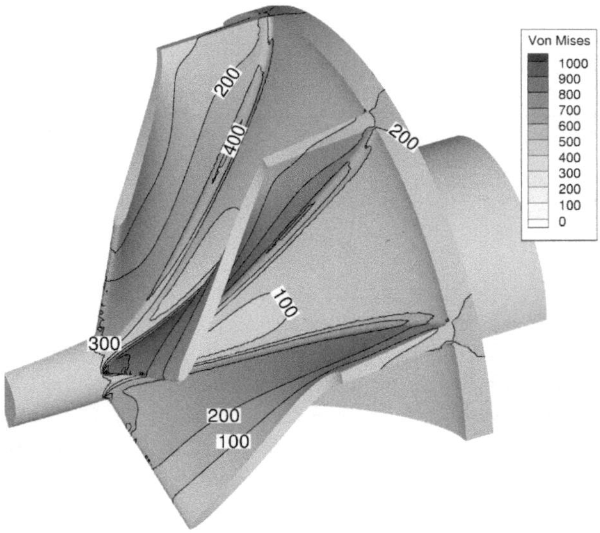

Fig. 6.31 von Mises stresses due to centrifugal loading in iteration 25

- the reduced blade height at the leading edge, resulting in lower centrifugal forces at the leading edge hub;
- the increase of blade thickness at the hub;
- the modified blade curvature.

The latter two result in a decrease of the bending stresses.

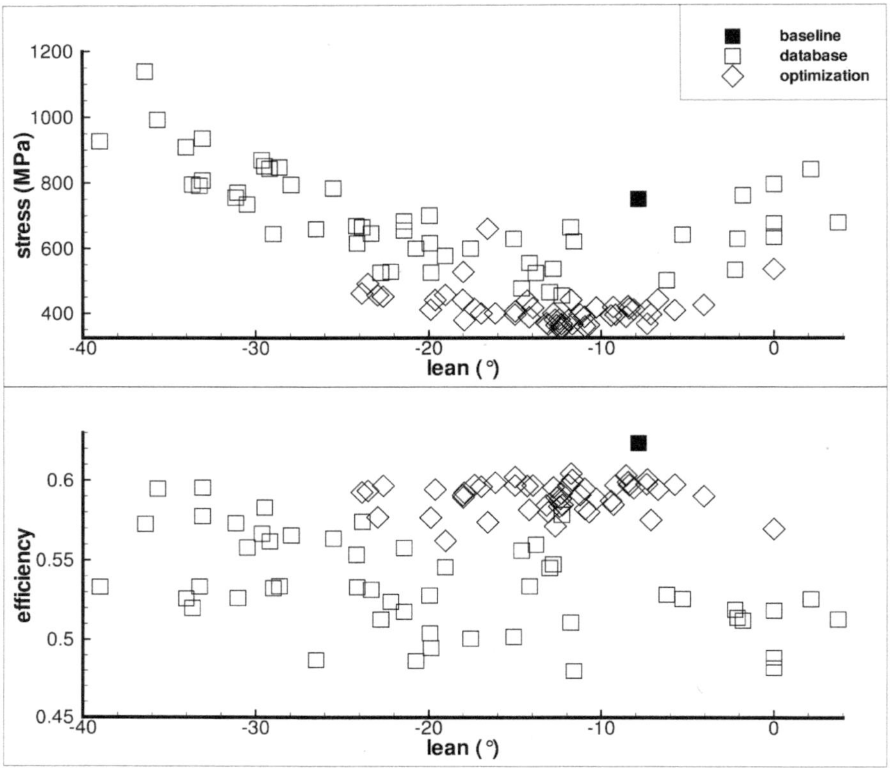

Fig. 6.32 Blade lean versus stress and efficiency for database and optimization geometries

The blade lean is positive in the direction of rotation (Fig. 6.7). It results from the integration of the β distribution at hub and shroud when the trailing edge rake is limited to 45.0°. It is often limited to a small arbitrary value for stress reasons. Iteration 49 on Table 6.5 shows that it can be as low as −15.0° without exceeding the maximum stress limit. Figure 6.32 confirms that minimum stresses are observed around −15.0°. Several geometries with good efficiency are found for lean angles between −40.0° to −5.0°. The decreasing efficiency for lean angles larger than −5.0° suggests that a limited negative lean may have a favorable effect on performance.

Figure 6.33 shows the impact of the leading edge blade height on the stress and efficiency. The radius at the LE shroud (see Fig. 6.22) can vary between 6.5 and 7.5 mm, resulting in a blade height of 4.25 and 5.25 mm, respectively. Values at 4.5 mm and 5.0 mm are database samples.

Shortening the blades lowers the stresses but the database samples suggest a small drop in efficiency. This explains the difficulty in maintaining a high efficiency when reducing the stress. However, the optimized geometries have shorter vanes and show a high efficiency. This indicates that the efficiency also

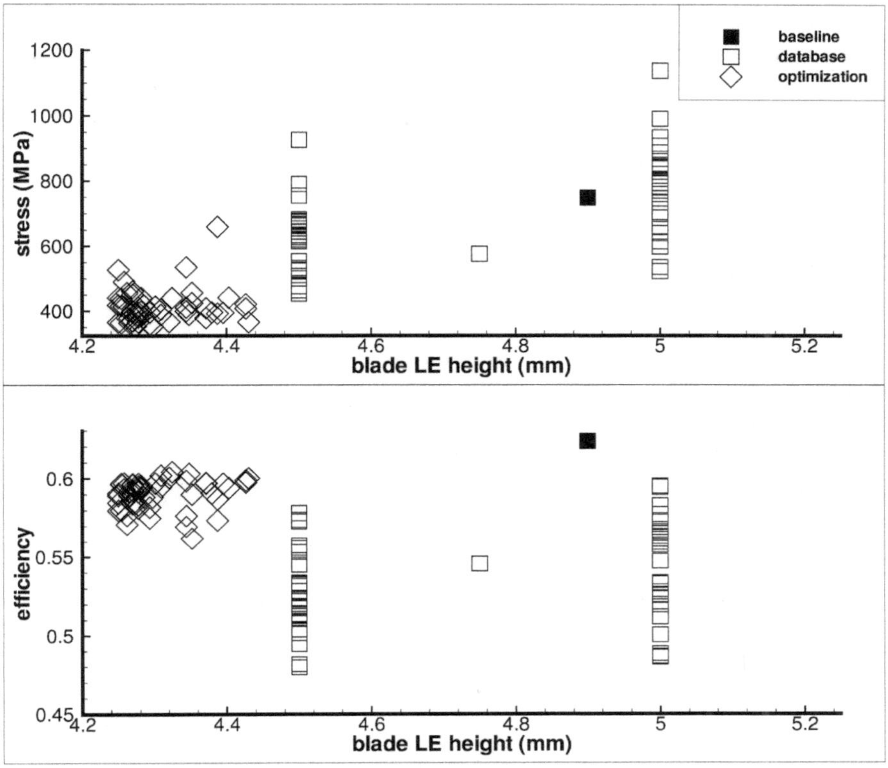

Fig. 6.33 Blade leading edge height versus stress and efficiency

depends on an optimum choice of other parameters. Although they have a less pronounced influence on stress and efficiency, a correct definition of their value is needed to reach the optimum. This illustrates the strongly coupled nature of the design problem and the need for an optimization tool.

6.7 Conclusions

It has been shown how a two-level optimization technique, an adequate parameter selection for the GA, the use of DOE for the definition of the database and an optimized learning technique for the ANN can considerably decrease the computational effort required by evolutionary theories. The proposed procedure is a self-learning system that makes full use of the expertise gained during previous designs.

The automated design method can be used with any flow solver and does not require the definition of a target pressure or Mach number distribution.

The Mach number-based criteria included in the Objective Function help enforce the convergence to the optimum and improve the off-design performance.

The reduction in computer effort makes the design of customized profiles and the multipoint optimizations affordable, as illustrated by the design of a transonic turbine blade and a Low Solidity Diffuser.

The use of a pseudo Objective Function to account for the mechanical and other constraints is presented, and the advantages and disadvantages are discussed. It is shown that the method is able to find the optimum combination of design parameters allowing a drastic reduction of the stresses with minimum penalty on efficiency.

The optimization algorithm provides the designer more insight into the multidisciplinary design problem. The main parameters that allow a reduction of the stresses are identified during the optimization process. This optimum combination may result in unexpected geometries that would not be accepted when using simplified stress criteria.

It has been shown how the use of computerized design techniques is a powerful tool to cope with the increasing complexity of advanced turbomachinery component design. However, the outcome of this valuable support still depends on the input of the designer in terms of a careful selection of design parameters, a clear definition of objectives and constraints as well as validation of results.

Acknowledgements The contributions by Dr. Stephane Pierret, Dr. Zuheyr Alsalihi and Dr. Tom Verstraete to the development of this method are gratefully acknowledged.

References

1. Aarts, E.H.L., Korst, J.H.M.: Simulated annealing in Boltzmann machines. Wiley Chichester (1987)
2. Alsalihi, Z., Van den Braembussche, R.A.: Evaluation of a design method for radial impellers based on artificial neural network and genetic algorithm. In: Proc. of ESDA 2002, 6th Biennial Conference on Engineering Systems Design and Analysis. Istanbul (2002)
3. Arnone, A.: Viscous analysis of three-dimensional rotor flow using a multigrid method. ASME Journal of Turbomachinery **116**, 435–445 (1994)
4. Bäck, T.: Evolutionary Algorithms in Theory and Practice. Oxford University Press, New York (1996)
5. Carroll, D.L.: FORTRAN genetic algorithm (GA) driver version 1.7.1a (2001). URL `URL:http:/cuaerospace.com/carroll/ga.html`
6. Cosentini, R., Alsalihi, Z., Van den Braembussche, R.A.: Expert system for radial compressor optimization. In: Proc. 4th European Conference on Turbomachinery. Firenze (2001)
7. Demeulenaere, A., Van den Braembussche, R.A.: Three-dimensional inverse design method for turbine and compressor blades. In: Design Principles and Methods for Aircraft Gas Turbine Engines, RTO MP-8 (1998)

8. Harinck, J., Alsalihi, Z., Van Buytenen, J.P., Van den Braembussche, R.A.: Optimization of a 3D radial turbine by means of an improved genetic algorithm. In: Proceedings of European Turbomachinery Conference. Lille (2005)
9. Kostrewa, K., Van den Braembussche, R.A., Alsalihi, Z.: Optimization of radial turbines by means of design of experiment. Tech. Rep. VKI-PR-2003-17, von Kármán Institute for Fluid Dynamics (2003)
10. Lichtfuss, H.J.: Customized profiles – the beginning of an area. In: ASME Turbo Expo 2004, Paper GT2004-53742 (2004)
11. Montgomery, D.C.: Design of Experiments. John Wiley & Sons, Inc. (1997)
12. Nursen, C., Van den Braembussche, R.A., Alsalihi, Z.: Analysis and multipoint optimization of low solidity vaned diffusers. Tech. Rep. VKI-SR-2002-31, von Kármán Institute for Fluid Dynamics (2002)
13. Pierret, S., Van den Braembussche, R.A.: Turbomachinery blading design using Navier Stokes solver and artificial neural network. ASME Journal of Turbomachinery **121**, 326–332 (1999)
14. SAMTECH group: SAMCEF FEA code. URL `www.samcef.com`
15. Siamion, J., Coton, T., Van den Braembussche, R.A.: Design and evaluation of a highly loaded LP turbine blade. In: Proc. 5th ISAIF Conference (International Symposium on Experimental and Computational Aerothermodynamics of Internal Flows). Gdansk (2001)
16. Thilmany, J.: Walkabout in another world. Mechanical Engineering **122**(11), 1–9 (2000)
17. Vanderplaats, G.N.: Numerical Optimization Techniques for Engineering Design. McGraw-Hill (1984)
18. Verstraete, T., Alsalihi, Z., Van den Braembussche, R.A.: Multidisciplinary optimization of a radial compressor for micro gas turbine applications. In: ASME Turbo Expo 2007, Paper GT2007-27484 (2007)
19. Verstraete, T., Alsalihi, Z., Van den Braembussche, R.A.: Numerical study of the heat transfer in micro gas turbines. Journal of Turbomachinery **129**(4), 835–841 (2007)
20. Volpe, G.: Geometric and surface pressure restrictions in airfoil design. In: Special Course on Inverses Methods for Airfoil Design for Aeronautical and Turbomachinery Applications, AGARD-R-780 (1990)

Chapter 7
CFD-based Optimization for Automotive Aerodynamics

Laurent Dumas

Abstract The car drag reduction problem is a major topic in the automotive industry because of its close link with fuel consumption reduction. Until recently, a computational approach of this problem was unattainable because of its complexity and its computational cost. A first attempt in this direction has been presented by the present author as part of a collaborative work with the French car manufacturer Peugeot Citroën PSA [4]. This article described the drag minimization of a simplified 3D car shape with a global optimization method that coupled a Genetic Algorithm (GA) and a second-order Broyden-Fletcher-Goldfarb-Shanno (BFGS) method. The present chapter is intended to give a more detailed version of this work as well as its recent improvements. An overview of the main characteristics of automotive aerodynamics and a detailed presentation of the car drag reduction problem are respectively proposed in Sects. 7.1 and 7.2. Section 7.3 is devoted to the description of various fast and global optimization methods that are then applied to the drag minimization of a simplified car shape discussed in Sect. 7.4. Finally in Sect. 7.5, the chapter ends by proposing the applicability of CFD-based optimization in the field of airplane engines.

Laurent Dumas
Université Pierre et Marie Curie
Laboratoire Jacques-Louis Lions
4, place Jussieu, 75230 Paris Cedex 05, France
(e-mail: `laurent.dumas@upmc.fr`)

7.1 Introducing Automotive Aerodynamics

7.1.1 A Major Concern for Car Manufacturers

In the past, the external shape of cars has evolved particularly for safety reasons, comfort improvement and also aesthetic considerations. Consequences of these guidelines on car aerodynamics were not of major concern for many years. However, this situation changed in the 70's with the emergence of the oil crisis. To promote energy conservation, studies were carried out and it was discovered that the amount of the aerodynamic drag in the fuel consumption ranges between 30% during an urban cycle and 75% at a 120 km/h cruise speed. Since then, decreasing the drag force acting on road vehicles and thus their fuel consumption, became a major concern for car manufacturers. Growing ecological concerns within the last decade further make this a critically relevant issue in the automotive research centers.

The process of drag creation and the way to control it was first discovered experimentally. In particular, it was found that the major amount of drag was due to the emergence of flow separation at the rear surface of cars. Unfortunately, unlike in aeronautics where it can be largely excluded from the body surface, this aerodynamic phenomenon is an inherent problem for ground vehicles and can not be avoided. Moreover, the associated three-dimensional flow in the wake behind a car exhibits a complex 3D behavior and is very difficult to control because of its unsteadiness and its sensitivity to the car geometry.

The pioneering experiments of Morel and Ahmed done in the late 70's on simplified geometries also called bluff bodies, are now described in Sects. 7.1.2 to 7.1.4.

7.1.2 Experiments on Bluff Bodies

Two major experiments have been done on bluff bodies, the first one by Morel in 1978 [18] and the second by Ahmed in 1984 [1]. The objective was to study the flow behavior around cars with a particular type of rear shape called hatchback or fastback. These experiments are even now used as a reference in many numerical studies [8, 10, 12, 13, 16].

The bluff body used by Ahmed, similar to the one used by Morel, is illustrated in Fig. 7.1. It has the same proportions as a realistic car but with sharp edges. More precisely, the ratio of length/width/height is equal to 3.33/1.5/1. In both cases, the rear base is interchangeable by modifying the slant angle denoted here as α. The Reynolds numbers are taken equal to 1.4×10^6 and 4.29×10^6 in the Morel and Ahmed experiments, respectively.

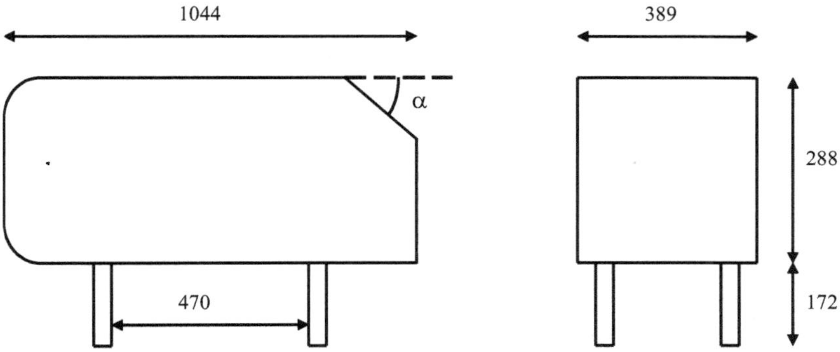

Fig. 7.1 The Ahmed bluff body

7.1.3 Wake Flow Behind a Bluff Body

The most difficult flow region to predict is located at the wake of the car where recirculation and separation occur. It is also the region which is responsible for most of the car drag (see Sect. 7.1.4).

In a time-averaged sense, two distinct regimes depending on the slant angle α, called Regime I and II, have been observed in the experiments done by Morel and Ahmed. The value of the critical angle α_c between both regimes is approximately equal to 30 degrees in each experiment but can slightly change depending on the Reynolds number and the exact geometry.

- Regime I ($\alpha_c < \alpha < 90°$): In this case, the flow exhibits a full 3D behavior with a separation area including the whole slant and base area. The recirculation zones, coming from the four parts of the car (roof, floor and the two base sides) gather and form a pair of horseshoe vortices situated one above another in a separation bulb (see zones A and B in Fig. 7.2). Vortices, coming off the slant side edges are also present (zone C in Fig. 7.2).
- Regime II ($0 < \alpha < \alpha_c$): For low values of α, the flow remains two-dimensional and separates only at the rear base. Two counter-rotating vortices appear from the roof and the floor similar to what happens around airfoils. When α increases up to α_c, the flow becomes three-dimensional because of the appearance of two longitudinal vortices issued from the side walls of the car.

The critical value of α_c corresponds to an unstable configuration associated with a peak in the drag coefficient (see Sect. 7.1.4). In this case, a slight change can generate a high modification of the wake flow. For these reasons, it is essential to avoid such angle value in the design of real cars.

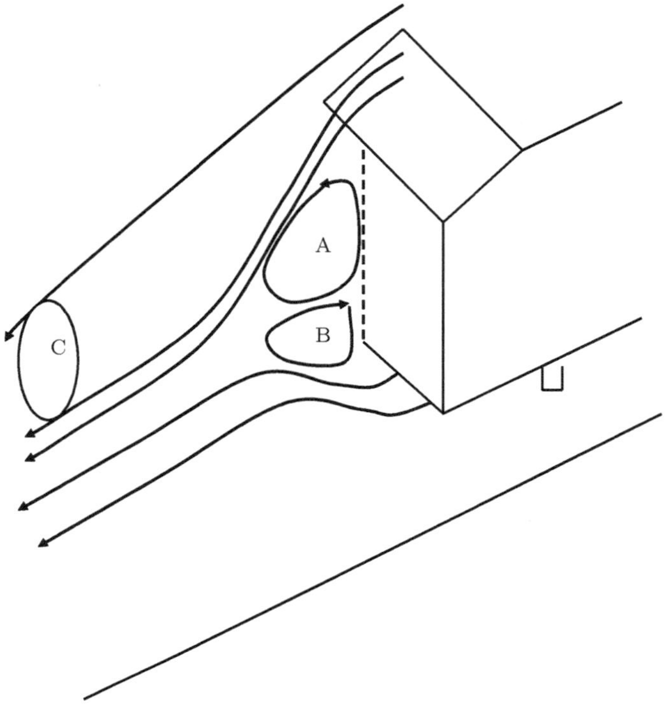

Fig. 7.2 Wake flow behavior behind Ahmed's bluff body

7.1.4 Drag Variation with the Slant Angle

A dimensionless coefficient, called drag coefficient and related to the drag force acting on the bluff body, is defined as follows:

$$C_d = \frac{F_d}{\frac{1}{2}\rho V_\infty^2 S} \ . \tag{7.1}$$

In this expression, ρ represents the air density, V_∞ is the freestream velocity, S is the cross section area and F_d is the total drag force acting on the car projected on the longitudinal direction. Note that the drag force F_d can be decomposed into a sum of a viscous drag force and a pressure drag force.

A first striking result observed by Morel is that the slant surface and the rear base are responsible for more than 90% of the pressure drag force. Moreover, the latter represents more than 70% of the total drag force. These observations have been confirmed by the Ahmed experiment where only 15% to 25% of the drag is due to the viscous drag. Such results can be explained by the analysis of the wake flow discussed in Sect. 7.1.3, that is, the large separation area will induce the major part of the total drag force.

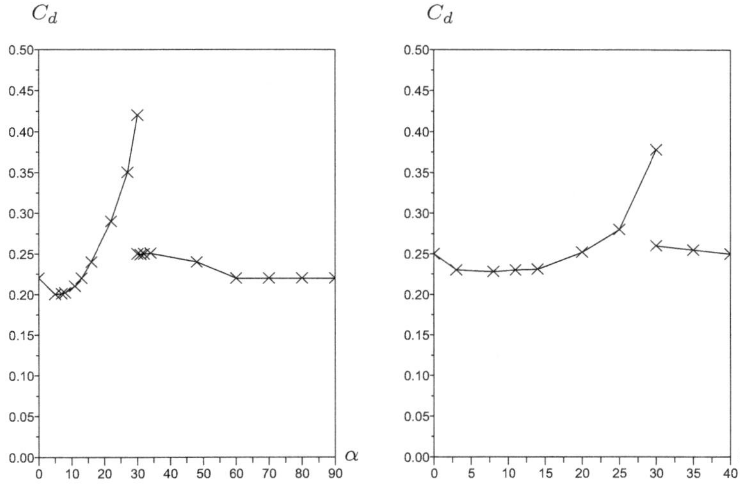

Fig. 7.3 Drag measured for the Morel (left) and Ahmed (right) bluff-body for various slant angles

The variations of the drag coefficient for the Morel and the Ahmed bluff body with respect to the slant angle α are displayed in Fig. 7.3.

Both graphs exhibit the same variations, in particular with a peak value at a critical angle α_c near 30 degrees, already introduced in the previous subsection. The shape of the curve can also be explained by referring to the wake flow behind the car: for small values of α for which the flow is two-dimensional, the drag is directly linked to the dimensions of the separated area. Then, when the flow becomes three dimensional, the separation bulb that appears absorbs a growing amount of the flow energy, thus leading to a large increase of the drag coefficient until α reaches α_c. Above this value, the airflow no longer feeds the vortex systems. Consequently, the static and total pressure experience a sudden rise at the base thus drastically reducing the drag coefficient almost to its value at a zero slant angle.

7.2 The Drag Reduction Problem

After the description in the previous section of the main features of automotive aerodynamics, the car drag reduction problem is now stated for real cars in Sect. 7.2.1. The numerical modelization of this problem in view of an automatic drag minimization is then presented in Sect. 7.2.2.

Table 7.1 Examples of drag coefficient of old or modern cars

Car	C_d
Ford T (1908)	0.8
Hummer H2 (2003)	0.57
Citroen SM (1970)	0.33
Peugeot 407 (2004)	0.29
Tatra T77 (1935)	0.212

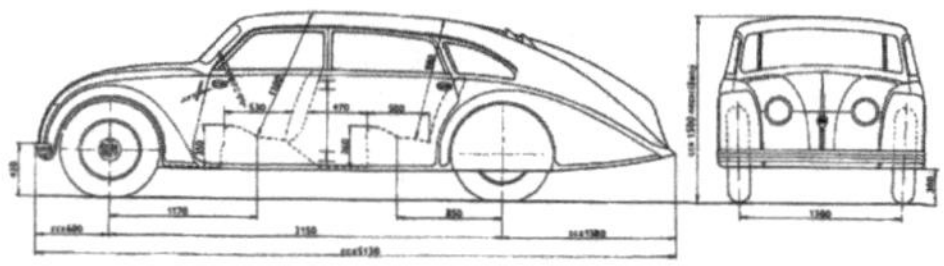

Fig. 7.4 Side and front view of Tatra T77 (1935, $C_d = 0.212$). With permission of Tatra Auto Klub, Slovakia

7.2.1 Drag Reduction in the Automotive Industry

The dimensionless drag coefficient C_d defined in Eq. (7.1), is the main coefficient for measuring the aerodynamic performance of a given car. Examples of values of C_d for past or existing cars are presented in Table 7.1.

It can be seen from this table that the drag coefficient of cars has been decreasing during the last century even though other considerations like comfort or aesthetics have also been taken into account to popularize a high drag model (see Hummer H2) or abandon a low drag model (see Tatra T77).

The last model of Tatra T77 had a remarkable low drag value of 0.212. A schematic side and front view of this car is illustrated in Fig. 7.4. The short forebody compared to the extended tailored rear shape conforms to the experimental observations stated in the previous section, saying that the separation zone at the slant and base area, largely reduced here, is responsible for the major part of C_d.

For a standard car, Table 7.2 displays the repartition of the relative contribution of various elements on the total drag.

It shows in particular that 70% of the drag coefficient depends on the external shape. Such a large value justifies the interest of a numerical modelization of the problem in order to find numerically innovative external shapes that will largely reduce the drag coefficient.

Table 7.2 Drag repartition on a realistic car

Position	percent of C_d
Upper surface	40%
Lower surface	30%
Wheels	15%
Cooling	10%
Others	5%

7.2.2 Numerical Modelization

7.2.2.1 The Navier-Stokes Equations

The incompressible Navier-Stokes equations govern the flow around the car shape. Denote S_c the car surface, G the ground surface and Ω a large volume around S_c and above G, then using the Einstein notation, the Navier-Stokes equations are written as follows:

- Incompressibility:

$$\frac{\partial u_i}{\partial x_i} = 0 \quad \text{on } \Omega \tag{7.2}$$

- Momentum ($1 \leq j \leq 3$):

$$\frac{\partial u_j}{\partial t} + \frac{\partial u_i u_j}{\partial x_i} = -\frac{1}{\rho}\frac{\partial p}{\partial x_j} + \frac{\partial}{\partial x_i}\left[\nu\left(\frac{\partial u_i}{\partial x_j} + \frac{\partial u_j}{\partial x_i}\right)\right] \quad \text{on } \Omega \tag{7.3}$$

where $u_j(t,x)$, $p(t,x)$ and ρ are respectively the flow velocity, pressure and density. The boundary conditions for the velocity are of Dirichlet type:

$$u_i = 0 \quad \text{on } S_c \cup G \quad \text{and} \quad u_i = V_\infty \quad \text{on } \partial\Omega \setminus (S_c \cup G)\ . \tag{7.4}$$

As the Reynolds number is very high in real configurations, usually more than 10^6, a turbulence model must be added. This model must be of reduced computational cost in view of the large number of simulations, more than 100, that need to be done during the optimization process. This explains why the Large-Eddy Simulation (LES) model can not be used here (see [8] and references herein for an example of the application of LES for a single computation around the Ahmed bluff-body).

A Reynolds-Averaged Navier-Stokes (RANS) turbulence model which consists of averaging the previous equations (7.2) and (7.3), is chosen here. Denoting $\bar{u}_i$ the averaged velocity, a closure principle for the term $\overline{u_i u_j}$ has to be defined. The most popular way to do it leads to the well known k–ε model. In this model, the averaged equations (7.3) are rewritten as:

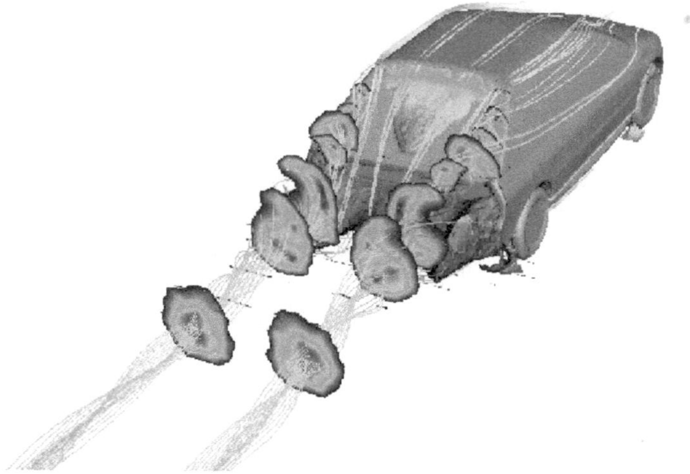

Fig. 7.5 Numerical flow field around a realistic car (Peugeot 206)

$$\frac{\partial \bar{u}_j}{\partial t} + \frac{\partial \bar{u}_i \bar{u}_j}{\partial x_i} = -\frac{1}{\rho}\frac{\partial \bar{p}}{\partial x_j} + \frac{\partial}{\partial x_i}\left[(\nu + \nu_t)\left(\frac{\partial \bar{u}_i}{\partial x_j} + \frac{\partial \bar{u}_j}{\partial x_i}\right) - \frac{2}{3}\delta_{i,j}k\right] \quad \text{on } \Omega \tag{7.5}$$

where ν_t is called the eddy viscosity and is related to the turbulent kinetic energy k and its rate of dissipation ε by

$$\nu_t = C_\mu \frac{k^2}{\varepsilon} \ . \tag{7.6}$$

In this expression, C_μ is a constant and the new variables k and ε are obtained from a set of two equations [17].

In the present configuration, it has been observed that the second-order closure model called Reynolds Stress Model (RSM) with an adequate wall function gives better results than the k–ε method [16]. This model will be preferred in the forthcoming simulations despite the small computational overcost on the order of 40% A first example of such flow computations around a realistic car is given in Fig. 7.5.

7.2.2.2 The Cost Function to Minimize

The cost function that will have to be minimized is the drag coefficient already introduced in (7.1). It can be rewritten in the following way after separating the pressure part and the viscous part:

$$C_d = \frac{\iint_{S_c}(p - p_\infty)\boldsymbol{n} d\sigma}{\frac{1}{2}\rho V_\infty^2 S} + \frac{\iint_{S_c}\boldsymbol{\tau}\cdot\boldsymbol{\nu} d\sigma}{\frac{1}{2}\rho V_\infty^2 S} \tag{7.7}$$

where $\boldsymbol{n}$ is the normal vector, $\boldsymbol{\tau}$ the viscous stress tensor and $\boldsymbol{\nu}$ is the projection of the velocity vector to the element of shape $d\sigma$.

The optimization will thus consist to reduce this drag coefficient by changing the car shape S_c and particularly its rear shape. Of course, two types of constraints will have to be added: geometric type (on volume, total length or cross section) and aerodynamic type (by fixing other aerodynamic moments for instance).

The numerical computation of the drag coefficient being very costly and very sensitive to the rear geometry explains why the numerical approach of the drag reduction problem has been for so long unattainable. The next section that presents fast and global optimization methods tries to make it possible.

7.3 Fast and Global Optimization Methods

There exists many methods for minimizing a cost function J defined from a set $\mathcal{O} \subset \mathbb{R}^n$ to $\mathbb{R}_+$. Among them, the family of evolutionary algorithms, including the well known methods of Genetic Algorithms (GAs) and Evolution Strategies (ES) whose main principles are recalled in the next subsection, has the major advantage to seek for a global minimum. Unfortunately, in view of the drag reduction problem that will be considered in Sect. 7.4, this type of method needs to be improved because of the large number of cost function evaluations that is needed. The hybrid optimization methods presented in Sect. 7.3.2 greatly reduce this time cost by coupling an evolutionary algorithm with a deterministic descent method. Another way to speed up the convergence of an evolutionary algorithm is described in Sect. 7.3.3 and aims at doing fast but approximated evaluations during the optimization process. All these methods are validated in Sect. 7.3.4 on classical analytic test functions.

7.3.1 Evolutionary Algorithms

The family of evolutionary algorithms gathers all stochastic methods that have the ability to seek for a global minimum of an arbitrary cost function. Among them, the population-based methods of GAs and ES are widely used in many applications and will serve as the core tool in the "real world" applications presented in Sects. 7.4 and 7.5, respectively. Their main principles are recalled in the next two paragraphs.

7.3.1.1 Genetic Algorithms (GA)

GAs are global optimization methods directly inspired from the Darwinian theory of evolution of species [11]. They require following the evolution of a certain number N_p of possible solutions, also called population. A fitness value is associated to each element (or individual) $x_i \in \mathcal{O}$ of the population that is inversely proportional to $J(x_i)$ in case of a minimization problem. The population is regenerated N_g times by using three stochastic principles called selection, crossover and mutation, that mimic the biological law of "survival of the fittest".

The GA that will be used in the drag reduction problem in Sect. 7.4 acts in the following way: at each generation, $N_p/2$ couples are selected by using a roulette wheel process with respective parts based on the fitness rank of each individual in the population. To each selected couple, the crossover and mutation principles are then successively applied with a respective probability p_c and p_m. The crossover of two elements consists in creating two new elements by doing a barycentric combination of them with random and independent coefficients in each coordinate. The mutation principle consists of replacing a member of the population by a new one randomly chosen in its neighborhood. A one-elitism principle is added in order to be sure to keep in the population the best element of the previous generation.

Thus, the algorithm can be written as:

- *Choice of an initial population* $P_1 = \{x_i^1 \in \mathcal{O}, 1 \leq i \leq N_p\}$
- $n_g = 1$. *Repeat until* $n_g = N_g$
- *Evaluate* $\{J(x_i^{n_g}), 1 \leq i \leq N_p\}$ *and* $m = \min\{J(x_i^{n_g}), 1 \leq i \leq N_p\}$
- 1-*elitism: if* $n_g \geq 2$ *&* $J(X_{n_g-1}) < m$ *then* $x_i^{n_g} = X_{n_g-1}$ *for a random* i
- *for* k *from* 1 *to* $N_p/2$
-     *Selection of* $(x_\alpha^{n_g}, x_\beta^{n_g})$ *with a roulette wheel process*
-     *with probability* p_c*: replace* $(x_\alpha^{n_g}, x_\beta^{n_g})$ *by* $(y_\alpha^{n_g}, y_\beta^{n_g})$ *by crossover*
-     *with probability* p_m*: replace* $(y_\alpha^{n_g}, y_\beta^{n_g})$ *by* $(z_\alpha^{n_g}, z_\beta^{n_g})$ *by mutation*
- *end for*
- $n_g = n_g + 1$.
- *Generate the new population* P_{n_g}.
- *Call* X_{n_g} *the best element.*

7.3.1.2 Evolution Strategies (ES)

Evolution Strategies (ES) have been first introduced by H.P. Schwefel in the 60's [2]. As it is the case for GAs, it requires following the evolution of a population of potential solutions through the same three stochastic principles, selection, recombination and mutation. However, unlike the GAs, the major process is the mutation process and the selection is made deterministic.

The Evolution Strategy that will be used in the application of Sect. 7.5 is based on the $(\mu+\lambda)$ selection principle and on the 1/5th rule for the mutation strength. An intermediate recombination with two parents is also included. The algorithm is thus written as:

- *Choice of an initial population of* μ *parents:* $P_1 = \{x_i^1 \in \mathcal{O}, \quad 1 \leq i \leq \mu\}$
- $n_g = 1$. *Repeat until* $n_g = N_g$
- *Creation of a population of* $\lambda \geq \mu$ *offsprings* O_{n_g} *by:*
-     *Recombination:* $y_i^{n_g} = \frac{1}{2}(x_\alpha^{n_g} + x_\beta^{n_g})$
-     *Normal mutation:* $z_i^{n_g} = y_i^{n_g} + \mathcal{N}(0, \sigma)$
- *Update of the mutation strength* σ *with the* 1/5*th rule*
- *Evaluate* $\{J(z_i^{n_g}), z_i^{n_g} \in O_{n_g}\}$.
- $n_g = n_g + 1$.
- *Selection of the best* μ *new parents in the population* $P_{n_g} \cup O_{n_g}$.
- *Call* X_{n_g} *the best element.*

7.3.2 Adaptive Hybrid Methods (AHM)

In order to improve the convergence of evolutionary algorithms for time-consuming applications like the drag reduction problem presented in Sect. 7.4, the idea of coupling a population-based algorithm with a deterministic local search, for instance a descent method, has been explored for many years (see e.g., [20]). However, the obtained gain can be very different from one function to another, depending on the level of adaptivity of the coupling and the way it is done.

The method presented here called Adaptive Hybrid Method (AHM) whose general principles are summarized in Fig. 7.6, tries to remedy these drawbacks by answering in a fully adaptive way the three fundamental questions in the construction of a hybrid method: when to shift from global to local, when to return to global and to which elements apply a local search. This method includes some criteria introduced in [7], and defines new ones as the reduced clustering strategy.

Note that this adaptive coupling can be implemented with any type of population-based global search methods (GA, ES, etc.) and any type of deterministic local search methods (steepest descent, BFGS, etc.).

From global to local

The shift from a global search to a local search is useful when the exploration ability of the global search is no longer efficient. With this aim, a statistical coefficient associated to the cost function repartition values is introduced. It is equal to the ratio of the mean evaluation of the current population to its

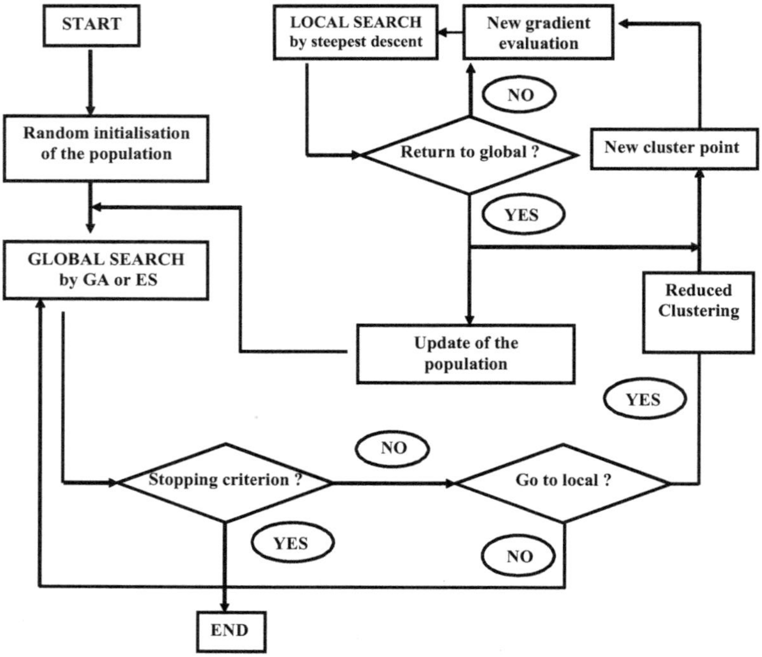

Fig. 7.6 General principle of the AHM

corresponding standard deviation computed with its variance:

$$CV = \frac{m}{\sigma} = \frac{\text{mean}\{J(x),\, x \in P_{n_g}\}}{\sqrt{\text{var}\{J(x),\, x \in P_{n_g}\}}} \tag{7.8}$$

and is named coefficient of variation CV. A local search will be utilized when this ratio increases within two consecutive generations of the evolutionary algorithm (either GA or ES).

From local to global

The local search is aimed at locally decreasing the cost function more efficiently than the random mutation. However, this gain must be counterbalanced after each evaluation with a characteristic gain of the global method. More precisely, the local search will continue here while:

$$G_{local} > G_{global}$$

where G_{local} is equal to the gain when passing from a point to the next one in the steepest descent algorithm and G_{global} is the gain of the last global phase evaluated with the decrease of m in formula (7.8). Both gains are scaled with the number of evaluations of the cost function needed to achieve them.

Reduced clustering

In order to spread as much as possible the local search in the whole domain, the population is divided into a certain number of sub-populations called clusters. To do so, a very classical and fast algorithm is used where each cluster is constructed such that all its associated elements are closer to its center of mass than to any other. After this preliminary step called clustering, the local search is applied to the best element (with respect to J) of each cluster.

A careful study of the appropriate number of clusters had never been done yet even though it appears to be rather important for the algorithm performance. To overcome the difficulty of choosing this number, a new method called reduced clustering has been proposed in [3] where the number of clusters is progressively decreased during the optimization process. It corresponds to the natural idea that the whole process will progressively focus on a reduced number of local minima. To do so, a deterministic rule of arithmetic decrease plus an adaptive strategy including the aggregation of too near clusters has been considered here and exhibits better results than any case with a fixed number of clusters as shown in Sect. 7.3.4.

7.3.3 Genetic Algorithms with Approximated Evaluations (AGA)

Another idea to speed up the convergence of an evolutionary algorithm when the computational time of the cost function $x \mapsto J(x)$ is high, is to take benefit of the large and growing data base of exact evaluations by making fast and approximated evaluations $x \mapsto \tilde{J}(x)$ leading to what is called surrogate or meta-models (see [9, 14, 15, 19]). In the present work, the chosen strategy is required to perform exact evaluations only for all the best fitted elements of the population (in the sense of $\tilde{J}$) and for one randomly chosen element. The new algorithm, called AGA is thus deduced from the algorithm of Sect. 7.3.1 by changing the evaluation phase into the following:

- *if* $n_g = 1$ *then make exact evaluations* $\{J(x_i^{n_g}), 1 \leq i \leq N_p\}$
- *elseif* $n_g \geq 2$
-     *for* i *from* 1 *to* N_p
-         *Make approximated evaluations* $\tilde{J}(x_i^{n_g})$.

- *if* $\tilde{J}(x_i^{n_g}) < J(X_{n_g-1})$ *then make an exact evaluation of* $J(x_i^{n_g})$
- *end for*
- *for a random* i*: make an exact evaluation of* $J(x_i^{n_g})$
- *end elseif*

The interpolation method chosen here comes from the field of neural networks and is called Radial Basis Function (RBF) interpolation [9]. Suppose that the function J is known on N points $\{T_i, 1 \leq i \leq N\}$, the idea is to approximate J at a new point x by making a linear combination of radial functions of the type:

$$\tilde{J}(x) = \sum_{i=1}^{n_c} \psi_i \Phi(||x - \hat{T}_i||) \tag{7.9}$$

where:

- $\{\hat{T}_i, 1 \leq i \leq n_c\} \subset \{T_i, 1 \leq i \leq N\}$ is the set of the $n_c \leq N$ nearest points to x for the euclidian norm $||.||$, on which an exact evaluation of J is known.
- Φ is a radial basis function chosen in the following set:

$$\Phi_1(u) = \exp(-\frac{u^2}{r^2}),$$

$$\Phi_2(u) = \sqrt{u^2 + r^2},$$

$$\Phi_3(u) = \frac{1}{\sqrt{u^2 + r^2}},$$

$$\Phi_4(u) = \exp(-\frac{u}{r}),$$

for which the parameter $r > 0$ is called the attenuation parameter.

The scalar coefficients $(\psi_i)_{1 \leq i \leq n_c}$ are obtained by solving the least square problem of size $N \times n_c$:

$$\text{minimize} \quad \text{err}(x) = \sum_{i=1}^{N} (J(T_i) - \tilde{J}(T_i))^2 + \lambda \sum_{j=1}^{n_c} \psi_j^2$$

where $\lambda > 0$ is called the regularization parameter.

In order to attenuate or even remove the dependence of this model to its attached parameters, a secondary global optimization procedure (a classical GA) has been over-added in order to determine for each x, the best values (with respect to $\text{err}(x)$) of the parameters n_c, $r \in [0.01, 10]$, $\lambda \in [0, 10]$ and $\Phi \in \{\Phi_1, \Phi_2, \Phi_3, \Phi_4\}$. As this new step introduces a second level of global optimization, it is only reserved to cases where the time evaluation of $x \mapsto J(x)$ is many orders of magnitude higher than the time evaluation of $x \mapsto \tilde{J}(x)$, as in the car drag reduction problem.

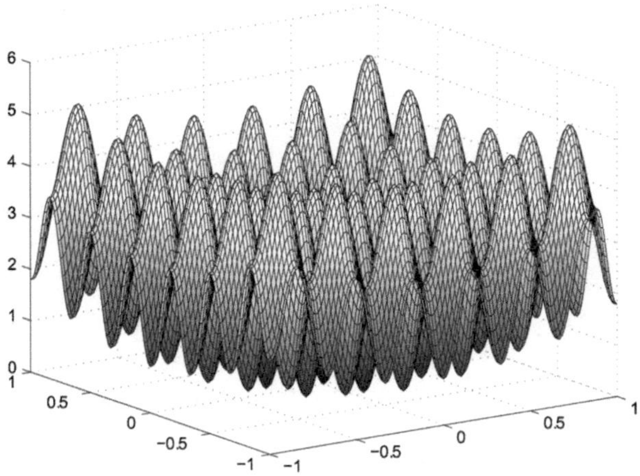

Fig. 7.7 The Rastrigin function $Rast_2$

7.3.4 Validation on Analytic Test Functions

Before applying them on real world applicative problems, all the previous global optimization algorithms have been tested and compared on various analytic test functions and among them the well-known Rastrigin function with n parameters:

$$Rast_n(x) = \sum_{i=1}^{n} \left(x_i^2 - \cos(2\pi x_i)\right) + n \tag{7.10}$$

defined on $\mathcal{O} = [-5, 5]^n$, for which there exists many local minima and only a global minimum located at $x_m = (0, ..., 0)$ and equal to 0 (see Fig. 7.7).

7.3.4.1 AHMs vs. Evolutionary Algorithms

A rather exhaustive comparison has been made between the classical evolutionary algorithm ES and the AHM introduced in Sect. 7.3.2. The statistical results are summarized in Table 7.3 for the Rastrigin function with 6 parameters. In this table, the success rate represents the rate of runs which were able to locate the correct attraction basin of the global minimum after a given number of evaluations of the cost function (respectively 500, 1000 and 2000). Note that any gradient evaluation counts for n evaluations of the cost function as it is the case in a finite-difference approximation.

Table 7.3 Comparison of ES and AHM for the Rastrigin function $Rast_6$

Method	Mean best (success rate) 500 evaluations	same after 1000 evaluations	same after 2000 evaluations
ES	6.47(0%)	3.46(0.5%)	0.46(80.5%)
AHM, 4 clust.	4.97(2.5%)	1.86(6%)	0.47(63.5%)
AHM, 8 clust.	3.54(7%)	1.77(11%)	0.34(67%)
AHM	3.52(10.5%)	1.51(17%)	0.34(68%)

As can be seen from this table, any AHM overperforms the ES at the early stage of the process in both performance criteria. Moreover, during this phase, the strategy of reduced clustering (last line in Table 7.3) significantly improves the results of a hybrid algorithm with a fixed number of clusters. When a large number of evaluations are done, the pure ES takes a slight advantage on the number of successes (but not on the mean best value) in this special case compared to any AHM.

These results can be summarized by saying that a hybrid algorithm will hasten convergence by enhancing the best elements in the population but on the other hand, such strategy can sometimes lead to a premature convergence. However, such drawback may not be too critical in a real applicative situation as it has only been observed with very special functions with a huge number of local minima like the Rastrigin function. Moreover, the main performance criterion of an algorithm for industrial purposes is its ability to achieve the best decrease of the cost function for a given amount of computational time.

7.3.4.2 AGA vs. GAs

Another statistical study has also been realized on the Rastrigin function with 3 parameters in order to compare the GA in Sect. 7.3.1 and the so-called AGA in Sect. 7.3.3. In order to achieve a quasi-certain convergence, the population number is fixed equal to $N_p = 30$ whereas the crossover and mutation probability are set to $p_c = 0.3$ and $p_m = 0.9$.

In this case, the average gain of an AGA compared to a classical GA is nearly equal to 4. It means that on average, the number of exact evaluations to achieve a given convergence level has been divided by a factor of 4.

In view of their promising results, both global optimization methods AHM and AGA are now used in the next two sections for solving realistic optimization problems.

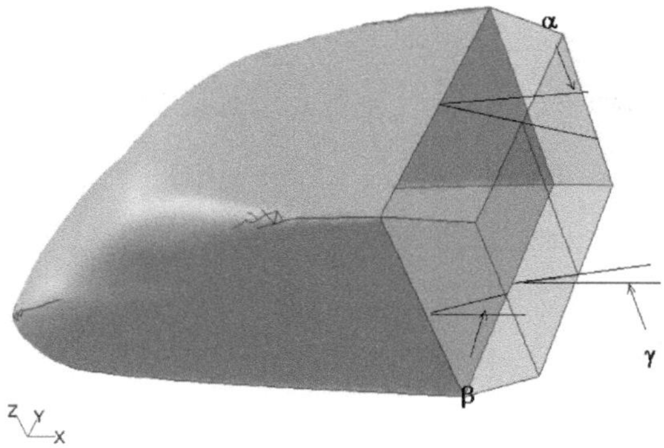

Fig. 7.8 3D car shape parametrized by its three rear angles α, β and γ

7.4 Car Drag Reduction with Numerical Optimization

In this section, the main results obtained on a car drag reduction problem are presented. All of them have been done in collaboration with two research engineers from Peugeot Citroën PSA, V. Herbert and F. Muyl, and have already been published in various journals [4, 5, 6].

7.4.1 Description of the Test Case

In order to test the fast and global optimization methods presented in Sect. 7.3 on a realistic car drag reduction problem, a simplified car geometry has been extensively studied. It comprises minimizing the drag coefficient, also called C_d and defined in Eq. (7.7), of a simplified car shape with respect to the three geometrical angles defining its rear shape (see Fig. 7.8): the slant angle (α), the boat-tail angle (β) and the ramp angle (γ). The forebody of the vehicle is fixed and closely resembles the shape of an existing vehicle, namely the Xsara Picasso from Citroën. The objective is thus to find the best rear shape that will reduce the total drag coefficient of the car, ignoring any aesthetic considerations. As it has been previously seen in Sect. 7.1 on the Ahmed bluff body, it is expected that modifying the rear shape will lead to a very important drag reduction.

7.4.2 Details of the Numerical Simulation

An automatic optimization loop has been implemented and is summarized in Fig. 7.9. This loop includes the following steps:

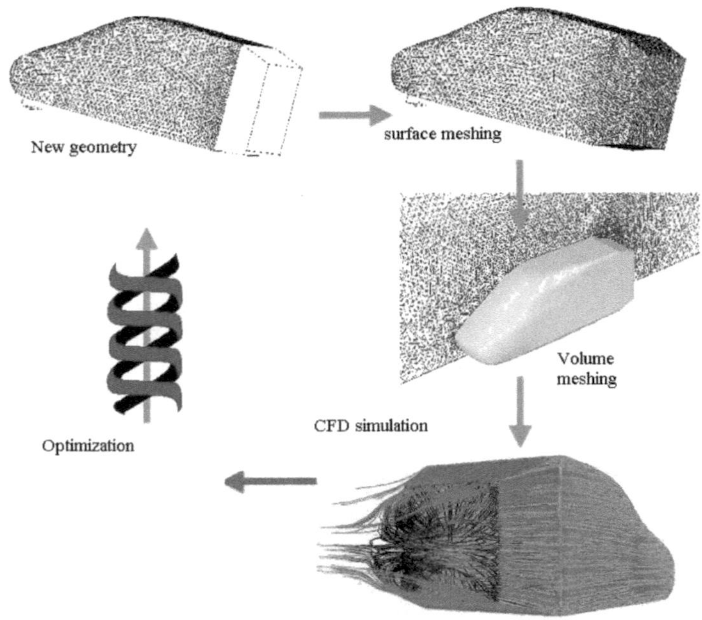

Fig. 7.9 General principles of the automatic optimization loop

(i) Car shape generation and meshing

In view of the experimental results obtained from using low drag car shapes presented in section 2, the three rear angles are sought in the following intervals (in degrees):

$$(\alpha, \beta, \gamma) \in [15, 25] \times [5, 15] \times [15, 25] \ .$$

For any given geometry, the 3D-mesh around the car shape is generated with the commercial grid generator Gambit. It contains a total of approximately 6 million cells that include both tetrahedrons and prisms. In order to simulate accurately the flow field behind the car which is responsible for the major part of the drag coefficient, the mesh is particularly refined in this region.

(ii) CFD simulation

The commercial CFD code Fluent is used for the computation of the flow field around the car. The Reynolds number based on the body length (3.95 m) and the velocity at infinity (40 m/s) is taken equal to 4.3×10^6 as in the Ahmed experiment [1]. The 7-equation RSM turbulence model is chosen as it gives better results in this case than the classical k–ε model (see [16]). The com-

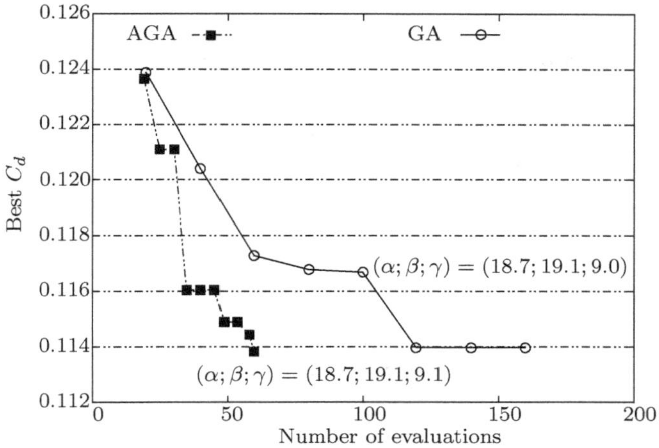

Fig. 7.10 Convergence history

putation is performed until a stationary state is observed for the main aerodynamic coefficients. This requires approximately 14 hours computational (CPU) time on a single-processor machine. In order to achieve a reasonable computational time, parallel evaluations on a cluster of workstations have been done.

(iii) Optimization method

Two different global optimization methods have been compared on this problem. The first one is a classical GA with a population number N_p equal to 20, a crossover and a mutation coefficient equaling to 0.9 and 0.6, respectively. The second method is similar to GA but with fast and approximated evaluations as presented in Sect. 7.3.3 (AGA). Note that the hybrid methods introduced in Sect. 7.3.2 have also been tested on this problem but are not presented here since AGA performs better. In contrast to these three global optimization methods, it is worth mentioning that a pure deterministic method like BFGS fails to find the global drag minimum at all (see [5] for a further comparison of all these methods).

7.4.3 Numerical Results

The convergence history of both optimization methods GA and AGA for the present drag reduction problem is depicted in Fig. 7.10. This figure shows in particular that both methods have nearly reached the same drag value,

Table 7.4 Four examples of characteristic shapes

shape	slant angle (α)	boat-tail angle (β)	ramp angle (γ)	C_d
(a)	21.1	24.1	14.0	0.1902
(b)	15.5	15.7	5.6	0.1448
(c)	16.9	17.7	11.3	0.1238
(d)	18.7	19.1	9.0	0.1140

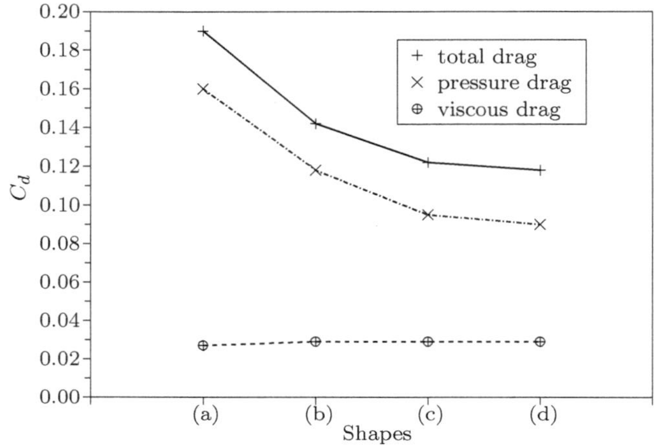

Fig. 7.11 Pressure and viscous part of drag for shapes (a) to (d)

namely 0.114, but with a different number of cost function evaluations. More precisely, the AGA algorithm has permitted to reduce the exact evaluation number by a factor of 2 compared to a classical GA, leading to the same proportional time saving.

The optimal angles obtained by both global optimization methods are nearly equal to $(\alpha, \beta, \gamma) = (18.7, 19.1, 9.0)$. These values have been experimentally validated in the Peugeot wind tunnel to be associated with the lowest drag value that can be reached with this particular forebody.

In order to understand in depth the complex phenomena involved in the variations of the drag coefficient, four characteristic shapes are presented in Table 7.4 and carefully studied.

The first shape (a) corresponds to a high-drag configuration whereas shapes (b), (c) and (d) correspond to low-drag configurations with an increasing value of slant and boat-tail angles. More precisely, shape (c) corresponds to the best shape obtained after the first generation of the GA whereas shape (d) is the best shape obtained in the whole range of admissible angles. Compared to shape (a), note that the value of the drag coefficient of shape (d) is almost divided by a factor of two which confirms the high dependence of C_d on the rear shape.

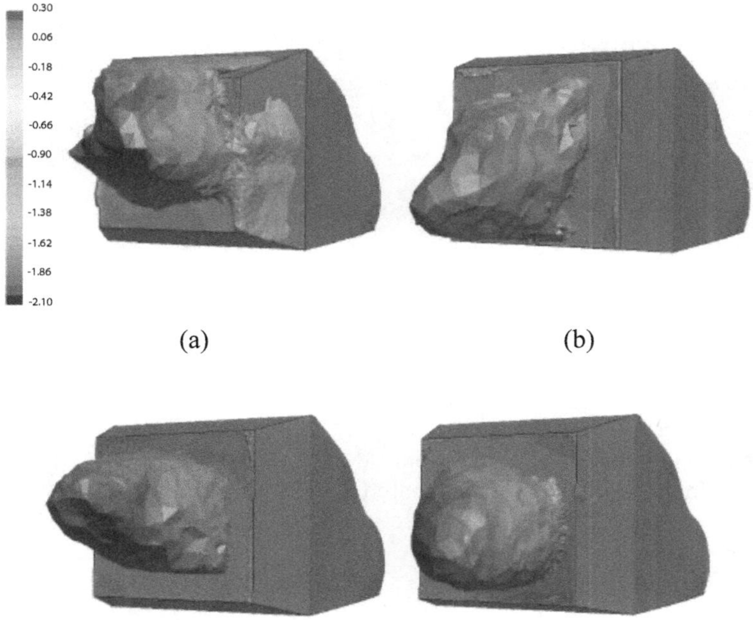

(c) (d)

Fig. 7.12 Isosurface of null longitudinal velocity colored with pressure coefficient for shapes (a) to (d)

Figure 7.11 displays for the four chosen configurations, the relative part of the pressure drag and the viscous drag in the value of the drag coefficient C_d. It is worth noticing that the pressure drag represents the major contribution to the total drag and also that the viscous drag remains almost constant for all cases. In particular, a higher pressure at the rear of the car will automatically reduce the drag. To see more precisely the topology and the pressure values at the wake flow, Fig. 7.12 depicts for shapes (a) to (d) the isosurface of null longitudinal velocity colored with the pressure coefficient. The latter corresponds to a dimensionless pressure value and is given by:

$$C_p = \frac{p - p_\infty}{\frac{1}{2}\rho V_\infty^2} \ . \tag{7.11}$$

It can be seen in particular on shapes (a) and (b) that respectively, either too high or too low slant and boat-tail angles will not generate a sufficient pressure level at the rear of the vehicle, thus increasing the pressure drag. On the contrary, intermediate and similar values of these two angles, coupled

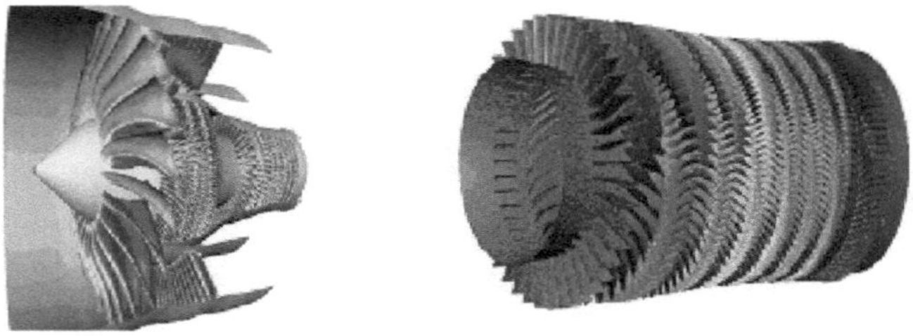

Fig. 7.13 Blades in the fan (left) and the high pressure compressor module (right)

with a low ramp angle as it is in the case for shapes (c) and (d), will improve the recompression at the car base. Note also that the optimal shape (d) is associated to a very narrow and regular recirculation bulb.

All these observations thus corroborate the qualitative trends well-known to car engineers since the experiments done by Morel [18] and Ahmed [1]. The numerical automatic tool presented here is now validated and appears to be very promising for car manufacturers to realistically design low drag car shapes in the near future.

7.5 Another Possible Application of CFD-O: Airplane Engines

In this section, another application of CFD-based optimization is given in the field of airplane engines. All the results presented here have been obtained in collaboration with two research engineers from Snecma-Moteurs (part of Safran group), B. Druez and N. Lecerf, and will be presented in more detail in [3].

7.5.1 General Description of the Optimization Case

In a turboreactor, the blades, which represent a big amount of an engine price (nearly 35%), are designed to create and control the aerodynamic flow through the engine (see Fig. 7.13).

The objective here is to optimize the design of the blades in the high pressure compressor module in order to minimize the mechanical efforts applied on them. Actually, it represents only a first step in the field of blade optimization since the main goal of a high pressure compressor designer is to increase the isentropic efficiency of the compressor. Nevertheless, this goal can not be achieved regardless of other engine features. Among them is the stall margin. This aerodynamic instability phenomenon consists in the stall of the flow

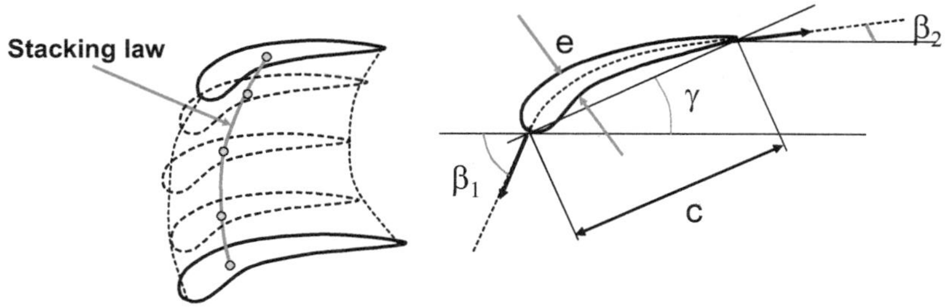

Fig. 7.14 Design parameters of a 3D blade

around the blades. This leads to backward flow inside the compressor and can result in engine shutdown, overtemperature in the low pressure turbine, high level of vibration or blade-out. To prevent such events, the designer will have to increase the compressor pressure ratio for low mass flow rates.

7.5.2 Details of the Computation

A 3D blade can be broken down into a set of several 2D airfoils profiles. The different airfoils are linked to the original blade through the stacking law (see Fig. 7.14). Each airfoil can then be described by a set of design parameters which reflect physical phenomenon that can be seized by the human designer. Figure 7.14 shows some common design parameters of the 2D profiles such as chord (c), maximum thickness value (e), upstream and downstream skeleton angles (β_1 and β_2), and stagger angle (γ). In the presented case, these parameters are kept fixed whereas the parameters to optimize, on the number of six, are all associated to the stacking law.

In order to minimize the mechanical efforts on the blade, the associated function to minimize is equal to the maximal value on 2D profiles of the von Mises constraints. Such a problem is highly non-linear, has a large number of constraints, many local minima and is also time consuming. A global, fast and robust method is thus needed.

7.5.3 Obtained Results

The AHM method presented in Sect. 7.3.2 has been used here to solve the blade optimization problem described above. Note in particular that constraints are handled with a penalization term whereas the gradients are approximated by finite differences.

Table 7.5 Convergence results on the blade optimization problem

Algorithm	number of evaluations	best obtained value
Evolutionary algorithm	460	163.5
AHM	480	158.6

The results are compared in Table 7.5 with those obtained with a classical evolutionary algorithm, here a simulated annealing. It can be seen that for the same simulation time (approximately 80 hours CPU), the AHM overperforms the simulated annealing. Indeed, even if the relative decrease obtained on the cost function appears to be small (3% approximately), it actually represents a significant improvement for the blade design.

7.6 Conclusion

In order to reduce fuel consumption, the minimization of the drag coefficient of cars has become a major research topic for car manufacturers. The development of fast and global optimization methods based either on hybridization of evolutionary algorithms with a local search process or on the use of surrogate models, has allowed only recently a first numerical and automatic approach for the drag reduction problem.

The results obtained from a simplified geometry called Ahmed bluff body are presented here. They confirm the experimental analysis saying that the drag coefficient is very sensitive to the rear geometry of the car due to the presence of separation and recirculation zones in this region, and thus can be largely reduced by shape optimization.

After this experimental validation, the numerical tool is now ready to be used by car designers for improving the drag coefficient of future car models.

References

1. Ahmed, S.R., Ramm, R., Faltin, G.: Some salient features of the time averaged ground vehicle wake. SAE Paper 840300 (1984)
2. Beyer, H.G., Schwefel, H.P.: Evolution Strategies. Kluwer Academic Publisher (2002)
3. Druez, N., Dumas, L., Lecerf, N.: Adaptive hybrid optimization of aircraft engine blades. Journal of Computational and Applied Mathematics, special issue in the honor of Hideo Kawarada 70th birthday (in press) (2007)
4. Dumas, L., Muyl, F., Herbert, V.: Hybrid method for aerodynamic shape optimization in automotive industry. Computers and Fluids **33**, 849–858 (2004)

5. Dumas, L., Muyl, F., Herbert, V.: Comparison of global optimization methods for drag reduction in the automotive industry. In: Lecture Notes in Computer Science, vol. 3483, pp. 948–957. Springer (2005)
6. Dumas, L., Muyl, F., Herbert, V.: Optimisation de forme en aérodynamique automobile. Mécanique et Industrie **6**(3), 285–288 (2005)
7. Espinoza, F., Minsker, B., Goldberg, D.E.: A self adaptive hybrid genetic algorithm. In: Proceedings of the Genetic and Evolutionary Computation Conference GECCO 2001, pp. 75–80. Morgan Kaufmann Publishers (2001)
8. Franck, G., Nigro, N., Storti, M., d'Elía, J.: Numerical simulation of the Ahmed vehicle model near-wake. Int. J. Num. Meth. Fluids (in press) (2007)
9. Giannakoglou, K.C.: Acceleration of ga using neural networks, theoretical background. GA for optimization in aeronautics and turbomachinery. In: VKI Lecture Series (2000)
10. Gillieron, P., Chometon, F.: Modelling of stationary three dimensional separated flows around an Ahmed reference model. In: ESAIM Proceedings. Third International Workshop on Vortex Flows and Related Numerical Methods, vol. 7, pp. 173–182 (1999)
11. Goldberg, D.E.: Genetic Algorithms in Search, Optimization and Machine Learning. Addison-Wesley (1989)
12. Han, T.: Computational analysis of three-dimensional turbulent flow around a bluff body in ground proximity. AIAA Journal **27**(9), 1213–1219 (1988)
13. Han, T., Hammond, D.C., Sagi, C.J.: Optimization of bluff body for minimum drag in ground proximity. AIAA Journal **30**(4), 882–889 (1992)
14. Jin, Y.: A comprehensive survey on fitness approximation in evolutionary computation. Soft Computing **9**, 3–12 (2005)
15. Jin, Y., Olhofer, M., Sendhoff, B.: A framework for evolutionary optimization with approximate fitness functions. IEEE Transactions on Evolutionary Computation **6**, 481–494 (2002)
16. Makowski, F.T., S.-E., K.: Advances in external-aero simulation of ground vehicles using the steady RANS equations. SAE Paper 2000-01-0484 (2000)
17. Mohammadi, B., Pironneau, O.: Analysis of the k–ε turbulence model. John Wiley & Sons (1994)
18. Morel, T.: Aerodynamic drag of bluff body shapes characteristic of hatch-back cars. SAE Paper 7802670 (1978)
19. Ong, Y.S., Nair, P.B., Keane, A.J., Wong, K.W.: Surrogate-assisted evolutionary optimization frameworks for high-fidelity engineering design problems. In: Knowledge Incorporation in Evolutionary Computation, Studies in Fuzziness and Soft Computing Series, pp. 307–332. Springer Verlag (2004)
20. Poloni, C.: Hybrid GA for multi objective aerodynamic shape optimisation, Genetic algorithms in engineering and computer science. In: G. Winter, J. Périaux, M. Galan, P. Cuesta (eds.) Genetic Algorithms in Engineering and Computer Science. John Wiley & Sons, Inc., New York, NY, USA (1995)

Chapter 8
Multi-objective Optimization for Problems Involving Convective Heat Transfer

Marco Manzan, Enrico Nobile, Stefano Pieri and Francesco Pinto

Abstract In this chapter, focused on Computational Fluid Dynamics (CFD)-based optimization for problems involving convective heat transfer, we present our approach for the multi-objective shape optimization of periodic wavy channels, representative of the repeating module of many heat exchangers.

The first problem is of fundamental nature and considers the geometric parametrization and shape optimization of two- and three-dimensional periodic wavy channels. The geometry of the channel is parametrized either by means of linear-piecewise profiles or by non-uniform rational B-splines. The second case, of industrial interest, illustrates the development and application of an automatic method for the design of gas turbine recuperators.

After a literature review of shape optimization in heat transfer, we describe in detail both aforementioned problems in terms of physical assumptions and mathematical formulation. In the *numerical methods* section we indicate the CFD codes used and describe the implementation of periodic boundary conditions. Thereafter in the *geometry parametrization* section, we illustrate the different types of numerical geometry representation used in the two problems, and the corresponding definition of the design variables whose variation leads to different shapes of the computational domain.

After a comprehensive classification and description of optimization methods and algorithms, we present the results obtained for the two different cases. For both problems the objectives considered are the maximization of heat transfer rate and the minimization of friction factor, with the additional objective of minimization of heat transfer surface for the recuperator module.

Marco Manzan · Enrico Nobile · Francesco Pinto
Dipartimento di Ingegneria Navale, del Mare e per l'Ambiente – DINMA, Sezione di Fisica Tecnica, Università di Trieste, Trieste, Italy
(e-mail: `manzan@units.it,nobile@units.it,fpinto@units.it`)

Stefano Pieri
Presently at Danieli & C. Officine Meccaniche Spa, Buttrio (UD), Italy
(e-mail: `s.pieri@danieli.it`)

Since there is no single optimum to be found, we use a multi-objective genetic algorithm and the so-called Pareto dominance concept.

The results obtained are very encouraging, and the procedure described can be applied, in principle, to even more complex convective problems.

8.1 Introduction

The common approach when solving thermofluid problems numerically is to first prescribe the geometry, boundary conditions, and thermophysical properties and then solve the governing equations for velocity, pressure, temperature, turbulent kinetic energy, etc. Problems of this type are referred to here as analysis problems. In design practice, the engineer usually tries different geometries, chooses other materials with different properties, and so on until satisfactory performance is obtained. Such cut-and-try method relies on the experience and skill of the designer to obtain any improvement but optimal performance is rarely achieved. Furthermore, this simple approach becomes impractical when the number of design variables is large, when there is more than one objective, and when there are several constraints to be satisfied. Therefore, in such cases, it is convenient, if not mandatory, to adopt an optimization strategy. However, the integration of numerical optimization techniques as part of the design process is still not very common today, particularly for complex heat transfer problems. In order to proceed in a systematic way, a basic requirement in shape optimization is to define the shape of the system to be optimized in terms of (known) functions and (unknown) parameters. This task is typically accomplished by means of a parametric computer-aided design (CAD) system, and by making use of an optimization procedure, automatic variations of the parameters associated with the geometric model lead to the creation of a variety of feasible shapes, which are then subjected to numerical analysis.

The aim of this article is to describe a general strategy for automatic, multi-objective shape optimization of heat exchanger modules. This represents a truly multi-objective optimization problem, since it is desired, from a design point of view, to maximize the heat transfer rate in order to, for example, reduce the volume of the equipment and to minimize the friction losses which are proportional to the pumping power required. These two goals are clearly conflicting, and therefore no single optimum can be found. For this reason we use a Multi-Objective Genetic Algorithm (MOGA), and the so-called Pareto dominance which allows us to obtain a design set rather than a single design.

Optimization of two-dimensional wavy channels is obtained by means of an unstructured finite-element (FE) solver, for a fluid of Prandtl number $\mathrm{Pr} = 0.7$, representative of air, assuming fully developed velocity and temperature fields, and steady laminar conditions. The geometry of the channels is parametrized by means of Non-Uniform Rational B-Splines (NURBS) and their control points represent the design variables. An alternative and simpler geometry is described by means of piecewise-linear profiles. An extension to 3D is made by considering channels obtained by extrusion at different angles of the 2D channels.

A similar strategy has been adopted for the multi-objective optimization of the periodic module of Cross-Corrugated (CC) compact recuperators. However, in this case, several widespread industrial codes are linked sequentially to obtain an automatic procedure for the recuperator design and optimization. The software tools used are CATIA for the geometric parametrization, ANSYS ICEM-CFD for the grid generation, and ANSYS-CFX for pre-processing, solution and post-processing. This study is focused on the development and validation of an automated calculation methodology for the thermo-fluid dynamic aspects of a recuperator design. Such a methodology correlates all the phases throughout the design process, i.e., the different mechanical, thermal and fluid dynamic aspects according to the concept of Multi Disciplinary Optimization (MDO). The final objective is to find optimum configurations for highly effective as well as tightly compact recuperators.

The described optimization procedure is robust and efficient, and the results obtained are very encouraging. In particular, we show that our approach is effective in performing genuine multi-objective shape optimizations that can deal with local minima and does not require knowledge of function gradients. There are no fundamental reasons, apart from computational costs and modeling accuracy, which prevent the application of the methodology to more complex geometries and more complex physics such as, for example, unsteady or turbulent flow regimes.

8.2 Literature Review

We limit our attention to optimization in heat transfer, and more specifically to shape optimization for heat transfer problems.

Gradient-based methods seem to be the most common approach for optimization of heat transfer. For example, Prasad and Kane [42] performed a three-dimensional design sensitivity analysis using a boundary element method. A two dimensional shape optimization for the Joule heating of solid bodies is described by Meric [28]. The sensitivity analysis was performed using the adjoint variable method and the material derivative technique, with a FE discretization of the non linear primary problem and the linear adjoint problem. Cheng and Wu [6], and Lan et al. [25], considered the direct design of shape for two-dimensional conductive bodies. In their approach the problem was discretized using a boundary-fitted Finite Volume method, coupled with a direct sensitivity analysis for minimization of the objective function.

More recently, Morimoto et al. [29] used the adjoint method to optimize counter-flow heat exchangers with oblique wavy walls.

Evaluation of the objective function in many engineering problems is costly and thus makes the optimization task prohibitively expensive. This is the case, for example, when using large scale CFD models [36]. A common approach in these situations is to construct, from a selected number of initial designs, a *surrogate* of the objective function, to be used for subsequent optimization. This strategy has been followed, among others, by Kim and Kim [23], who performed the optimization of a two-dimensional channel with periodic ribs using a response surface method, the latter representing an analytical surface which either interpolate or approximate the initial set of designs. The particular technique used by Kim and Kim, the *D-optimal design*, offers the opportunity to construct the response surface method with a small number of design points with considerable savings in computational resources. A CFD-based optimization of a plate-fin heat sink, i.e., minimization of pressure loss under a required temperature rise, has been recently described by Park and Moon [35]. The optimization has been performed with three geometric design variables using a progressive quadratic response surface method (PQRSM).

Traditional optimization algorithms based on sensitivity analysis have a series of drawbacks which make them not always reliable for engineering problems. The most important and relevant to the problem considered in this paper, is the risk of getting trapped in local minimum since the optimum obtained from these methods becomes a global one only if the objective and constraints are differentiable and convex functions.

Recently, new algorithms were used to overcome the difficulties arising from traditional optimization methods. They are known as genetic algorithms (GA), and they mimic the evolution of living organisms in nature. Their applicability is broad and for a general introduction and a review of applications, the reader can refer to [4] and [43], respectively. For heat transfer problems, Queipo et al. [44] used a GA to optimize the cooling of electronic components while the optimization of a two-dimensional polynomial fin profile was considered by Fabbri [15] where a GA was applied to an FE representation of the conduction problem. The same author later used GA for optimization, i.e., maximization of the heat transfer rate, of the shape and spacing of the fins of a heat sink cooled by laminar flow [16]. The optimization of two-dimensional corrugated channels under laminar steady conditions was also considered by Fabbri in [17]. In this latter paper, the channel consisted of flat-insulated wall and a corrugated one described by a fifth order polynomial. The objective was to maximize the Nusselt number for a given volume (material) of the thermally active wall or for a given pressure drop. An application of GA for the optimization (minimization of overall thermal resistance) of a stacked micro-channel heat sink is described by Wei and Joshi [54], while the application of GA for the inverse geometric problem of detection of subsurface cavities, using thermographic techniques, has been reported by Divo et al. [13]. In this case, the inverse problem has been solved as a single-objective optimization problem. Finally, a very recent application of evolution strategy (ES), for the shape optimization of thermoelectric bodies exchanging heat by convection and radiation, has been reported by Białecki el al. [3]. Like GAs, ES belong to the general category of Evolutionary Algorithms [4], but differ from the former mainly in the use of real-valued vectors of variables instead of bit-strings. Moreover, in contrast to GA which relies heavily on *recombination* to explore the design (search) space, ES use mutation as the main search operator. An interesting result from this study was the presence of very different shapes having practically the same value of the objective function. As shown in the results section, we also found a similar trend, i.e., two families of shapes characterized by the same performance figures.

This short review of the available literature is followed by some remarks. The first is that in the majority of cases, the authors have used in-house and/or problem specific methods – solvers and in particular optimization algorithms – that, while capable of guaranteeing high accuracy and computational efficiency, however lack generality and robustness. Hence, they could hardly be applied to the complex optimization tasks found in industrial applications. In fact, this type of problems are frequently computationally demanding, are affected by noise, and are characterized by several conflicting objectives as well as numerous constraints.

The second remark is relative to the fact that in all cited works, the problem was to optimize a single performance metric. In other words, it was considered a *single-objective* optimization problem.

When there were several objectives such as in [23] and [29], these objectives were incorporated into a single function using suitable weighting factors, thereby reducing the problem into one of single-objective optimization again. This approach, however, has several drawbacks, the first being that weights must be provided *a priori*, which can influence the solution to a large degree. Moreover, if the objectives are inherently very different such as in cost and thermal efficiency, it can be difficult to define a single all-inclusive objective function. Finally, the user should even take care of normalization, and this is not always a simple task since the range of variation of each objective may be unknown. True multi-objective optimization techniques overcome these problems by keeping the objectives separate during the optimization process, bearing in mind that there will frequently be no single optimum in cases with opposing objectives, since any solution will be a compromise. This is the case for the problem considered in this paper, where the two objectives, maximization of heat transfer rate and minimization of pressure losses, are clearly conflicting. We will show how to identify the solutions which lie on the *trade-off* curve known as the Pareto Frontier (named after the Italian-French economist, Vilfredo Pareto). These solutions, also known as *non-inferior*, *non-dominated* or *efficient* solutions, all have the characteristic that none of the objectives can be improved without prejudicing another. The advantage of the use of the Pareto dominance concept is that, as it will be shown in the result section, all shape-design alternatives become visible *at once* for the design engineer, and the set of optimized solutions is obtained in one optimization run, in contrast to classical single-objective methods.

The use of the Pareto dominance concept has been illustrated, for heat transfer optimization problems, by Park et al. [33, 34]. In both studies, there were two variables to be minimized: the pressure drop and the thermal resistance of a plate exchangers [34], and of a heat-sink [33]. However, the multi-objective optimization problems were treated as a single objective problem, using weight coefficients. By performing a series of optimizations with varying weighting factors, the authors obtained an approximate Pareto front.

8.3 Problem Statement

The problems considered in this work are: I – the multi-objective shape optimization of two-dimensional (2D) and three-dimensional (3D) convective wavy channels; II – the multi-objective optimization of Cross-Corrugated channels. The former is of fundamental nature and is of interest for heat exchangers and other heat transfer devices. The second problem has a practical significance since it represents the building block of many gas turbine recuperators [26].

For both problems, the study is limited to a single module at fully developed flow and heat transfer conditions. In such a circumstance, channels of periodic cross section form can be considered periodic in the flow and thermal fields as well. Therefore, the computational domain of interest becomes a single periodic module of the entire geometry as depicted in Figs. 8.1 and 8.2 for the 2D and 3D channels, respectively. CC heat exchanger is formed by stacking undulated plates at different inclination angles as reported in Fig. 8.3(a). The geometry obtained is presented in Fig. 8.3(b) while the repeating module is depicted in Fig. 8.3(c). It can be noticed that, unlike other authors, both hot and cold fluid domains are included [26, 27].

We have limited the study to the steady, laminar flow regime which is found in many practical circumstances. The Reynolds number chosen for the simulation is $\mathrm{Re} = 200$ for both two-dimensional channel and CC module which corresponds to the typical value found in microturbine recuperators, while the Prandtl number is assumed as $\mathrm{Pr} = 0.7$, representative of air and other gases. For computational convenience, a lower value of the Reynolds number, $\mathrm{Re} = 100$, has been selected for the 3D wavy-channels.

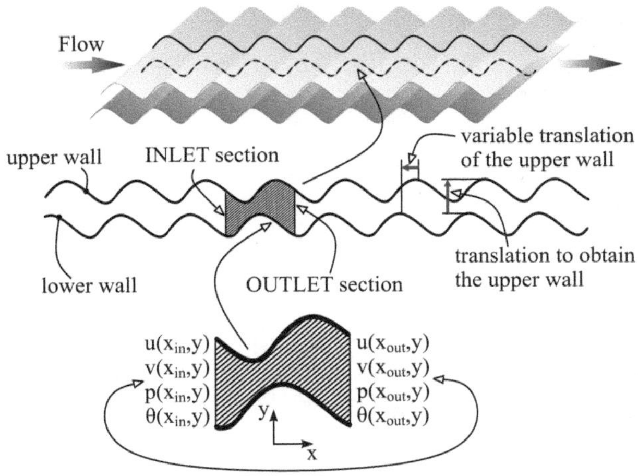

Fig. 8.1 Periodic module of the two-dimensional wavy channel

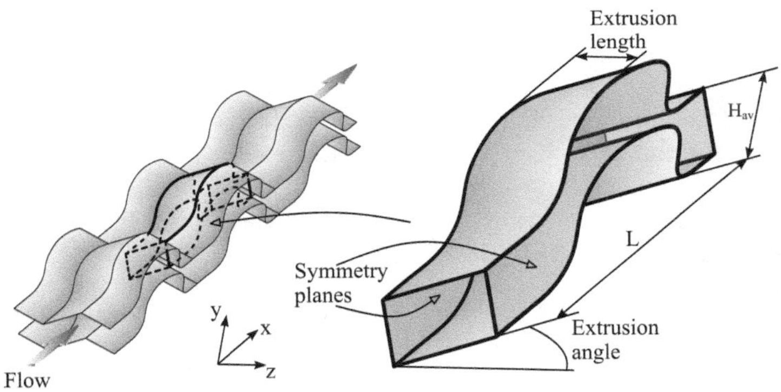

Fig. 8.2 Periodic module of the three-dimensional wavy channel

8.3.1 Governing Equations

The flow is assumed to be incompressible, Newtonian, and of constant thermophysical properties. Moreover, we refer to a stationary and laminar forced convection with negligible viscous dissipation and buoyancy contribution. In these conditions, continuity, Navier-Stokes, and energy equations are written as follows:

$$\nabla \cdot \boldsymbol{u} = 0 \tag{8.1}$$

$$\nabla \cdot (\rho \boldsymbol{u}\boldsymbol{u}) = \nabla \cdot (\mu \nabla \boldsymbol{u}) - \nabla p \tag{8.2}$$

$$\nabla \cdot (\rho c_p \boldsymbol{u} T) = \nabla \cdot (\lambda \nabla T) \ . \tag{8.3}$$

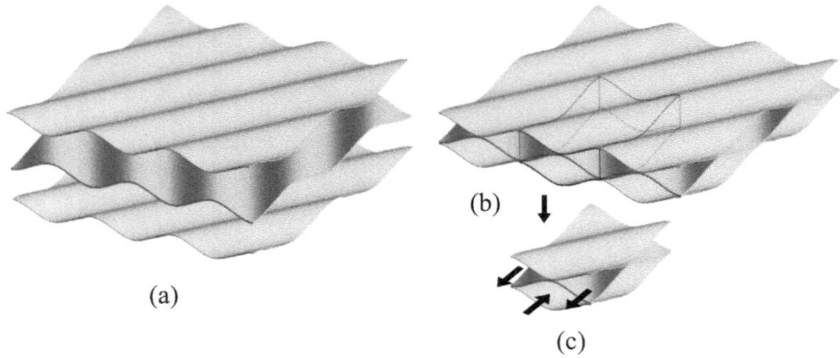

Fig. 8.3 Computational domain for the CC heat exchanger: **a** undulated plates to be stacked; **b** final geometry of CC channel; **c** periodic module

Before describing the boundary conditions, it is useful to focus on the meaning of the pressure term along with the mean flow direction, x, of the channel. Omitting any discussion about the entrance and exit regions of the channel, in the fully developed regime zone the velocity profiles can be considered periodic (of the same period as the channel). On the other hand, the effect of the pressure field is to allow the fluid flow, acting against friction forces due to the viscous behavior of the mean. The dissipative work is negligible and its contribution is not included in the energy equation, yet a pressure drop is present along the streamwise direction. The pressure field can be split into two contributions [49]:

1. a *linear decaying*, which counteracts friction forces;
2. a *periodic* term related to the detailed local motion.

It follows that the pressure assumes the following expression:

$$p(x, y, z) = -\beta \cdot x + \tilde{p}(x, y, z) \tag{8.4}$$

where $\tilde{p}(x, y, z)$ is the periodic part of the pressure. Therefore, the momentum equation in the x direction can be rewritten as:

$$\nabla \cdot (\rho \boldsymbol{u} \boldsymbol{u}) = \nabla \cdot (\mu \nabla \boldsymbol{u}) - \frac{\partial \tilde{p}}{\partial x} + \beta \tag{8.5}$$

where β becomes a volume force term whose value, influencing the Reynolds number obtained, will be adjusted in the solution procedure.

8.3.2 Fluid Dynamic Boundary Conditions

The condition of periodic velocity profiles can be expressed in the inlet and outlet boundaries, as illustrated in Fig. 8.1 for the 2D wavy-channel, as follows:

$$\boldsymbol{u}_{in} = \boldsymbol{u}_{out} \tag{8.6}$$

and, in general:

$$\boldsymbol{u}(x, y, z) = \boldsymbol{u}(x + L, y, z) \tag{8.7}$$

where L is the length of the repeating module. After the split of the total pressure, the newly introduced periodic pressure, $\tilde{p}$, can be handled as the velocity field:

$$\tilde{p}_{in} = \tilde{p}_{out} \tag{8.8}$$

and, in general:

$$\tilde{p}(x, y, z) = \tilde{p}(x + L, y, z) \ . \tag{8.9}$$

Finally, standard *no-slip* conditions are imposed at the walls.

8.3.3 Temperature Boundary Conditions

The temperature field in a heat exchanger or in a regenerator is not periodic since it changes continuously along the channel in the mean flow direction. However, a region of fully developed thermal condition can be identified and appropriate boundary conditions can be imposed for a single module. The thermal boundary conditions are different for the two cases presented in this work so they are treated separately. For the wavy channel case a constant-temperature boundary condition is used as representative of e.g., automotive radiators with negligible wall thermal resistance at high liquid flow rates. On the other hand, the CC channel is used in gas-gas recuperators where the wall temperature is not uniform. For this case, a flux boundary condition has been developed.

8.3.3.1 Wavy Channel

Shah and London [49] discuss fully developed flow in parallel plates for many wall-boundary condition cases. By *fully developed thermal field*, it is meant that the Nusselt number is constant along the flow.

As in Patankar et al. [37], the concept of the thermally developed regime is generalized to the *periodic* thermally developed one. The condition for a constant cross-sectional channel writes as follows:

$$\frac{\partial \theta_p}{\partial x} = 0 \tag{8.10}$$

where θ_p is the local non-dimensional temperature

$$\theta_p(x,y) = \frac{T(x,y) - T_w}{T_b(x) - T_w} \tag{8.11}$$

and $T_b(x)$ is the streamwise varying bulk temperature. In the case of periodic wavy channel, a weak condition, between two sections a period length apart, can be imposed:

$$\theta_p(x,y) = \theta_p(x+L,y) \; . \tag{8.12}$$

The use of the periodic temperature field, defined in Eq. (8.11), leads to a volume force term, in the energy equation, which depends on the x streamwise coordinate. So, another equation must be introduced [37], but this is cumbersome to implement on the triangular unstructured grids that COMSOL, the FE package employed for wavy channels, uses. For this reason, another strategy has been used to tackle the problem.

A fixed arbitrary reference value has been adopted to make the temperature non-dimensional. The value chosen, for convenience, is the bulk temperature at the inlet boundary. So the periodicity condition, Eq. (8.12), changes into:

$$\theta(x_{in},y) = \theta(x_{out},y) \cdot \sigma \tag{8.13}$$

where

$$\theta(x,y) = \frac{T(x,y) - T_w}{T_{b,in} - T_w} \tag{8.14}$$

and σ is the ratio between the inlet and outlet temperature differences:

$$\sigma = \frac{T_{b,in} - T_w}{T_{b,out} - T_w} \; . \tag{8.15}$$

In place of the adjoint equation introduced by Patankar [37], an iterative numerical procedure based on an energy balance has been introduced to reach fully developed conditions. This will be explained in Sect. 8.4.

8.3.3.2 CC Channel

In gas microturbine regenerators, the hot and cold streams are in countercurrent flow arrangement and are characterized by similar heat flow capacities. With these conditions the heat transferred between the fluids remains almost constant from repetitive module to module, so temperature and pressure can be treated in a similar way. The temperature is expressed as the sum of a periodic and a linear component driven by the gradient in the mean flow direction γ:

$$T(x,y,z) = \gamma\, x + \tilde{T}(x,y,z) \tag{8.16}$$

$$\tilde{T}(x,y,z) = \tilde{T}(x+L,y,z) \; . \tag{8.17}$$

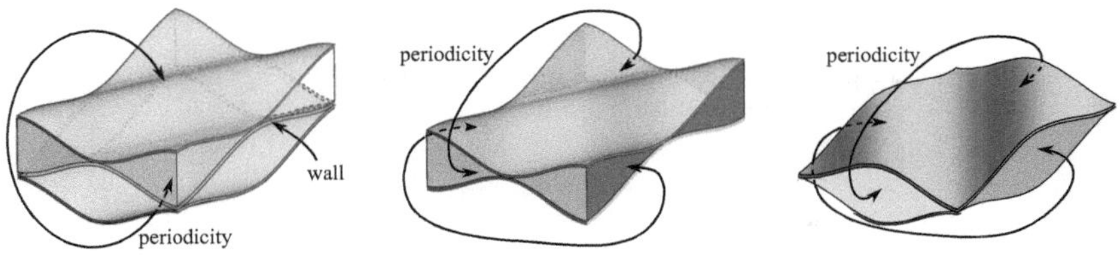

Fig. 8.4 Periodicity conditions for the CC heat exchanger: (**a**) assembled domain periodicity; (**b**) hot domain periodicity; (**c**) cold domain periodicity

Equation (8.16) can be substituted into the energy equation, Eq. (8.3), to obtain the transport equation for the periodic component:

$$\nabla \cdot \left(\rho\, c_p \boldsymbol{u}\, \tilde{T}\right) = \nabla \cdot \left(\lambda\, \nabla \tilde{T}\right) + \gamma\, u\, \rho\, c_p\, . \tag{8.18}$$

Equation (8.18) can be solved with appropriate boundary conditions as in [51] where a uniform wall flux was imposed and to satisfy the energy balance, the temperature gradient was computed as:

$$\gamma = \frac{\phi}{\dot{m}\, c_p L} \tag{8.19}$$

where L is the length of the periodic module in the mean flow direction, and ϕ is the overall heat flux specified at the walls. While the uniform flux is a practical solution for imposing a thermal boundary condition, it does not correctly describe the heat transfer in a recuperator module. Indeed the fluid flows in the furrows of the CC ducts with complicated three-dimensional patterns, strongly affecting the local heat transfer rates.

In this work the elementary periodic module is doubled to take into account both hot and cold fluids as shown in Fig. 8.3, while the periodicity conditions for both fluids are reported in Fig. 8.4. Equation (8.18) is applied to both domains and no boundary condition has to be imposed on the interface wall. A similar approach has been used by Morimoto et al. [29] to solve the heat transfer in a periodic channel, but with a different geometry. To satisfy the global energy balance the same flux has been applied to the hot and cold domains:

$$\phi_h = -\phi_c \tag{8.20}$$

using Eq. (8.19) to compute the last term of Eq. (8.18). Temperatures of the hot and cold domains are automatically adjusted during the iterative computation to satisfy automatically the overall energy balance:

$$\phi = U\, A\, \Delta T \tag{8.21}$$

where ΔT is the temperature difference between hot and cold fluids:

$$\Delta T = (T_{bi})_h - (T_{bo})_c = (T_{bo})_h - (T_{bi})_c \tag{8.22}$$

U is the global heat transfer coefficient, and A is a reference heat transfer surface, that has been taken as twice the projection of the interface wall on the horizontal plane, $A = 2 \cdot A_h$. This choice allows the comparison of heat transfer coefficients for different surface geometries.

8.4 Numerical Methods

The numerical solution of the 2D and 3D wavy channel case was carried out by means of the COMSOL software package [9]. Full details of the methodology have been previously published elsewhere [30] and are summarized hereafter. For the numerical simulation of the CC heat exchanger module, the ANSYS-CFX CFD package has been employed. Hexahedral structured grids have been generated by means of the ICEM-CFD grid generation package.

8.4.1 Fluid Dynamic Iterative Solution

For the wavy and CC channels the fluid dynamic equations are solved in a similar way. As shown in Eq. (8.5), a forcing term β appears, due to the pressure splitting introduced in Eq. (8.4). Since the Reynolds number is a given constant of the problem, the non-dimensional mean velocity in the channel must be unitary. It is, therefore, necessary to find the correct value of β that ensures this condition. Other authors [31] have used proportional-integrative iterative controls to reach the correct value of the pressure gradient, starting from a trial value. At first, a similar approach has been attempted, but it proved to be not very efficient in converging to the correct value of β, and this can be a limiting factor for CPU-intensive optimization studies.

By the definition of the friction factor f as the non-dimensional surface shear stress [49], it is easy to show by applying the second Newton's law that:

$$f = \beta \frac{D_h}{\frac{1}{2}\rho U_{av}^2} \ . \tag{8.23}$$

Recall that, for internal flows in laminar regime, the friction factor is proportional to the inverse of the Reynolds number. From the definition of Re, it follows that:

$$\beta \propto U_{av} \ . \tag{8.24}$$

This relation is not strictly valid in channels with varying cross section, but nevertheless it is expected a proportionality law such as:

$$\beta \propto U_{av}^{m} \tag{8.25}$$

to hold, with a value of the exponent m close to 1. In these conditions, evaluating the flow field with a test value for β, taking into account that the desired average velocity is unitary, and updating iteratively the pressure gradient as follows:

$$\beta_{n+1} = \frac{\beta_n}{U_{av,n}} \tag{8.26}$$

it is expected to reach the exact value of β in few steps. This has been verified during this study, where only 3 to 6 steps are required to reach a value of β which gives an error on Re below 0.1%.

8.4.2 Thermal Field Iterative Solution

The thermal field solution for the two cases reported here is quite different since for the wavy channel a uniform temperature boundary condition has been adopted while for the CC channel a constant flux has been considered. Therefore, the algorithms used in the two cases will be described independently.

8.4.2.1 Wavy Channels

After the velocity field has been obtained, the thermal field is computed with an iterative approach. This is based on the fact that the heat flux at the wall has to be balanced by the enthalpy difference between inlet and outlet. The task is to find a value of σ, Eq. (8.15), that ensures this balance:

$$\dot{m}c_p \left(T_{b,in} - T_{b,out}\right) = \int_w -k\frac{\partial T}{\partial n}\mathrm{d}s \; . \tag{8.27}$$

Assembling the dimensional terms and remembering Eq. (8.15), one is left with

$$1 - \frac{1}{\sigma} = \frac{2}{\mathrm{Re}\,\mathrm{Pr}} \int_w -\frac{\partial \theta}{\partial n}\mathrm{d}s \; . \tag{8.28}$$

It can be recognized that there is a relation between σ and the heat flux at the wall and in particular, an incorrect value of σ leads to a generation term on the outlet boundary which has no physical meaning. So, starting from a tentative value of σ and updating it iteratively by means of (8.28), it converges towards the correct solution.

Once the correct thermal field has been calculated, the mean Nusselt number, Nu, is obtained as:

$$\mathrm{Nu} = \frac{1}{2L} \int_w \mathrm{Nu}_x \mathrm{d}s \tag{8.29}$$

$$\mathrm{Nu}_x = \frac{h_x D_h}{k} \tag{8.30}$$

where Nu_x and h_x are, respectively, the local values of the Nusselt number and heat transfer coefficient. From an energy balance, again in dimensional form, at a generic section of the channel, one has:

$$\dot{m}\, c_p \,\mathrm{d}T_b = -h_x (T_b - T_w) \mathrm{d}s \; . \tag{8.31}$$

Multiplying each side by $(\mu\, k\, D_h)$, expressing the mass flow in its components and integrating, one finally obtains

$$\mathrm{Nu} = \ln \sigma \frac{D_h}{4L} \mathrm{Re}\, \mathrm{Pr} \; . \tag{8.32}$$

A grid independence study was performed for different geometric configurations of the channel (designs) in order to ensure the accuracy of the results presented.

8.4.2.2 Cross-Corrugated Channels

A first simulation of the CC channel, with a grid similar to those used in this optimization study, has been conducted for comparison with literature data and the results are presented elsewhere [26]. The geometry considered is similar to the base one used for the optimization. The channel has been generated using sinusoidal profiles as in [10] with values of $W/a = 12$, channel height $b = 2a$ and inclination angle $\theta = 90°$. The meaning of W, a, and b is given in Sect. 8.5.2. The generated structured grid is composed by 155×10^3 hexahedral elements, selected as a trade-off between accuracy and computational costs. Computations with refined grids up to 371×10^3 elements have been carried out to check for grid independence. It has been verified that the coarsest mesh used in this work gives an estimated error of 1% for the Nusselt number, and 0.4% for the friction factor.

Most literature results were obtained using uniform temperature boundary conditions for the energy equation. Thus, this same condition has also been implemented in ANSYS-CFX [27]. Nevertheless, in the present work, the optimization has been carried out using the constant flux boundary condition, since it better describes the real working conditions for a microturbine recuperator.

The functionalities of industrial codes may be extended by means of script functions or user-written routines. To implement the constant flux condition in ANSYS-CFX, the flexibility of the Ansys CCL (CFX Command Language) has been exploited. The value of the pressure gradient β of Eq. (8.5) has been added as a source term to the momentum equation and updated using Eq. (8.26) by means of CCL expression to obtain the desired velocity. To solve the thermal field, Eq. (8.19) is computed at each time step using a CCL expression, and the gradient γ is introduced in the last term of Eq. (8.18) to compute the source term of the energy equation for both hot and cold fluid streams.

The average Nusselt number can be easily obtained by inspecting Eq. (8.21). The global heat transfer coefficient, neglecting wall resistance, can be written as

$$U = \left(\frac{1}{\bar{h}_h} + \frac{1}{\bar{h}_c} \right)^{-1} \tag{8.33}$$

where $\bar{h}_h$ and $\bar{h}_c$ are the mean heat transfer coefficients for the hot and cold side, respectively. Since the fluids, the flow heat capacities and the geometry for the hot and cold ducts are equal, a unique mean heat transfer coefficient can be introduced $\bar{h}_h \equiv \bar{h}_c \equiv h$, and Eq. (8.21) can be simplified as:

$$\phi = \frac{h}{2} \cdot A \cdot \Delta T \tag{8.34}$$

from which the mean Nusselt number can be obtained as:

$$\overline{\mathrm{Nu}} = \frac{\phi \ D_h}{\Delta T \ A_h \ k} \tag{8.35}$$

where $D_h = 2b$ is the hydraulic diameter of the channel [26]. The local Nusselt number can be derived from (8.35) as:

$$\mathrm{Nu} = \frac{2 \ \phi''(x) \ D_h}{\Delta T \ k} \tag{8.36}$$

where $\phi''(x)$ is the local specific heat flux at the wall.

8.5 Geometry Parametrization

Different geometry parametrizations have been used for the wavy channels and the CC module and are described next.

8.5.1 Wavy Channels

The shape of two-dimensional convective channel is represented either by linear-piecewise profiles, or by NURBS. The two different geometric profiles,

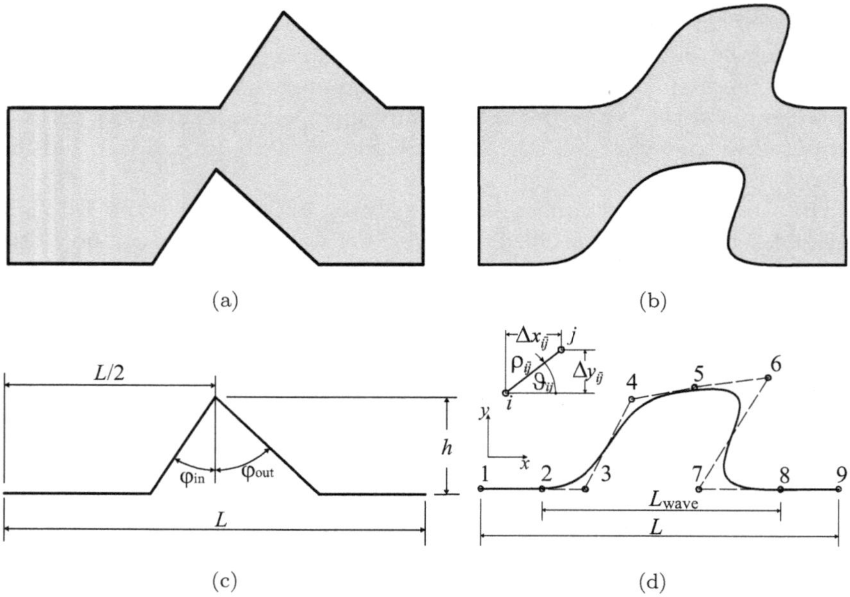

Fig. 8.5 Periodic channels and geometrical parametrization. (**a**) Linear piecewise channel; (**b**) NURBS channel; (**c**) Linear piecewise parametrization; (**d**) NURBS parametrization and control points

the linear-piecewise and the NURBS, are depicted in Figs. 8.5(a) and 8.5(b), respectively.

8.5.1.1 2D Linear Piecewise Parametrization

The linear piecewise wall profile of the channel is characterized by a small number of *design variables* or Degrees Of Freedom (DOFs). Their exact meaning will be described in Sect. 8.6. Here it is sufficient to say that they represent the independent variables that may be adjusted, during the optimization process to achieve the desired goal(s). The variables introduced in this wall parametrization are presented in Fig. 8.5(c) and summarized in Table 8.1. There are four DOFs requested for describing the wall profile. Realized by defining a set of points and connecting them with straight lines, the profile is constructed only for convenience in order to have the edge of the corrugation centered on the wall.

Table 8.1 Variables defining the linear piecewise channel

Variable	Symbols	Range
module length	L	$[0.8; 2.0]$
corrugation height	h	$[0.0; 0.5]$
forward edge angle	φ_{in}	$[10°; 60°]$
backward edge angle	φ_{out}	$[10°; 60°]$
translation of upper wall	$transl$	$[-0.5L; 0.5L]$

Table 8.2 Control points for the NURBS-profile and parameters required

Point	x	y	DOFs
1	0	0	0
2	$(L - L_{wave})/2$	0	0
3	$x_2 + \Delta x_{\overline{23}}$	0	1
4	$x_5 + \rho_{\overline{54}} \cos \vartheta_{54}$	$y_5 + \rho_{\overline{54}} \sin \vartheta_{54}$	2
5	$x_1 + \Delta x_5$	$y_1 + \Delta y_5$	2
6	$x_5 + \rho_{\overline{56}} \cos \vartheta_{56}$	$y_5 + \rho_{\overline{56}} \sin \vartheta_{56}$	2
7	$x_8 - \Delta x_{\overline{87}}$	0	1
8	$x_2 + L_{wave}$	0	1
9	L	0	1

8.5.1.2 2D NURBS Parametrization

In order to generate the NURBS channel during the optimization process, let's start by defining first the wall profile. As a good compromise between the number of DOFs and the geometrical complexity, a 9 control-point periodic cubic NURBS has been chosen, to ensure the periodicity of the channel itself. A large number of DOFs allows to describe minutely the profile, but if this number is excessive, it would make the optimization process quite difficult and expensive. For this reason, it has been also decided to fix to a unitary value the curve's weights. This, together with the uniform knots distribution we adopted, makes our NURBS curve practically equivalent to a B-Spline curve [18, 38].

As depicted in Fig. 8.5(d), both the first and the last three aligned control points are needed to maintain the entrance and the exit of the profile parallel to the x direction. The remaining ones give freedom to the wavy section. The parameters required for the definition of the lower profile are explained in Table 8.2 and numbered as in Fig. 8.5(d). The symbol Δ means the difference between the coordinates of two points while $\rho_{\overline{ij}}$ is the module and ϑ_{ij} the phase of a polar-coordinate system centered on the point i. The phase is positive counterclockwise with respect to the positive direction of the x axis. The profile is again constructed, for convenience, in order to have the wavy part centered on the wall. Therefore, the value of parameter x_2 is equal to $(L - L_{wave})/2$. The number of DOFs is now 10.

8.5.1.3 Channel Construction

In Figs. 8.5(a) and 8.5(b) both types of channel are depicted, i.e., linear piecewise channel and NURBS channel, and they are constructed in the same way. Once the lower wall profile has been obtained, the upper one, as illustrated in Fig. 8.1, is made first by a simple translation in y-direction of the former in order to obtain the height of the channel, followed by a translation in x-direction. This guarantees the realistic desire to construct the channels of e.g., a finned heat exchanger, by simple juxtaposition of identical wavy plates. This action introduces in both cases another DOF which defines the x translation of the upper profile, variable *transl* in Tables 8.1 and 8.4. In order to avoid linear-dependent channels, the translation range is bounded to one half-period in both the positive and negative x-direction. The y-translation is instead fixed: in this way the average height of the channel is set to 0.5, that is half of the non-dimensional hydraulic diameter (D_h). Finally the periodic duct module is cut by two straight lines, representing the inlet and the outlet section, as sketched in Fig. 8.1. Overall, we are left with 5 DOFs for the linear piecewise model, and 11 DOFs for the NURBS channel.

8.5.1.4 Three-dimensional Extension

Once the two-dimensional results have been obtained, it has been tried to encourage further mixing in the flow, and therefore an augmentation of the heat transfer rate by forcing a non-zero value of the z-component of the velocity vector [47]. Three-dimensional analysis has been performed by simple extrusion from two-dimensional modules. For this purpose, the best two-dimensional optimized channels in terms of low friction factor and high Nusselt number have been selected first, and one new additional variable was introduced such as the extrusion angle. The constructing procedure is described in Fig. 8.2. Due to the exploratory nature of this study, the extrusion length is fixed to the same value as the channel height.

8.5.2 CC Module

The CAD package CATIA [11] has been utilized to describe the heat transfer surface geometry, thanks to its key feature of allowing the generation of parametric drawings. In Fig. 8.6, the parameters required to describe the geometry of a CC surface are presented. Points P1, P2, P3 and P4 define the

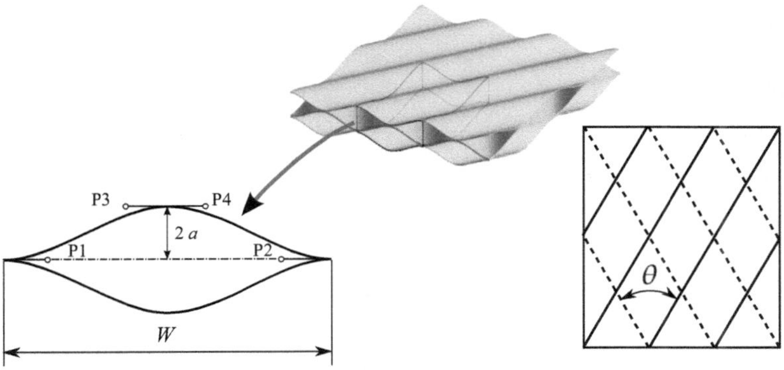

Fig. 8.6 Geometric parametrization for the CC module using *Catia*

Bézier curves, which give the shape of both extruded virtual profiles forming the flow channel. The half-height of the channel is controlled by parameter a, while its width is defined by parameter W. The corrugation angle between upper and lower plates is given by θ. By changing the parameter values, various different geometries can be easily produced.

The geometry generated for a set of the parameters can be exported and subsequently loaded by the software ANSYS ICEM-CFD [2], which generates the grid. In this work, because of their superior accuracy for Finite Volume discretization, only structured hexahedral grids have been used, but alternative possibilities exist for more complicated geometries. The generated grid is imported into the pre-processor *cfx5pre* of ANSYS CFX [1] where fluid characteristics and boundary conditions are automatically defined. The case file exported by the *cfx5pre* module is processed by the *cfx5solve* that solves fluid and thermal equations. Finally the results are post-processed by the *cfx5post* code to obtain the non-dimensional synthetic data for performance evaluation.

8.6 Optimization Methods

In design, construction, and maintenance of any engineering system, technological and managerial decisions have to be taken at several stages. The objective of such a decision process is either to maximize the desired benefit or to minimize the required effort in terms of time, money, materials, energy, environmental impact, etc. Such a process can be abstracted from the engineering field and applied to whatsoever situation in which human choice is present.

A decision-making environment can be linked to the concept of *system*. A system is an entity, a set of logical or physical connections that gives determinate outputs, when it undergoes through certain inputs. Whoever wants to look into a system, she/he has to, first of all, build a model of it: the simplest possible representation, yet bearing the most important features of the system itself. By means of its model, a system can be studied and improved using mathematical tools, whenever a quantitative relation between inputs and outputs can be established. Thus, a system can be seen as a *black box* acting on inputs to produce outputs, as in the following relation:

$$\boldsymbol{O} = \left\{ \begin{matrix} O_1 \\ \dots \\ O_m \end{matrix} \right\} = f(\boldsymbol{X}) = f \left\{ \begin{matrix} X_1 \\ \dots \\ X_n \end{matrix} \right\} \tag{8.37}$$

where $\boldsymbol{O}$ is a set of outputs and f is a generic relation linking outputs to inputs set $\boldsymbol{X}$.

Optimization is the act of obtaining the best solution under given circumstances, as Rao states in [45]. From a system point of view, this means one is searching a maximum, or respectively a minimum, for function f, depending on the desired goal. Without loss of generality, noting that the maximum of f coincides with the minimum of its opposite $-f$, an optimization problem can be taken as either a minimization or a maximization one.

The existence of optimization methods can be traced back to the beginning of differential calculus that allows minimization of functionals, in both unconstrained and constrained domains. But, in real problems, the function f is unlikely to be a simple analytical expression, in which case the study of the function by classical mathematical analysis is sufficient. It is rather a usually unknown relation that might lack continuity, derivativeness, or connectedness. Differential calculus, therefore, is of little help in such circumstances.

When a relation is unknown, a *trial and error* methodology is the oldest practice, and no further contributions to optimization techniques has been provided until the advent of digital computers, which have made implementation of optimization procedures a feasible task. From an optimization point of view inputs and outputs in Eq. (8.37) can be renamed after their conceptual meanings. Inputs are usually known in literature as *design variables*, while outputs, being the goal of an optimization process, are known as *objective functions* or simply *objectives*. In many practical problems, design variables cannot be chosen arbitrarily, but they have to satisfy specified requirements. These are called *design constraints*. Even the objectives could undergo restrictions. They are called *functional constraints*. In addition to Eq. (8.37), these two kind of constraints can be formally expressed as:

$$g\left(\boldsymbol{X}\right) \leq 0 \tag{8.38}$$

$$m\left(f\left(\boldsymbol{X}\right)\right) \leq 0 \tag{8.39}$$

where g and m are two general applications. Equality relations are easily obtained replacing the symbol "≤" with "=". Optimization problems can be classified in several ways that depend on different aspects of the problems themselves. In [45], the following classifications are highlighted.

A first classification can be based on:

1. *Existence of constraints.* As stressed earlier, problems can be classified constrained or unconstrained. Constraint handling is not a trivial task for most optimization techniques.
2. *Nature of design variables.* f can be a function of a *primitive* set of variables depending on further parameters, thus becoming *trajectory optimization problems* [24].
3. *Physical structure of the problem.* Depending on the structure of the problem, *optimal control* theory can be applied, where a global cost functional is minimized to obtain the desired solution.
4. *Nature of the relations involved.* When known or at least well guessed, the nature of the equations governing the model of the system under study can address the choice to the most efficient among a set of optimization methods. Linear, quadratic geometric and nonlinear programming are examples.
5. *Permissible values of design variables.* Design variables can be real value or discrete.
6. *Deterministic nature of the variables.* The deterministic or stochastic nature of the parameters is a criterion to classify optimization problems. In particular, the concept of *Robust Design* or *Robust Optimization* has recently gained popularity [32].
7. *Separability of the functions.* A problem is considered separable if f functions can be considered a combination of functions of single design variables $f_1(X_1)$, $f_2(X_2), \ldots, f_2(X_n)$ and f becomes:

$$f(\boldsymbol{X}) = \sum_{i=1}^{n} f_i(X_i) \ .$$

   The advantage of such a feature is that in nonlinear problems, nonlinearities are mathematically independent [22].
8. *Number of objective functions.* Depending on the number of objective functions, the problem can be single- or multi-objective. This is an outstanding distinction, since in multi-objective optimizations, objectives are usually conflicting. No single optimum exists, but rather a set of designs the decision maker has to choose from. This is one of the motivations that have led to the birth of *Evolutionary Multi-Objective Optimization* (EMOO) [8].

Depending on the characteristics of optimization problems, many techniques have been developed to solve them. These can be roughly divided into two categories.

1. *Traditional mathematical programming techniques.* They require a certain knowledge of the relation between objectives and design variables, and they are usually best suited for single-objective optimizations.

2. *Evolutionary Algorithms* (EA). They are heuristic methods that use some mechanisms inspired by biological evolution: reproduction, mutation, recombination, natural selection and survival of the fittest. Their most important feature is the applicability to almost all types of problems, because they do not make any assumption about the system under study as classical techniques do. They can be used when relation f is a completely unknown function. In their multi-objective version, *Multi-Objective Evolutionary Algorithms* (MOEA), being part of the recently cited research area known as evolutionary multi-objective optimization (EMOO), are shown to be capable of dealing with truly multi-objective optimizations [7].

Differential calculus is the first example of traditional programming techniques, but there is a widely developed literature [45] on the subject. Linear programming, quadratic programming and nonlinear programming are just some of the examples. To perform the optimization processes described in this chapter, evolutionary techniques have been used that are available in the modeFRONTIER© optimization program [14]. modeFRONTIER©, used throughout, is a state-of-the-art optimization package that includes most instruments relevant to data analysis and single- and multi-objective optimizations.

8.6.1 Design of Experiment

Heuristic evolutionary techniques do not make any assumption on the relation between objectives and design variables, thus providing an analogy with experimental dataset analysis. A good initial sampling, which allows an initial guess on the relations between inputs and outputs, is of great relevance in reducing optimization effort and improving results [40].

Design Of Experiments (DOE) is a methodology applicable to the design of all information-gathering activities where variation of decisional parameters (design variables) is present. It is a technique aimed at gaining the most possible knowledge within a given dataset. The first statistician to consider a formal mathematical methodology for the design of experiments was Sir Ronald A. Fisher, in 1920.

Before the advent of DOE methodology, the traditional approach was the so-called *One Factor At a Time* (OFAT). Each factor (design variable) which would influence the system used to be moved within its interval, while keeping the others constant. In as much as it would require a usually large number of evaluations, such a process could be quite time consuming. By contrast, Fischer's approach was to consider all variables simultaneously, varying more than one at a time, so as to obtain the most relevant information with minimum effort. It is a powerful tool for designing and analyzing data: it eliminates redundant observations, thus reducing time and resources in experiments, and giving a clearer understanding of the influence of design variables. Three main aspects must be considered when choosing a DOE:

1. The number of design variables (i.e., domain space dimension);
2. The effort of a single experiment;
3. The expected complexity of the objective function.

The *modeFRONTIER* package includes several algorithms for the DOE selection [14]. In this work, we have mainly used *full factorial* and *Sobol* ones.

Full Factorial

Full factorial (FF) algorithm sample each variable span for n values called *levels* and evaluates every possible combination. The number of total experiments is

$$N = \prod_{i=1}^{k} n_i \tag{8.40}$$

where n_i is the number of levels for the i-th variable and k is the number of design variables. Full factorial provides a very good sampling of the variables domain space, giving complete information on the influence of each parameter on the system. The higher the number of levels, the better the information. However, this algorithm bears an important drawback, namely that the number of samples increase exponentially with the number of variables. This makes the use of FF unaffordable in many practical circumstances.

Sobol

Sobol algorithm creates sequences of n points that fill the n-dimensional space more uniformly than a random sequence does. These types of sequences are called *quasi-random* sequences. This term is misleading since there is nothing random in this algorithm. The data in this type of sequence are chosen as to avoid each other, filling in a uniform way the design space. This is illustrated in Fig. 8.7(b).

8.7 Optimization Algorithms

Optimization algorithms investigate the behavior of a system, seeking for design variable combinations that give optimal performances. In terms of objective function values, an optimal performance means the attainment of extrema. Extrema are points in which the value of the function is either minimum or maximum. Generally speaking, a function might present more than

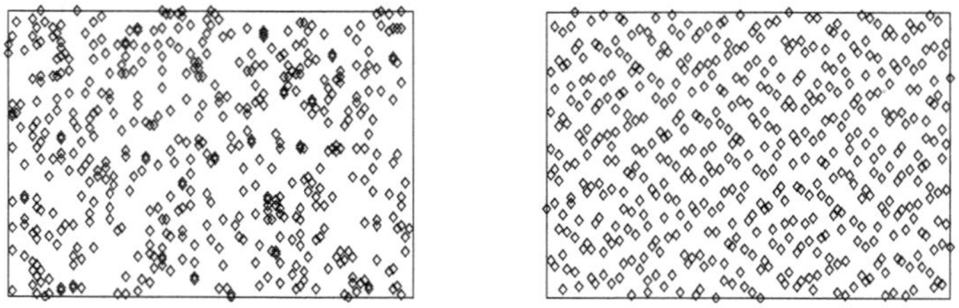

(a) random sampling (b) Sobol sampling

Fig. 8.7 Random sampling vs. Sobol (quasi-random) sampling

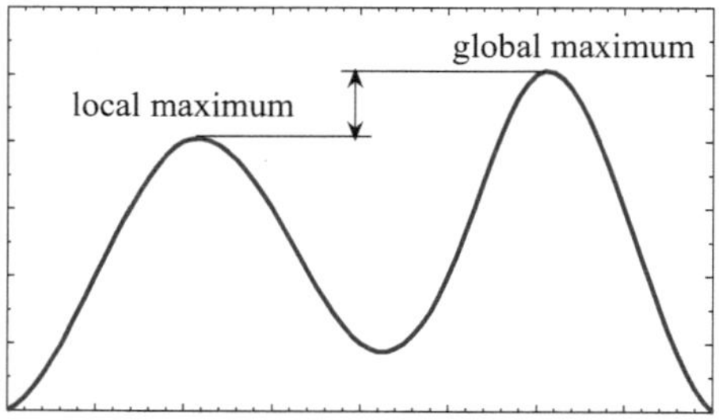

Fig. 8.8 Multiple extrema points

one extreme point, called *local extrema*, see Fig. 8.8. It is of great importance for an algorithm to be capable of finding the *global extremum* of an objective with the minimum effort. There are three main characteristics that distinguish and classify the efficiency of an optimization algorithm:

- **Robustness.** Robustness is the capability of reaching a global optimum point without being stuck in local extrema, or blocked for lack of useful data. This is the most important feature in measuring the efficiency of an optimization technique. The more an algorithm is robust, the higher the chance to reach a global optimum or sub-optimum, i.e., a point close to the global optimum.
- **Accuracy.** Accuracy is the ability of an algorithm to reach the actual extrema, either global or local, around its proximity. Usually, accuracy and robustness are conflicting attributes, so that robust algorithms are not accurate and vice versa.
- **Convergence rate.** Convergence rate is a measure of the effort an algorithm has to carry on to reach its goal. Again, robust algorithms are usually slow. However, fast but not robust ones might not reach the goal at all.

In this work, evolutionary techniques have been used that are usually robust but neither accurate nor fast. However, as already stressed, their most important attribute as well as reason for their popularity is the applicability

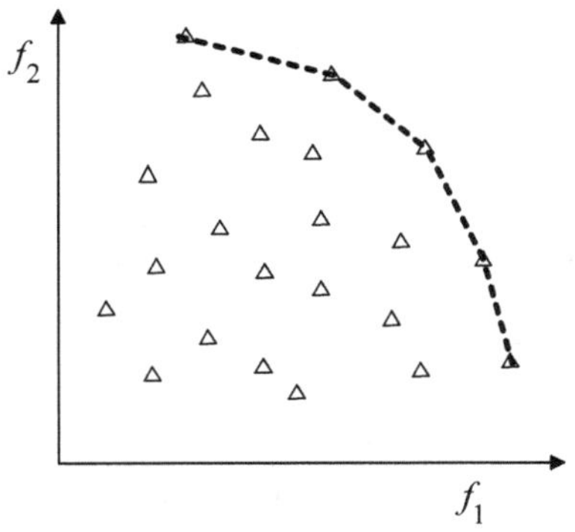

Fig. 8.9 Pareto front for a two-objective optimization

to almost any single- or multi-objective optimization problem of whichever complexity. In EA heuristic processes, most of the current research aims at proving actual convergence [7, 46]. Nevertheless, the wide literature on successful applications of evolutionary optimizations sets EMOO as a promising field [54].

Pareto Optimality

As soon as there are many, possibly conflicting, objectives, it is rarely the case that there is a single design variables combination that simultaneously optimizes all the objective functions in a multi-objective optimization problem. Rather, there usually exists a whole set of possible solutions of equivalent quality. This can be the case with heat exchangers and in particular of their building modules studied in this chapter: the desired objectives might be to maximize heat transfer rate per unit volume, to minimize the pumping power, to minimize cost and weight, and to minimize the performance degradation due to fouling. These goals are clearly conflicting and, therefore, there is no single optimum to be found. For this reason, the so-called *Pareto dominance* or *Pareto optimality* concept must be used, according to which *design a dominates design b* if and only if:

$$(\forall i\, f_i(a) \geq f_i(b)) \cap (\exists j : \, f_j(a) > f_j(b)) \tag{8.41}$$

where f_i is the *i-th* objective function, and for simplicity it is assumed that we are considering the maximization of the n objectives. Expression (8.41) means that at least in one purpose design a is better than design b, while in the others they might be equal. Let us consider, as an example, a problem where it is desired to maximize two objective functions f_1 and f_2. Each design evaluation, during the optimization, produces an $[f_1, f_2]$ couple, and all these

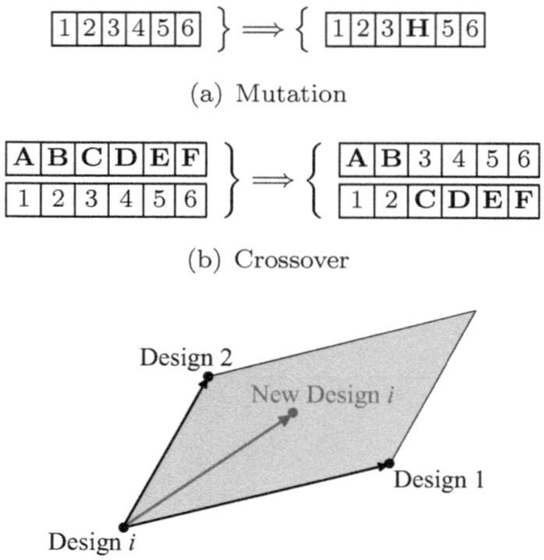

(a) Mutation

(b) Crossover

(c) Directional Crossover

Fig. 8.10 Evolutionary operators of a GA

data can be represented graphically as in Fig. 8.9. The relations (8.41) allows the determination of a design set – those joined by a dashed line – which is called *Pareto front* or *Pareto optimal set* and whose dimension is equal to $n-1$. In this example $n=2$, and the front is a line.

8.7.1 Genetic Algorithm

Genetic algorithm (GA) is the most popular type of EA. The basic idea underlying the method comes from the behavior of living organisms in nature. An initial set of *individuals*, called *initial population*, undergoes a natural selection process. So each individual can be seen as a DNA string. Parental populations give birth to offsprings. GAs work on individuals as coded bit strings, thus they need discrete variable intervals. The new generations are created following a series of genetic rules:

- **Selection.** Selection operator randomly shifts a defined number of individuals to the next generation, keeping them unchanged. The probability for an individual to undergo a process of selection is weighed on the fitness[1] value of each design. The better the fitness, the higher the probability to be selected for the new population.

[1] *fitness* is the measure of how a design variable set *fits* the goal of an optimization. Its quantitative definition depends on the chosen algorithm.

$$P(\boldsymbol{X}_i) = \frac{F(\boldsymbol{X}_i)}{\sum_{j=1}^{n} F(\boldsymbol{X}_j)}$$

where $P(\boldsymbol{X}_i)$ is the probability of selection for the i-th individual with fitness $F(\boldsymbol{X}_i)$ of the n-sized population.
- **Mutation.** Mutation operator consists of the random substitution of some bits (nucleotides) in the numeric string representing an individual. The role of mutation is to enhance the probability of exploring untouched areas of the design space avoiding premature convergence. Mutation generally involves less than 10% of the individuals.
- **Crossover.** Crossover is a genetic recombination between two individuals, whose strings are randomly cut and re-assembled.

The version of GA in modeFRONTIER© implements a fourth operator called *directional crossover*. It assumes that a direction of improvement can be detected comparing the fitness values of two reference individuals. This operator usually speeds up the convergence process, though it reduces robustness.

Directional cross over works as follows:

1. Select an individual i;
2. Select reference individuals i_1 and i_2;
3. Create the new individual as:

$$\boldsymbol{X}_{i_{new}} = \boldsymbol{X}_i + s \cdot \text{sign}(F_i - F_{i_1})(\boldsymbol{X}_i - \boldsymbol{X}_{i_1}) + t \cdot \text{sign}(F_i - F_{i_2})(\boldsymbol{X}_i - \boldsymbol{X}_{i_2})$$

where s and t are two random parameters. The genetic operator behavior is illustrated in Fig. 8.10. Each operator can be applied with a certain probability. Different combinations of operator probabilities may lead to different levels of robustness, accuracy and convergence rate.

8.7.2 Multi-objective Approaches

When there is a single-objective function, the definition of a metric for evaluating design *fitness* is straightforward. Pareto optimal set reduces to a single point and the best design is a global extreme. On the other hand, the introduction of multiple objectives tangles things a bit. It has been already stressed that in multi-objective optimization, there exists a set of solutions that present equal quality or effectiveness. These are the non-dominated designs as defined in Eq. (8.41). The problem arises as to how to compare and judge designs in order to get a scalar evaluation scale for their fitness.

MOEA approaches can be roughly divided into three categories: *Aggregating Functions*, *Population-based Approaches*, and *Pareto-based Approaches*.

8.7.2.1 Aggregating Functions

The most straightforward approach in handling multiple objectives is the use of an arithmetical combination of all the objectives. Thus, the resulting single function can be studied with any of the single-objective algorithms either evolutionary or classical. Aggregating approaches are the oldest mathematical programming methods found in literature [45].

Applied to EA, the aggregate function approach does not require any change to the basic search mechanism. Therefore, it is efficient, simple, and of easy implementation. It can be successfully used on simple multi-objective optimizations that present continuous convex Pareto fronts. An example of this approach is a linear sum of weights of the form:

$$\min \left(\sum_{i=1}^{m} w_i f_i(\boldsymbol{X}) \right)$$

where the weight w_i represent the relative importance of the m-th objective function. The weighting coefficients are usually assumed to sum at 1:

$$\sum_{i=1}^{m} w_i = 1 \ .$$

Aggregate function may be linear as in the previous example or nonlinear. Both types of function have been used with evolutionary algorithms but, generally speaking, aggregating methods have some limitations in generating complex Pareto fronts, though nonlinear aggregating function do not necessarily present such limitations [8]. Aggregating functions are widely used in *Multi-Criteria Decision Analysis* (MCDA) or *Multi-Criteria Decision Making* (MCDM) [5]. MCDA is a discipline aimed at supporting decision makers who are faced with making numerous and conflicting evaluations. MCDM is frequently used by EA users for *a posteriori* analysis of optimization results, but the discipline can be applied in a much more sophisticated way for *a priori* analysis. MCDM is briefly introduced in Sect. 8.7.3.

8.7.2.2 Population-based Approaches

In these techniques the population of an EA is used to diversify the search, but the concept of Pareto dominance is not directly incorporated into the selection process.

The first approach of this kind is the *Vector Evaluation Genetic Algorithm* (VEGA) introduced by Schaffer [48]. At each generation this algorithm performs the selection operation based on the objective switching rule, i.e., selection is done for each objective separately, filling equal portions of mating pool (the new generation) [12]. Afterwards, the mating pool is shuffled, and crossover and mutation are performed as in basic GAs. Solutions proposed by VEGA are locally non-dominated, but not necessarily globally non-dominated. This comes from the selection operator which looks for optimal individuals for a single objective at a time. This problem is known as *speciation*. Groups of individuals with good performances within each objective function are created, but non-dominated intermediate solutions are not preserved.

8.7.2.3 Pareto-based Approaches

Recognizing the drawbacks of VEGA, Goldberg [20] proposed a way of tackling multi-objective problems that would become the standard in MOEA for several years.

Pareto-based approaches can be historically divided into two generations. The first is characterized by *fitness sharing* and *niching* combined with the concept of *Pareto ranking*. Keeping in mind the definition of non-dominated individual, the rank of a design corresponds to the number of individuals by which it is dominated. Pareto front element has a rank equal to 1. The most representative algorithms of the first generation are *Nondominated Sorting Genetic Algorithm* (NSGA), proposed by Srinivas and Deb [50], *Niched-Pareto Genetic Algorithm* (NPGA) by Horn et al. [21], and *Multi-Objective Genetic Algorithm* by Fonseca and Fleming [19].

The second generation of MOEAs was born with the introduction of *elitism*. Elitism refers to the use of an external population to keep track of non-dominated individuals. In such a way, viable solutions are never disregarded in generating offsprings.

The second generations of multi-objective algorithms based on GA are:

1. **MOGA-II.** MOGA-II uses the concept of Pareto ranking. Considering a population of n individuals, if at generation t, individual x_i is dominated by $p_i^{(t)}$ designs of the current generation, its rank is given by:

   $$rank(x_i, t) = 1 + p_i^{(t)} \ .$$

   All non-dominated individuals are ranked 1. Fitness assignment is performed by:

   a) Sort population according to rank;
   b) Assign fitness to individuals by interpolating from the best to the worst;
   c) Average the fitness of individuals with the same rank. In this way, the global fitness of the population remains constant, giving quite a selective pressure on better individuals.

   The modeFRONTIER© version of MOGA-II includes the following smart elitism operator [39]:

a) MOGA-II starts with an initial population P of size n and an empty elite set $E = \emptyset$
b) For each generation, compute $P' = P \cup E$
c) If cardinality of P' is greater than cardinality of P, reduce P' randomly removing exceeding points
d) Generate P'' by applying MOGA algorithm to P'
e) Calculate P'' fitness and copy non-dominated designs to E
f) Purge E from duplicated and dominated designs
g) If cardinality of E is greater than cardinality of P randomly shrink the set
h) Update P with P'' and return to step (b)

2. **NSGA-II.** NSGA-II employs a fast non-dominated sorting procedure and uses the crowding distance (which is an estimate of the density of solutions in the objective space) as a diversity preservation mechanism. Moreover, NSGA-II has an implementation of the crossover operator that allows the use of both continuous and discrete variables. The NSGA-II does not use an external memory as MOGA does. Its elitist mechanism consists of combining the best parents with the best offspring obtained.

8.7.3 Multi-Criteria Decision Making (MCDM)

As already stated, it is impossible to find out a unique best solution in a multi-objective optimization process, but rather a whole group of designs that dominate the others: the *Pareto front* or *Pareto optimal set.* All Pareto optimal solutions can be regarded as equally desirable in a mathematical sense. But from an engineering point of view, the goal is a single solution to be put into practice at the end of an optimization. Hence, the need for a decision maker (DM) who is able to identify the most preferred one among the solutions. The decision maker is a person who is able to express preference information related to the conflicting objectives. Ranking between alternatives is a common and difficult task, especially when several solutions are available or when many objectives or decision makers are involved.

Decisions have taken over a limited set of good alternatives mainly due to the experience and competence of the single DM. Therefore, the decision stage can be described as *subjective and qualitative* rather than *objective and quantitative.* Multi-Criteria Decision Making (MCDM) refers to the solving of decision problems involving multiple and conflicting goals, coming up with a final solution that represents a good compromise that is acceptable to the entire team. As already underlined, when dealing with a multi-objective optimization, the decision making stage can be done in three different ways [53]:

- **Decision-making and then search (a priori approach).** The preferences for each objective are set by the decision-makers and then, one or various solutions satisfying these preferences have to be found.
- **Search and then decision-making (a posteriori approach).** Various solutions are found and then, the decision-makers select the most adequate. The solutions chosen should represent a trade-off between the various objectives.
- **Interactive search and decision-making.** The decision-makers intervene during the search in order to guide it towards promising solutions by adjusting the preferences in the process.

modeFRONTIER© implementation of MCDM allows the user to classify all the available alternatives through pair-wise comparisons on attributes and designs. Moreover, modeFRONTIER helps decision makers to verify the coherence of relationships. Thus it is an *a posteriori* or *interactive* oriented tool.

To be coherent, a set of relationships should be both rational and transitive. To be rational means that if the decision maker thinks that solution A is better than solution B, then solution B is worse than solution A. To be transitive means that if the decision maker says that A is better than B and B is better than C, then solution A should be always considered better than solution C. In this way, the tool allows the correct grouping of outputs into a single utility function that is coherent with the preferences expressed by the user. Four algorithms are implemented in modeFRONTIER:

- **Linear MCDM.** When the number of decision variables is small;
- **GA MCDM.** This algorithm does not perform an exact search so it can be used even when the number of decision attributes is large;
- **Hurwicz criterion.** Used for the uncertain decision problems;
- **Savage criterion.** Used for the uncertain decision problems where both the decision states and their likelihoods are unknown.

8.7.4 Optimization Process

The parameters that influence the performances of a heat exchanger, and in particular of the single repeating module under study, are its overall heat transfer coefficient and the pumping power required to supply a desired mass flow. In a dimensionless perspective, these two parameters are represented by the friction factor and the Nusselt number, respectively. Both quantities are function of the geometric variables:

$$[f, \mathrm{Nu}] = g(\boldsymbol{X}) \tag{8.42}$$

where vector $\boldsymbol{X}$ is constituted by the degrees of freedom of the chosen parametrization, either linear piecewise, Bézier or NURBS. As stated before,

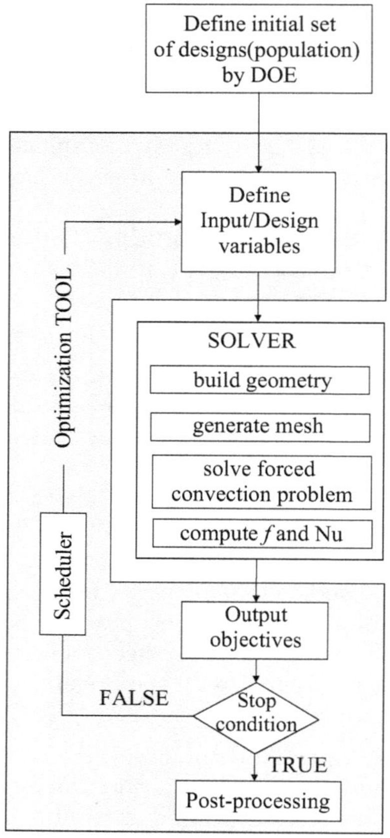

Fig. 8.11 Optimization work-flow

there is an ample variety of numerical techniques to perform optimization tasks. EAs are the most robust ones, and have been used in this optimization process. In particular, MOGA-II algorithm, has been employed.

The optimization process, sketched in Fig. 8.11, follows these tasks:

- Automatically the optimization software generates a set of numbers, i.e., the geometrical design variables, representing the shape of a channel (called *individual*).
- These variables are written in an input file, which is sent to the CFD solver. This, in turn, computes the flow and thermal fields, and from these, it evaluates the friction factor and the Nusselt number, which represent the objective functions.
- The numerical values of the objective functions are sent back to the optimizer, which generates another set of geometrical parameters.

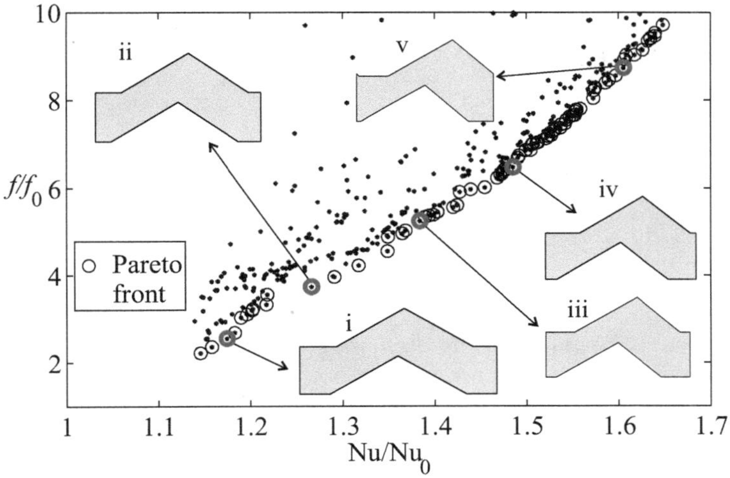

Fig. 8.12 Pareto front for the 2D linear piecewise optimization process

The results are expressed in terms of the ratios Nu/Nu_0 and f/f_0, where Nu_0 and f_0 are the Nusselt number and the friction factor for a parallel plate channel, taken as reference geometry.

8.8 Results and Discussion

8.8.1 Linear Piecewise Optimization

The first analysis has been conducted on the linear piecewise geometry type. The variables are small in number and clearly related to the shape of the channel: the depth of the asperity, the forward side angle, the backward side angle, the length of the channel and the translation of the upper wall.

The Full Factorial algorithm [14] with three levels has been used (DOE) to sample the design space. This method gives the best homogeneous distribution of the samples. The number of Full Factorial samples is 243 and the parameters ranges are given in Table 8.1. In a GA optimization, the size of the population within the design space affects the convergence ratio. After having performed the numerical simulation on this set, its Pareto front has been chosen as the initial population of multi-objective optimization with GA. This preliminary Pareto front is a good selection in order to limit the number of starting channels. The optimization has been performed with MOGA-II along 30 generations of 20 individuals each: so the total number of designs evaluated is 600. The results are sketched in Fig. 8.12 where the Pareto front

Table 8.3 Values of the design variables for the selected linear piecewise channels

	ID i	ID ii	ID iii	ID iv	ID v
L	2.00	1.64	1.54	1.58	1.47
h	0.400	0.400	0.383	0.400	0.400
φ_{in}	60.00°	59.04°	60°	59.04°	60.00°
φ_{out}	60.00°	56.04°	50.16°	51.24°	49.32°
$transl$	0.066	0.102	0.202	0.236	0.298

is highlighted. Due to the simplicity of the parametrization, the dominant set can probably be taken as the limit performance of this kind of geometry. In fact, further optimization process has not given appreciable improvements on design objectives.

From the analysis of the shape of the channels along the Pareto front, sketched for convenience in the same figure, no sensible fluctuation is evident for variables like φ_{in} and h, which remain close to 60° and 0.4, respectively. What makes the difference is the translation of the upper profile, responsible for the development of the separation bubble induced by the corrugation. In Table 8.3, the values of the design variables required for the definition of the channels, marked in Fig. 8.12 for illustrative purpose, are presented.

8.8.2 NURBS Optimization

As already stated, the increased number of degrees of freedom causes a more expensive optimization task. In contrast with linear piecewise optimization, the Full Factorial algorithm has not been used as first exploration of the design space. Since there are 11 degrees of freedom, the number of individuals to be computed would have risen to 3^{11}, that means 177, 147, with a three-level Full Factorial. The Sobol algorithm [14] has been chosen to define an initial population of 50 individuals. The optimization algorithm chosen was again MOGA-II. The optimization has started allowing great freedom to the design variables, and no constraint has been imposed. In Table 8.4, the summary of design variable ranges and the number of steps (basis), which notches them to discrete form, are presented. The choice of the variables and their range might dramatically affect the convergence rate to a good solution. In our case, the generation of input strings leading to incoherent geometries has to be avoided as much as possible. One strategy is to scale the x-direction Cartesian variables to the length of the channel L. Therefore, all parameters are proportional to each other. Figure 8.13(a) shows the Pareto front after this first optimization stage. Five channels, representative of different combination of the two objectives, are highlighted. From Fig. 8.13(a), it is clear that all the individuals selected have in common the presence of closing bends

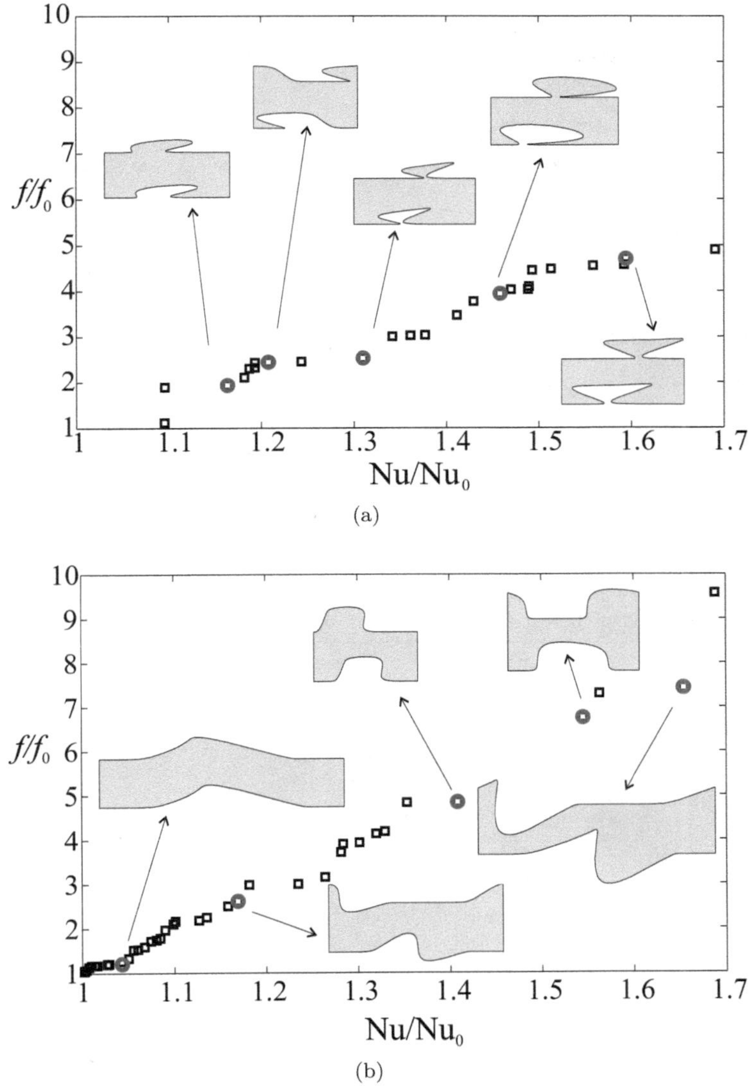

Fig. 8.13 Pareto front comparison between different optimization processes for the 2D NURBS channels: (**a**) first MOGA; (**b**) first MOGA with constraint; see also Fig. 8.14

in wall profile. The same happens for almost all the other channels. Taking into account the industrial feasibility, though only from a methodological point of view, these channels would be very difficult to realize, for example, in a forming process. Therefore, a sort of *fabricability check* has been implemented, and its aim was to discard those geometries whose wall profile

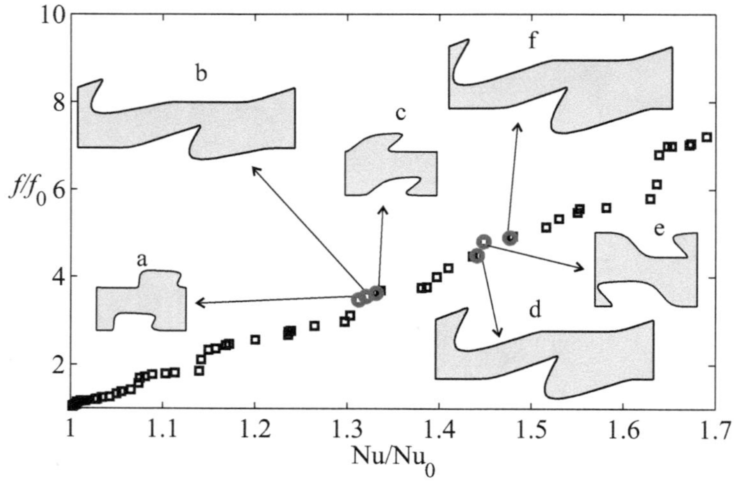

Fig. 8.14 Pareto front comparison between different optimization processes for the 2D NURBS channels: last Pareto front; see also Fig. 8.13

Table 8.4 Ranges of the variables for the first NURBS optimization

Parameter	Range	Basis
L	$[1.00; 2.50]$	1001
L_{wave}	$[0.15\,L; 0.85\,L]$	1001
Δx_{23}	$[0.01\,L; 0.20\,L]$	1001
ρ_{54}	$[0.05; 1.20]$	1001
ϑ_{54}	$[70°; 300°]$	501
Δx_5	$[0.30\,L; 0.75\,L]$	1001
Δy_5	$[0.00; 0.50]$	1001
ρ_{56}	$[0.05; 1.10]$	1001
ϑ_{56}	$[-120°; 150°]$	501
Δ_{87}	$[0.01\,L; 0.80\,L]$	1001
$transl$	$[-0.500\,L; 0.500\,L]$	1001

could not be obtained by moulding. This constraint can be summarized as follows. During the construction procedure, after the lower wall profile has been drawn, it is *brushed* along x direction by straight lines which progressively change their slope ($[-45°; +45°]$ with respect to y axis), Fig. 8.15. The channel is declared feasible if and only if, there exists at least one direction at which all the lines encounter the wall profile no more than once. The simulation has been restarted with the same parameters, but introducing the fabricability check just described. Results are depicted in Fig. 8.13(b). The presence of such a constraint makes the optimization process closer to the channel shapes more common in practice. Its observation reveals the decreas-

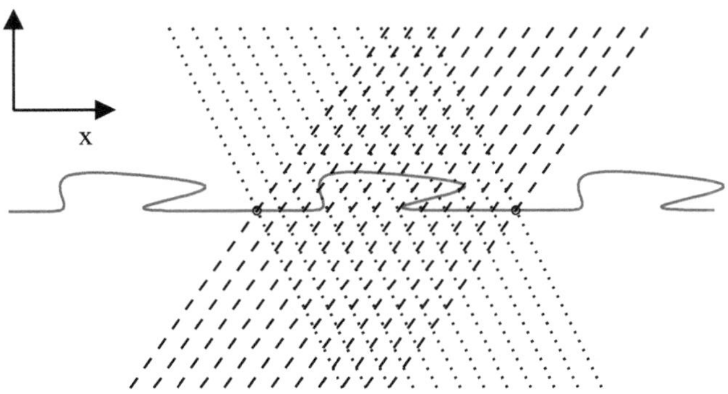

Fig. 8.15 Fabricability check

ing performances of the second set, as it comes out from a comparison between Figs. 8.13(a) and 8.13(b). As already anticipated the MOGA algorithm is well suited for truly multi-objective problems. It is robust, albeit somehow slow for increasing number of objective functions or design variables.

Once a high quality Pareto front has been obtained, one can choose slightly different strategies in order to improve the fitness of the channels. One way is to transform the multi-objective problem into a single-objective one by means of a weighted function, involving objectives. The mono-objective optimization, performed using MOGA-II and Simplex [45] algorithms, makes the process faster. Having two starting objectives, imposing different relations between designs, a different weight distribution on f and Nu could be given and a different ranking to each alternative would be assigned. Using MCDM, two kind of utility functions were created: the first privileges the increase of Nusselt number, whereas the other is more focused toward the reduction of the friction factor. In this way, the results summarized in Fig. 8.14 have been obtained after evaluating 2,500 designs.

In the same figure two sequences of three channels, each one having almost the same performance metrics, are marked. They represent different arrangements of geometries in the Pareto front. After the NURBS optimization process it has been recognized that two different families of channel shapes belong to the part of the Pareto front characterized by high values of f and Nu, and they are shuffled.

Although various kind of corrugations are present, the main difference between the two types, the one called S for short (ID a, ID c, ID e), the other L for long (ID b, ID d, ID f), is the length of the module. The average length of S-channels is 1, while for L-channels it is 2, so the ratio between length of channels S and channels L is 0.5. This is an important feature of the optimization because it shows the non-univocity of the solution, i.e., similar performances can be reached by very different geometries [41]. In Fig. 8.16, the design database has been filtered and divided into two categories. Pareto

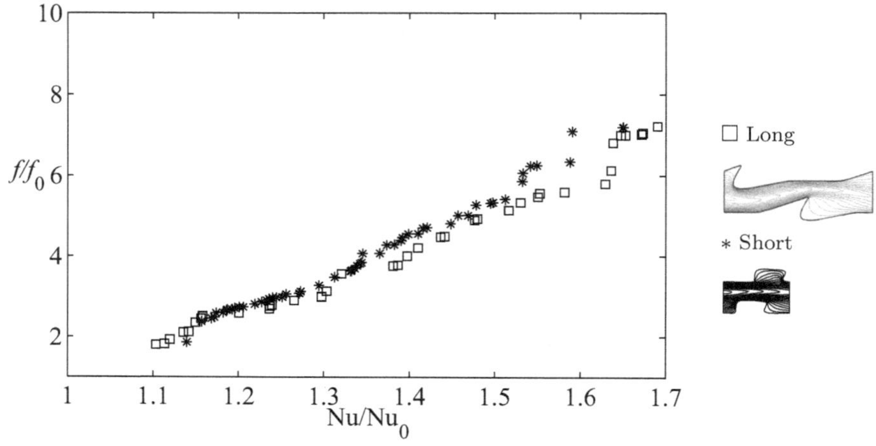

Fig. 8.16 Two distinct Pareto front for the different type of geometries

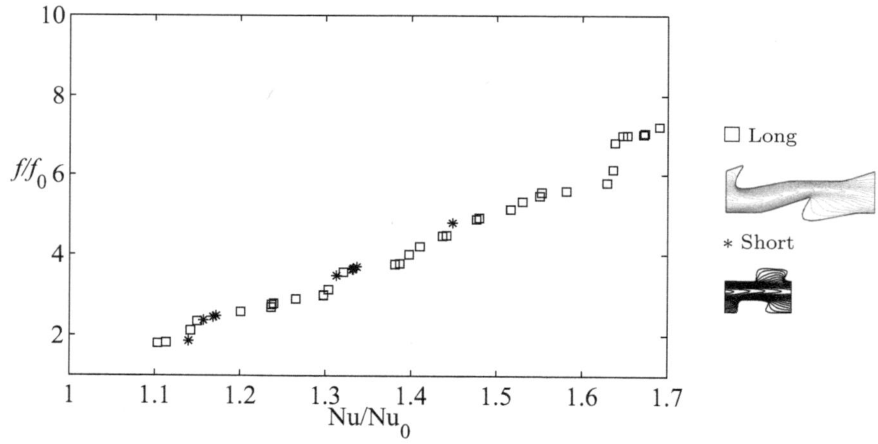

Fig. 8.17 Overlapping the solutions, leading to a single Pareto front

dominance applied to both subsets of design shows there are clearly two fronts made of completely different geometries that overlap. From Fig. 8.17, it is clear that the unfiltered front is mainly made by long geometries with superpositions of short channels. Figure 8.14 shows two examples of this phenomenon. It should be noted that at low values of f and Nu, the shape of designs is close to parallel plate channel, so the different length does not affect the form of the wave.

The six channels marked in Fig. 8.14 are shown in Fig. 8.18, together with the streamlines and non-dimensional temperature field. They are scaled with the same average height in order to be directly compared. In Table 8.5, the

Table 8.5 Control points for the selected NURBS-based channels

	ID a	ID b	ID c
(x_1, y_1)	$(0, 0)$	$(0, 0)$	$(0, 0)$
(x_2, y_2)	$(0.160, 0)$	$(0.354, 0)$	$(0.179, 0)$
(x_3, y_3)	$(0.207, 0)$	$(0.457, 0)$	$(0.252, 0)$
(x_4, y_4)	$(0.190, 0.220)$	$(1.183, 0.251)$	$(0.409, 0.161)$
(x_5, y_5)	$(0.410, 0.199)$	$(0.196, 0.267)$	$(0.612, 0.207)$
(x_6, y_6)	$(1.348, 0.198)$	$(0.893, -0.234$	$(0.965, 0.224)$
(x_7, y_7)	$(0.768, 0)$	$(1.830, 0)$	$(0.580, 0)$
(x_8, y_8)	$(0.554, 0)$	$(1.950, 0)$	$(0.843, 0)$
(x_9, y_9)	$(0.870, 0)$	$(2.304, 0)$	$(1.023, 0)$
transl	0.290	−1.039	−0.204
	ID d	**ID e**	**ID f**
(x_1, y_1)	$(0, 0)$	$(0, 0)$	$(0, 0)$
(x_2, y_2)	$(0.341, 0)$	$(0.113, 0)$	$(0.347, 0)$
(x_3, y_3)	$(0.487, 0)$	$(0.171, 0)$	$(0.480, 0)$
(x_4, y_4)	$(1.149, 0.248)$	$(-0.105, 0.249)$	$(1.068, 0.209)$
(x_5, y_5)	$(1.152, 0.268)$	$(0.208, 0.279)$	$(1.111, 0.262)$
(x_6, y_6)	$(0.827, -0.362)$	$(0.551, -0436)$	$(0.696, -0.440)$
(x_7, y_7)	$(1.853, 0)$	$(0.685, 0)$	$(1.890, 0)$
(x_8, y_8)	$(1.880, 0)$	$(0.909, 0)$	$(1.916, 0)$
(x_9, y_9)	$(2.221, 0)$	$(1.023, 0)$	$(2.263, 0)$
transl	−0.920	−0.268	−0.994

values of control points and of upper wall translation for the six channels are listed.

8.8.3 Linear Piecewise versus NURBS

In Fig. 8.20, the results sets obtained by the linear piecewise and NURBS optimization are compared. The ones marked by stars, represent channels with linear piecewise wall profile. The others marked by squares, represent channels with smooth NURBS wall profile. The higher geometrical complexity and computational costs of NURBS channels is well counterbalanced by the significant performance improvement over the simpler channels with linear piecewise walls. The greater number of variables makes the convergence slower and an asymptotic limit of the front was not reached during the optimization of NURBS profile channel.

Channel (a) $f = 0.417$ Nu $= 9.90$

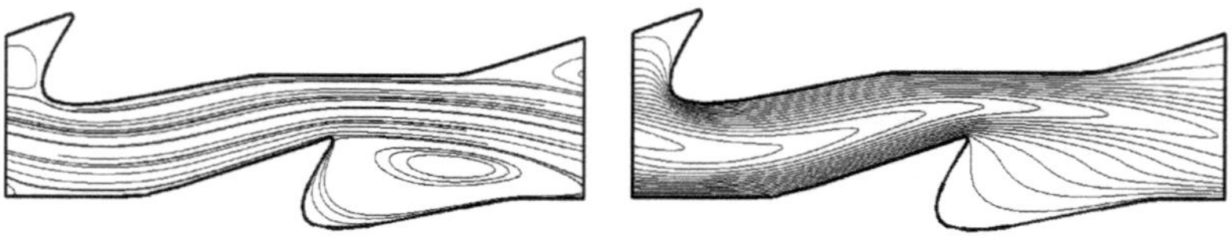

Channel (b) $f = 0.427$ Nu $= 9.96$

Channel (c) $f = 0.436$ Nu $= 10.04$

Fig. 8.18 Selected NURBS-based channels: fluid-dynamic (left) and thermal (right) fields. These three channels correspond to designs marked in Fig. 8.14; see also Fig. 8.19

8.8.4 Three-dimensional Analysis

This part of the work is a proof-of-concept in 3D applications. A true optimization has not been performed on NURBS channels but, due to time limitation and computing resources available, a parametric analysis has been done in order to verify the applicability of the method to 3D problems. For the same reasons, the Reynolds number has been reduced from 200 to 100, in order to guarantee an adequate level of numerical accuracy. Therefore, the 2D Pareto fronts have been recomputed at this reduced value of the Reynolds number, in order to compare the performances of the 2D channels with the 3D ones. These have been obtained, as indicated in Fig. 8.2, by extrusion, at different angles, of selected 2D channels. In Figs. 8.21(a) and 8.21(b), the results obtained are presented. In particular, the results for the simpler channels with linear piecewise walls are shown in Fig. 8.21(a), while the objective functions for the NURBS-based channels are shown in Fig. 8.21(b). From the figures it is evident that, for both type of channels, the Nusselt number in

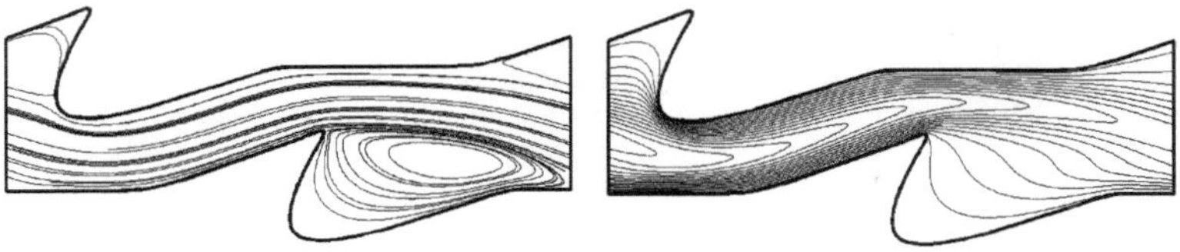

Channel (d) $f = 0.539$ Nu $= 10.87$

Channel (e) $f = 0.577$ Nu $= 10.92$

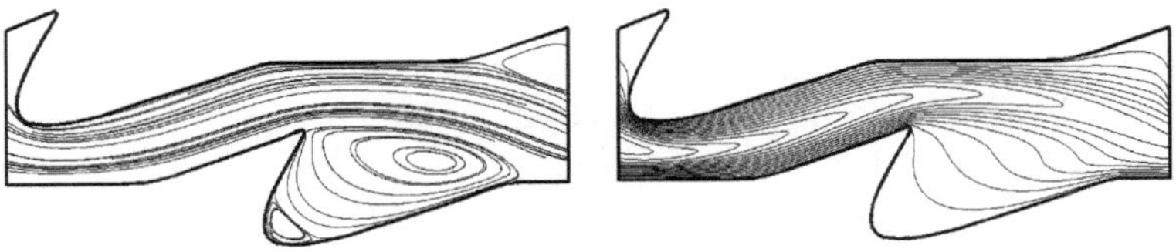

Channel (f) $f = 0.587$ Nu $= 11.13$

Fig. 8.19 Selected NURBS-based channels: fluid-dynamic (left) and thermal (right) fields. These three channels correspond to designs marked in Fig. 8.14; see also Fig. 8.18

Table 8.6 Variation of friction factor and Nusselt number versus extrusion angle for one design

f/f_0	Nu/Nu_0	Extrusion angle
3.807	1.307	0°
3.932	1.366	20°
3.937	1.508	30°
3.909	1.530	40°

general tends to increase, without a sensible increment in the friction factor, for the 3D channels when the extrusion gives rise to a 3D flow pattern.

This trend is due to the presence of secondary motions that enhance the heat transfer at the walls. In Table 8.6, the results for one of the NURBS profile extrusions are reported, and the positive effect of secondary motions

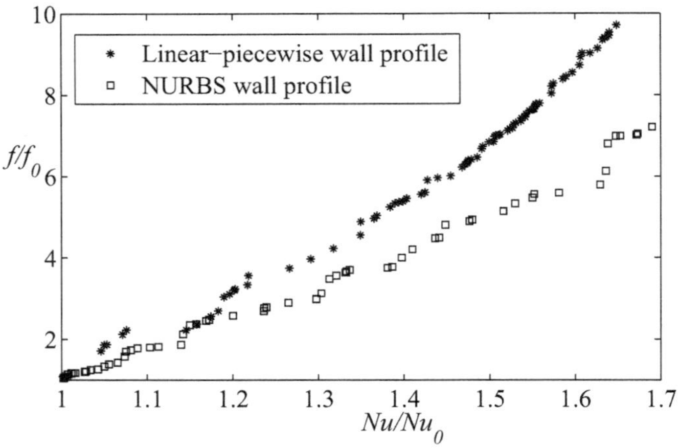

Fig. 8.20 Pareto front comparison between linear piecewise and NURBS-based channels

and longitudinal steady vortices on the heat transfer rate is again clear. For illustrative purposes, the secondary flow pattern for two channels obtained with a 20° and 40° extrusion respectively, are depicted in Fig. 8.22.

Finally, a MOGA optimization has been started on extruded linear piecewise channels for a short number of generations. The Pareto front of this optimization is compared in Fig. 8.23 with the 2D linear piecewise and 2D NURBS fronts. Though the 3D linear piecewise front is rather sparse because of a small number of individuals processed, the heat transfer augmentation due to secondary motions is clearly visible.

8.8.5 CC Module

Again with MOGA-II, an original design was used for the DOE when optimization was first attempted. The design had sinusoidal wall profile (design 0), characterized by a corrugation angle $\theta = 60°$, and 35 designs have been chosen with Sobol method. The value of the angle θ has been chosen, following Stasiek et al. [52], as a good compromise in terms of heat transfer rate and friction factor. The six design variables were, with reference to Fig. 8.6, P1, P2, P3, P4, W and θ. The three objectives to be minimized were the pressure gradient β, the temperature difference between the two fluids ΔT and the heat transfer surface area. The first objective has been selected to reduce the pressure drop across the regenerator. The second one drives the heat transfer performance of the regenerator. For instance, based on inspection of Eq. (8.35), a lower ΔT leads to a higher mean Nusselt number. The last objective has been selected to reduce the overall cost of the

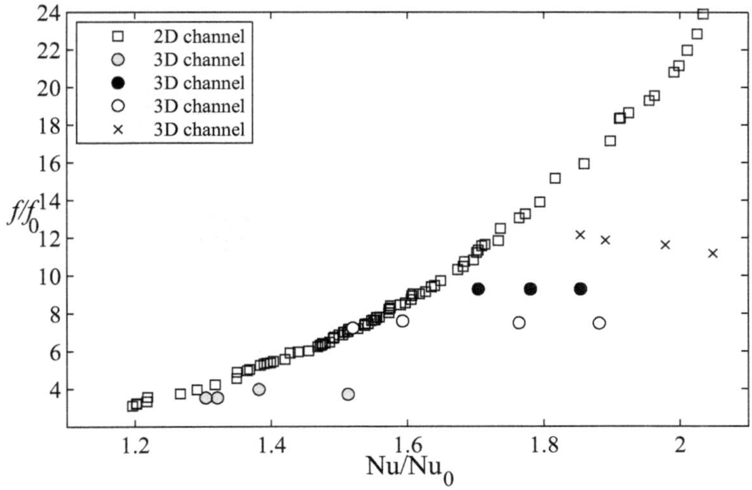

(a)

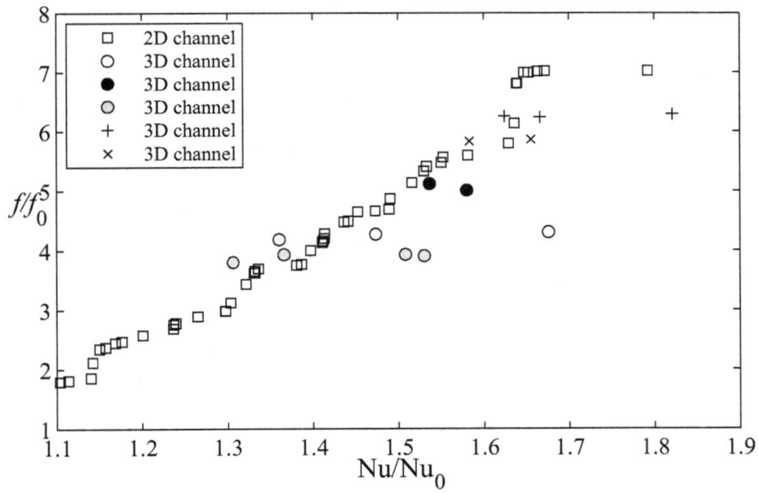

(b)

Fig. 8.21 Pareto front comparison: (**a**) linear piecewise profile; (**b**) NURBS profile

material. All three objectives were constrained to be smaller or at least equal to those for the reference design 0. This preliminary attempt, however, was quite unsuccessful, with most of the computed designs considered *unfeasible* (outside the constraints domain).

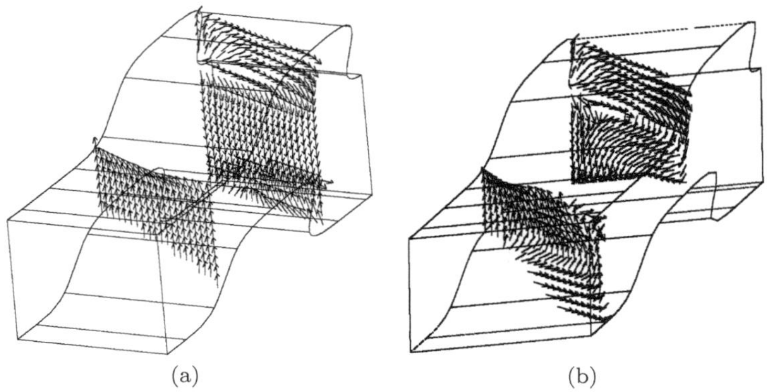

Fig. 8.22 Example of secondary vortices for the 3D NURBS channels: (**a**) extrusion angle of 20°; (**b**) extrusion angle of 40°

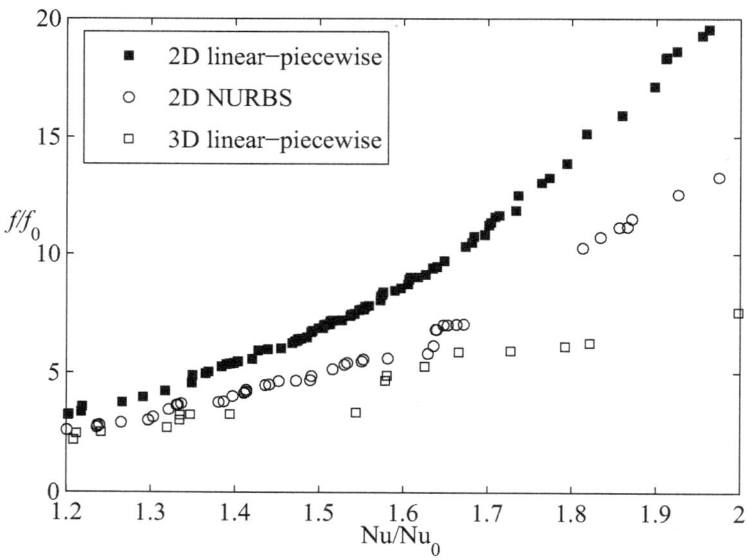

Fig. 8.23 Pareto fronts: 2D-3D linear piecewise and 2D smooth NURBS profile

Therefore, a second, simpler optimization has been carried out using 5 design variables, by fixing W. Moreover the objective ΔT has been transformed to a constraint, in order to simplify the optimization and to favour module geometries with lower surface area and lower pressure drop. The optimization has been carried out for 332 designs. At this point the process was practically converged. The Pareto front of this optimization is reported in Fig. 8.24 where it can be seen that the designs with the same inclination angle θ are rather perfectly aligned. A decrease of the inclination angle leads to lower

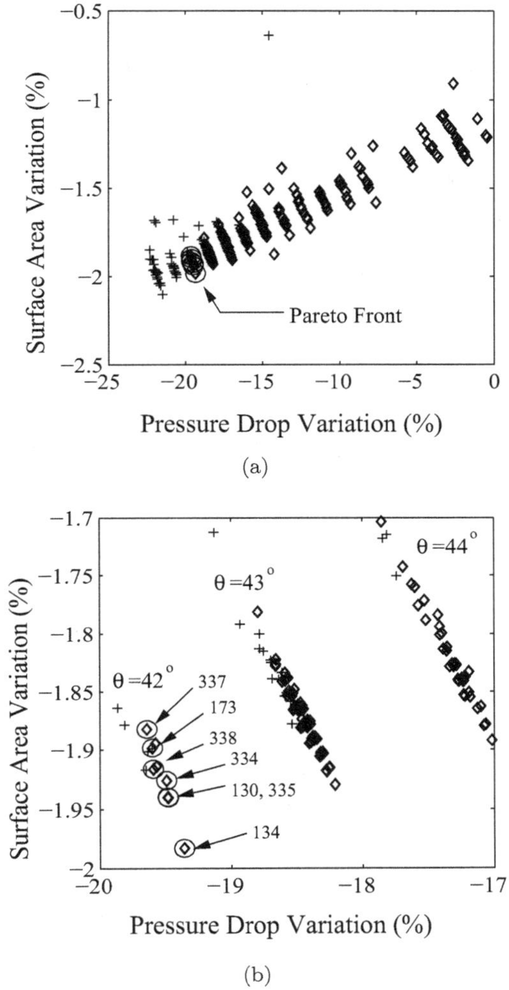

Fig. 8.24 Pareto fronts for the CC channel: ◇ feasible designs, + unfeasible designs, ○ Pareto front, (**a**) Pareto fronts, (**b**) design numbers

pressure drop and surface area as shown in Fig. 8.24(b) while maintaining the same heat transfer characteristics. This trend is maintained up to an angle of 42°, while for lower values the same heat transfer characteristics could not be maintained. These correspond to unfeasible results, since the value of the mean Nusselt number is lower than that of the original design 0.

Among the designs that belong to the Pareto front, the *design 134* has been chosen for illustrative purposes. In fact it has less than 19% pressure gradient β, and a surface area 2% lower than that of the original one while

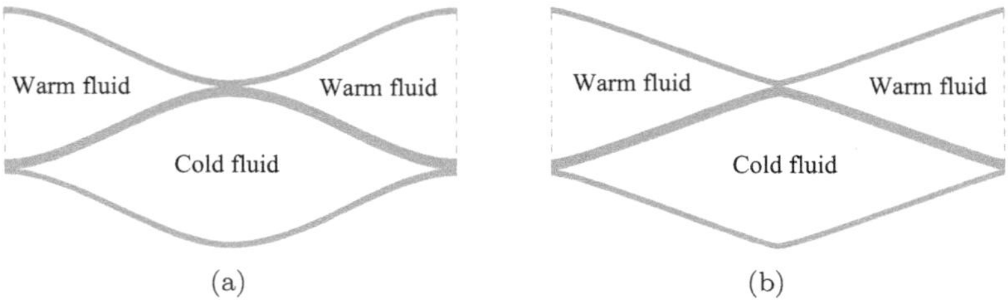

Fig. 8.25 (**a**) Original design 0, with $\theta = 60°$; (**b**) optimized design 134, with $\theta = 42°$

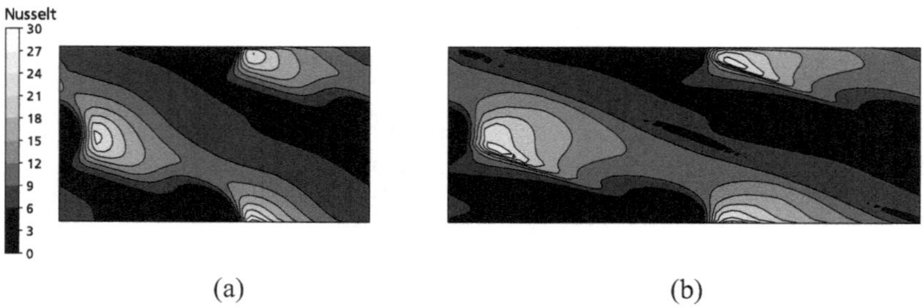

Fig. 8.26 (**a**) Heat flux on the wall for the original design; (**b**) heat flux on the wall for the optimized design

maintaining the same ΔT. These results have been duly verified by running grid-independence tests at several grid resolutions, up to 5×10^5 cells. The newly optimized geometry has a sharper shape of the walls and a lower value of the corrugation angle, i.e., $\theta = 42°$, which means a major length of the repeating module. The original and optimized profiles are given in Fig. 8.25, while the local Nusselt number distribution on the interface wall, for the original and optimized design, are presented in Fig. 8.26.

8.9 Concluding Remarks

In this chapter, we have described our approach for the multi-objective shape optimization of convective periodic channels, which represent a fundamental building block of many heat exchangers. The optimization is performed for a fluid of Prandtl number 0.7, representative of air and other gases, assuming fully developed velocity and temperature fields, and steady laminar conditions.

We considered first the optimization of two-dimensional channels, described either by linear piecewise corrugated walls, or by wavy NURBS-based profiles. It has been observed that simpler linear piecewise channels, though easier to optimize, cannot provide the same performances obtained by NURBS channels. It has been shown that, as in similar studies, very different channel shapes offer almost the same flow and heat transfer performance, i.e., non-uniqueness of the shape optimization problem. The 3D channels have been obtained by extrusion at variable angles of linear piecewise and NURBS channels. The former has been optimized while for the latter, a parametric analysis has been done. In both cases, the presence of secondary motions, and in particular steady longitudinal vortices, leads to a significant increase of the heat transfer rate in comparison with the 2D channels.

The optimization of the CC periodic module has been carried out considering both hot and cold fluid streams, without the necessity of imposing artificial constant-temperature or constant-flux boundary conditions. In this case, a major effort was the proper linking of the different software packages and additional utilities. The first results are very encouraging since one of the optimized geometries leads to almost the same heat transfer performance of the original design, but with a significant pressure drop reduction of about 20% and without increase of the heat transfer surface.

The procedure described has been proven robust and efficient, and in principle, could be applied to even more complex problems.

Acknowledgements Financial support for this research is provided by MIUR, PRIN 2004, *Surface Optimization for Heat Transfer Problems*, and PRIN 2005, *Study and Optimization of Buoyancy-controlled Thermal Systems*, and is gratefully acknowledged.

References

1. ANSYS, Inc.: ANSYS-CFX, Release 10.0. Canonsburg, PA, USA (2005). See also URL `http://www.ansys.com/products/cfx.asp`
2. ANSYS, Inc.: ANSYS ICEM-CFD Documentation. Canonsburg, PA, USA (2005). See also URL `http://www.ansys.com/products/icemcfd.asp`
3. Białecki, R.A., Burczyński, T., Długosz, A., Kuś, W., Ostrowski, Z.: Evolutionary shape optimization of thermoelastic bodies exchanging heat by convection and radiation. Comp. Methods Appl. Mech. Engrg. **194**(17), 1839–1859 (2005)
4. Biethahan, J., Nissen, V. (eds.): Evolutionary Algorithms in Management Applications. Springer, Berlin (1995)
5. Bouyssou, D., Marchant, T., Pirlot, M., Tsoukiàs, A., Vincke, P.: Evaluation and decision models with multiple criteria: Stepping stones for the analyst, 1st edn. International Series in Operations Research and Management Science, Volume 86. Springer, Boston (2006)
6. Cheng, C.H., Wu, C.Y.: An approach combining body-fitted grid generation and conjugate gradient methods for shape design in heat conduction problems. Numerical Heat Transfer **37, part B**, 69–83 (2000)
7. Coello Coello, C.A.: Recent Trends in Evolutionary Multiobjective Optimization: Theoretical Advances and Applications, chap. Recent Trends in Evolutionary Multiobjective Optimization, pp. 7–32. Springer-Verlag, London (2005)
8. Coello Coello, C.A., Van Veldhuizen, D.A., Lamont, G.B.: Evolutionary Algorithms for Solving Multi-Objective Problems. Kluwer Academic Publishers, New York (2002)

9. COMSOL AB: COMSOL 3.3 Documentation. Burlington, MA, USA (2006). See also URL http://www.comsol.com
10. Croce, G., D'Agaro, P.: Numerical analysis of forced convection in plate and frame heat exchangers. Int. J. Numer. Methods Heat Fluid Flow **12**(6), 756–771 (2002)
11. Dassault Systemes: CATIA Documentation (2004). See also URL http://www.3ds.com/products-solutions/plm-solutions/catia/
12. Dias, A., Vasconcelos, J.: Multiobjective genetic algorithm applied to solve optimization problems. IEEE Trans. Magnetics **38**, 1133–1136 (2002)
13. Divo, E., Kassab, A., Rodriguez, F.: An efficient singular superposition technique for cavity detection and shape optimization. Numerical Heat Transfer, Part B: Fundamentals **46**(1), 1–30 (2004)
14. ESTECO s.r.l: modeFRONTIER version 3 Documentation. Trieste, Italy (2006). See also URL http://www.esteco.com
15. Fabbri, G.: A genetic algorithm for fin profile optimization. Int. J. Heat Mass Transfer **40**, 2165–2172 (1997)
16. Fabbri, G.: Optimization of heat transfer through finned dissipators cooled by laminar flow. Int. J. Heat Mass Transfer **19**, 644–654 (1998)
17. Fabbri, G.: Heat transfer optimization in corrugated wall channels. Int. J. Heat Mass Transfer **43**, 4299–4310 (2000)
18. Farin, G.: Curves and Surfaces in Computer-Aided Geometric Design: A Practical Guide, fifth edn. Academic Press, San Francisco, CA (2002)
19. Fonseca, C.M., Fleming, P.J.: Genetic algorithms for multiobjective optimization: Formulation, discussion and generalization. In: Genetic Algorithms: Proceedings of the Fifth International Conference, pp. 416–423. Morgan Kaufmann (1993)
20. Goldberg, D.E.: Genetic Algorithms in Search, Optimization, and Machine Learning. Addison-Wesley Professional, Reading, MA (1989)
21. Horn, J., Nafpliotis, N., Goldberg, D.E.: A Niched Pareto Genetic Algorithm for Multiobjective Optimization. In: Proceedings of the First IEEE Conference on Evolutionary Computation, IEEE World Congress on Computational Intelligence, vol. 1, pp. 82–87. IEEE Service Center, Piscataway, New Jersey (1994)
22. Jensen, P.A., Bard, J.F.: Operations Research Models and Methods. Wiley and Sons (2003)
23. Kim, H.M., Kim, K.Y.: Design optimization of rib-roughened channel to enhance turbulent heat transfer. Int. J. Heat Mass Transfer **47**(23), 5159–5168 (2004)
24. Kirk, D.E.: Optimal control theory: an introduction. Prentice-Hall, Englewood Cliffs, NY (1970)
25. Lan, C.H., Cheng, C.H., Wu, C.Y.: Shape design for heat conduction problems using curvilinear grid generation, conjugate, and redistribution methods. Numerical Heat Transfer **39, part A**, 487–510 (2001)
26. Manzan, M., Micheli, M., Pieri, S.: Automatic integration in the design of a microturbine compact recuperator. In: Procs. GT2006 ASME Turbo Expo 2006. Barcelona, Spain (2006)
27. Manzan, M., Pieri, S., Nobile, E.: Thermal periodic boundary conditions for automatic design process of cross corrugated channels. In: Procs. XXIV UIT Conf. Napoli, Italy (2006)
28. Meric, R.A.: Shape design sensitivity analysis and optimization for nonlinear heat and electric conduction problems. Numerical Heat Transfer **34, part A**(2), 185–203 (1998)
29. Morimoto, K., Suzuki, Y., Kasagi, N.: Optimal shape design of counter-flow primary surface recuperators. In: Procs. Fifth International Conf. on Enhanced, Compact and Ultra-Compact Heat Exchangers: Science, Engeneering and Technology, pp. 124–131. Hoboken, NJ, USA (2005)
30. Nobile, E., Pinto, F., Rizzetto, G.: Multiobjective shape optimization of two-dimensional convective periodic channels. Numerical Heat Transfer **Part B**(50), 425–453 (2006)

31. Nonino, C., Comini, G.: Finite-element analysis of convection problems in spatially periodic domains. Numerical Heat Transfer **Part B**(34), 361–378 (1998)
32. Padovan, L., Pediroda, V., Poloni, C.: Multidisciplinary Methods for Analysis Optimization and Control of Complex Systems, *Mathematics in Industry*, vol. 6, chap. Multi Objective Robust Design Optimization of Airfoils in Transonic Field, pp. 283–295. Springer, Berlin, Heidelberg (2005)
33. Park, K., Choi, D.H., Lee, K.S.: Numerical shape optimization for high performance of a heat sink with pin-fins. Numerical Heat Transfer, Part A **46**, 909–927 (2004)
34. Park, K., Choi, D.H., Lee, K.S.: Optimum design of plate heat exchangers with staggered pin arrays. Numerical Heat Transfer, Part A **45**, 347–361 (2004)
35. Park, K., Moon, S.: Optimal design of heat exchangers using the progressive quadratic response surface model. Int. J. Heat Mass Transfer **48**(11), 2126–2139 (2005)
36. Parry, J., Bornoff, R.B., Stehouwer, P., Driessen, L.T., Stinstra, E.: Simulation-based design optimization methodologies applied to CFD. In: IEEE Trans. Compon., Packag. and Manufact. Technol., vol. 27, pp. 391–397 (2004)
37. Patankar, S.V., Liu, C.H., Sparrow, E.M.: Fully developed flow and heat transfer in ducts having streamwise-periodic variations of cross-sectional area. ASME J. Heat Transfer **99**, 180–186 (1977)
38. Piegl, L., Tiller, W.: The NURBS Book, second edn. Springer-Verlag, Berlin (1997)
39. Poles, S.: MOGA-II an improved multi-objective genetic algorithm. Tech. Rep. 003-006, ESTECO, Trieste, ITALY (2003)
40. Poles, S., Fu, Y., Rigoni, E.: The effect of initial population sampling on the convergence of multi-objective genetic algorithms. In: 7th international conference on multi-objective programming and goal programming (MOPGP'06). Tours, France (2006)
41. Poloni, C., Pediroda, V.: GA coupled with computationally expensive simulations: tools to improve efficiency. In: D. Quagliarella, J. Périaux, C. Poloni, G. Winter (eds.) Genetic algorithms and evolution strategy in engineering and computer science: recent advances and industrial applications, chap. 13, pp. 267–288. John Wiley & Sons, Chichester, UK (1996)
42. Prasad, K.G., Kane, J.H.: Three-dimensional boundary element thermal shape sensitivity analysis. Int. J. Heat Mass Transfer **35**, 1427–1439 (1992)
43. Quagliarella, D., Périaux, J., Poloni, C., Winter, G. (eds.): Genetic algorithms and evolution strategy in engineering and computer science: recent advances and industrial applications. John Wiley & Sons, Chichester, UK (1996)
44. Queipo, N., Devarakonda, N., Humphrey, A.C.: Genetic algorithms for thermosciences research: application to the optimized cooling of electronic components. Int. J. Heat Mass Transfer **37**, 893–908 (1994)
45. Rao, S.: Engineering Optimization, Theory and Practice, third edn. John Wiley & Sons, Inc., New York (1996)
46. Rudolph, G.: Convergence of evolutionary algorithms in general search spaces. In: International Conference on Evolutionary Computation, pp. 50–54 (1996)
47. Sawyers, D.R., Sen, M., Chang, H.C.: Heat transfer enhancement in three-dimensional corrugated channel flow. Int. J. Heat Mass Transfer **41**, 3559–3573 (1998)
48. Schaffer, J.D.: Multiple objective optimization with vector evaluated genetic algorithms. In: Proc. of the International Conference on Genetic Algorithms and Their Applications, pp. 93–100. Pittsburgh, PA (1985)
49. Shah, R.K., London, A.L.: Laminar Flow Forced Convection in Ducts. Academic Press, New York (1978)
50. Srinivas, N., Deb, K.: Multiobjective optimization using nondominated sorting in genetic algorithms. Evolutionary Computation **2**(3), 221–248 (1994)
51. Stalio, E., Nobile, E.: Direct numerical simulation of heat transfer over riblets. Int. J. Heat Fluid Flow **24**, 356–371 (2003)
52. Stasiek, J., Collins, M.W., Ciofalo, M.: Investigation of flow and heat transfer in corrugated passages – I. experimental results. Int. J. Heat Mass Transfer **39**(1), 165–192 (1996)

53. Steuer, R.E.: Multiple Criteria Optimization: Theory, Computation and Application. John Wiley, New York (1986)
54. Wei, X., Joshi, Y.: Optimization study of stacked micro-channel heat sinks for microelectronic cooling. In: IEEE Trans. Compon., Packag. and Manufact. Technol., vol. 26, pp. 55–61 (2003)

Chapter 9
CFD-based Optimization for a Complete Industrial Process: Papermaking

Jari Hämäläinen, Taija Hämäläinen, Elina Madetoja and Henri Ruotsalainen

Abstract Development of tailored software tools based on coupling of Computational Fluid Dynamics (CFD) with optimization is presented in this paper. In papermaking, industrial applications deal with fluid dynamics at the wet-end of a paper machine as well as in the entire papermaking process.

First, the CFD tools being developed for optimal shape design and optimal control problems at the wet-end (where the paper web is formed) are presented. Different levels of complexity of CFD modeling and a dimension reduction technique are considered in this paper. The reduced CFD model used is validated with a complete model.

In addition, optimization of the whole papermaking process being modeled with different modeling techniques is considered. Our approach is based on interactive multi-objective optimization because the papermaking process as well as the produced paper require multiple criteria to be optimized simultaneously. Typically, the objectives are conflicting which means that compromises need to be done. This is illustrated with numerical examples.

Finally, a completely new design of decision support tools based on multiobjective optimization and multiphysical modeling of large industrial systems is discussed.

9.1 Introduction

Optimal shape design is the best-known example of model-based optimization in which the model is described as a system of partial differential equations. In a sense, optimal shape design is an inverse problem, that is, the

Jari Hämäläinen · Taija Hämäläinen · Elina Madetoja · Henri Ruotsalainen
Department of Physics, University of Kuopio, P.O.Box 1627, FI-70211 Kuopio, Finland
(e-mail: jari.hamalainen@uku.fi, taija.hamalainen@uku.fi, elina.madetoja@uku.fi, henri.ruotsalainen@uku.fi)

shape of a domain is not known *a-priori*. Instead, a shape is sought such that the cost function is minimized (or maximized), and the state equation and possible constraints are fulfilled. The cost function, also known as the objective function, measures the quality of the shape. Typically, it is formulated such that its minimum (or maximum) value corresponds to the best possible shape (optimal shape). The state equation is a model describing the physical phenomenon to be studied, as in the case of the Navier-Stokes equations, for example. The constraints ensure that the optimal shape is reasonable, for instance, by setting limits to total material usage in solid mechanical problems, minimum and maximum material thicknesses. Optimal control is a special case of shape optimization. Instead of the shape of the domain, the boundary data, for example, are now *a-priori* unknown. A typical optimal control problem related to CFD is the search for an optimal velocity profile at an inlet or optimal heat flux through a wall.

Optimal shape design is said to have originated with Hadamard in 1910 [9], who first supplied a formula for a partial differential equation in order to evaluate the change due to a boundary modification of the domain. The first engineering applications were in solid mechanics. CFD emerged through potential flows, Euler equations, and finally viscous Navier-Stokes equations including turbulence models. For further information, see [18, 29, 32] and the bibliographical studies cited there.

In traditional optimal shape design or optimal control problems, there is only one objective (also known as cost function) to be optimized. However, everyday engineering problems typically involve several conflicting objectives that should be achieved simultaneously. A single objective function can be derived from multiple functions, as a weighted sum, for example, but then the practical relevance of the cost function values may be lost. Instead, it is more reasonable to handle multiple objectives as they arise naturally from engineering problems, by using methods of multi-objective optimization [26, 34]. The importance of multi-objective optimization becomes even more significant when dealing with large industrial processes and taking into account consecutive unit-processes, raw material, energy consumption, and end-product quality and price, each of which have their own objectives. Applications of multi-objective optimization are not only necessary to handle engineering problems, but will play an important role also in decision support systems in the future.

Computing capacity is always limited which leads to compromises between accuracy, scope and computing time as illustrated in Fig. 9.1. Naturally, there is a need for detailed small-scale modeling such as Direct Numerical Simulation (DNS) which requires high performance computing even without any optimization. But the detailed models cannot be coupled with optimization if total computational (CPU) time is to be kept reasonable. When a new product is being designed, a relatively long CPU time, days or even weeks, is considered acceptable. Therefore, the optimal shape design can be used based on a complex CFD model. But when a fast response, in seconds or

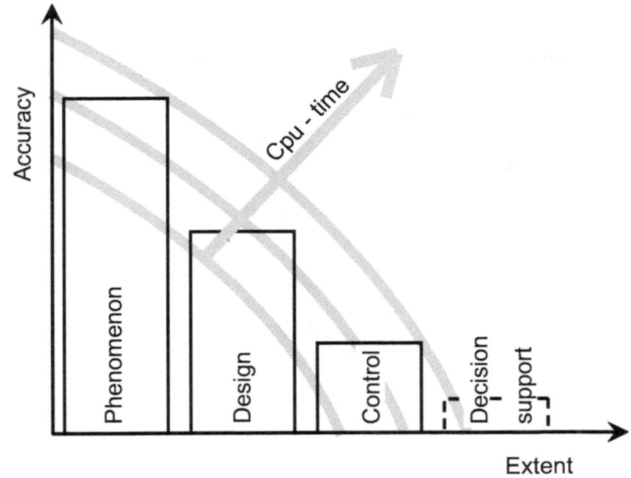

Fig. 9.1 Compromise between accuracy and extent of CFD when coupled with optimization

minutes, is required, the cost function evaluation has to be carried out extremely quickly. This is possible only with the use of reduced CFD models. Moreover, when the scope extends to a large system consisting of numerous unit-process models coupled with multi-objective optimization, models need to be reduced even further. Such is the case when modeling the entire papermaking process. Evidently, CFD and different CFD based optimization applications involve varying demands for accuracy, complexity and CPU time.

CFD-based optimization tools developed for the paper industry are discussed in this paper. Applications vary from headbox optimal shape design (Sect. 9.2) and optimal control (Sect. 9.3) to multi-objective optimization (Sect. 9.4) and decision support systems (Sect. 9.5) designed for the whole paper machine, all having basis in reduced CFD models. In this paper, one such model, the depth-averaged Navier-Stokes equations, has been validated in a contracting channel.

9.2 Optimal Shape Design of the Tapered Header

A traditional design problem in the paper machine headbox is the tapering of the header that allows even outlet flow rate distribution across the full width of the paper machine [1, 24, 35]. The headbox is the first part of the paper machine from where the fiber-water mixture starts the papermaking process by going through the header as the first flow passage (Fig. 9.2). The fiber-

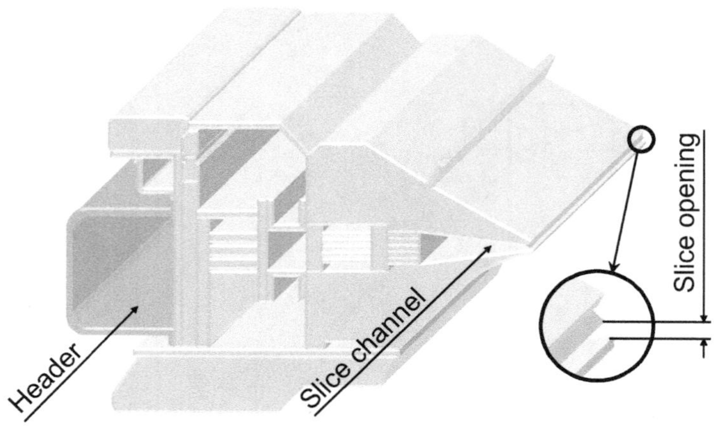

Fig. 9.2 Paper machine headbox

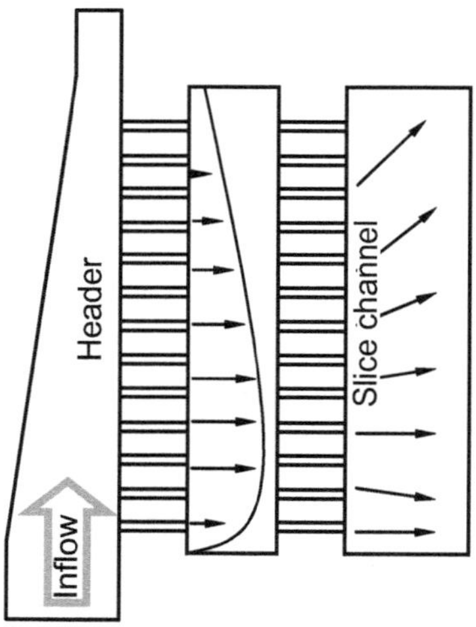

Fig. 9.3 Cross-directional flows inside a headbox due to uneven flow distribution coming from the tapered header

water mixture coming from a pump is turned 90° into the machine direction and distributed inside the header as illustrated in Fig. 9.3. An uneven flow rate distribution may cause an uneven basis weight profile in the paper web produced. Inside the headbox, a non-optimal header may also be a potential source of cross-directional flows which in turn causes misaligned fibers in the paper since fibers approximately follow the velocity vectors at the outlet jet of the headbox. Optimization is utilized to minimize these disturbances.

The design of the tapered header has been identified as an optimal shape design problem. A cost function describes how even the flow rate distribution after the header should be and design variables define the location of the back

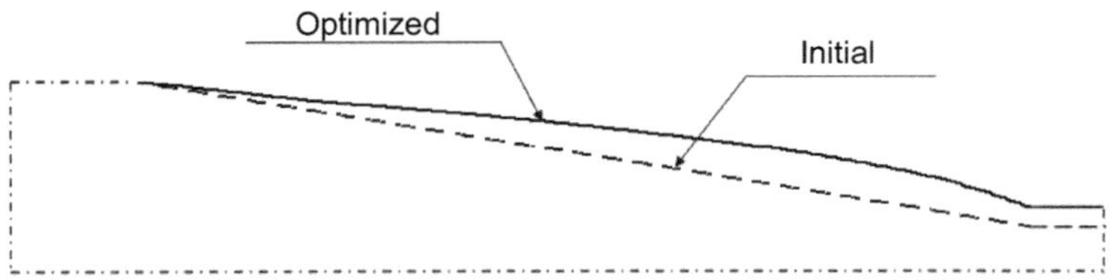

Fig. 9.4 Initial and optimized back wall of the tapered header

wall of the header. The first numerical experiments in header optimization were reported in the early 1990's [10, 11].

Modeling of the headbox flows includes certain special features. First, the tube bundles (see Figs. 9.2 and 9.3) consist of hundreds of small tubes. They cannot be included in detail in CFD but are taken into account as an effective porous medium. A three-dimensional (3D) CFD model would also be too time-consuming for optimization. Hence, specific two-dimensional (2D) models have been developed for the headbox applications [12]. The header model is derived from a 3D, incompressible, k–ε turbulence model by averaging the equations in the vertical direction, which results in a non-standard 2D flow model. A similar approach has also been studied in [33] for open-channel problems.

In addition to model reductions, optimization methods also have a significant influence on computing efficiency. In general, gradient-based methods have proven to be more efficient than gradient-free methods for the optimization problems introduced in this paper. The most critical step in gradient-based optimization algorithms is the evaluation of the cost function gradient. Finite difference approximation is easy to obtain, even for complex models, but on the other hand, it is inefficient. The so-called adjoint state technique has also been studied for the tapered header [13] but only for laminar or algebraic turbulence models. Evaluation of the cost function gradient can be avoided by using genetic algorithms [4, 14]. Nevertheless, at least in papermaking applications, gradient-based methods are much faster than genetic algorithms, even when the gradient is approximated by finite differences. In genetic algorithms, a large set of uninteresting solutions has to be calculated in order to find the interesting ones.

The designing software tool for optimization of the tapered header was introduced in the industry in 1995. The design procedure takes place only once a week, and thus, a CPU time of several hours is acceptable. It is also justified to utilize a two-equation turbulence model, together with finite difference approximation of the cost function gradient, instead of a fast solver based on an algebraic turbulence model and the adjoint state technique.

Optimal shape design of the header has been utilized in the design process by the industry for a decade [15]. One example of the initial and optimized

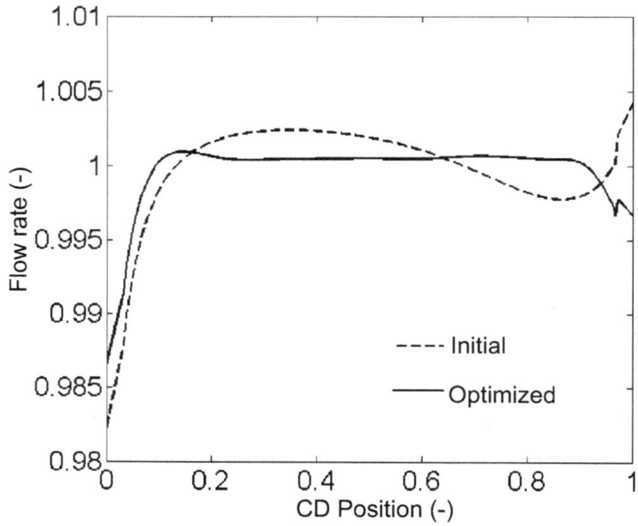

Fig. 9.5 Initial and optimized flow rate distribution of the header

shape of the header, and its resulting outflow velocity profile, is illustrated in Figs. 9.4 and 9.5, respectively. As can be seen, the optimized velocity profile is notably even across the width of the machine, except for some minor boundary layer effects at the edges.

In design processes, relatively complex and computationally expensive models can be coupled with optimization, as shown in this example. In the header optimization tool, a 2D turbulence model has been used, but when faster response times are required, more reduced models are needed. An example of such a control problem is presented in the next section.

9.3 Optimal Control of the Fiber Orientation in the Slice Channel

9.3.1 On Modeling Fiber Orientation

Modeling of fiber suspension flows involves numerous challenges. Wood fibers are non-spherical particles with a length-to-diameter aspect ratio typically of the order of 100. Having approximately the same density as water, they tend to align with the velocity direction but, because of their flexibility, they also form accumulations called "fiber flocs". Moreover, fibers interact with the carrying phase and turbulence as well. Thus, we face conflicting modeling challenges: the CFD model should be as simple as possible when coupled with optimization, despite that the complete 3D turbulence model is insufficiently accurate to predict all detailed fluid flow phenomena occurring in the headbox. Therefore, the basic phenomena affecting fiber orientation need to be studied by developing elaborated fluid flow models while separate reduced models are used in the following optimization.

When designing the header (Sect. 9.2), several hours of CPU time is acceptable. But when developing tools for the engineers whose task is to go round the paper mills and optimize the papermaking process, a much faster CFD-based optimization tool is required. To fulfill these demands, a simulator called Headbox Optimization Control Simulator (HOCS) Fiber [16] has been developed. It solves an optimal control problem in the order of one minute. Efficiency is obtained by a reduced CFD model, that is, a laminar depth-averaged Navier-Stokes equation, and by using adjoint state technique in a gradient-based optimization algorithm. However, simplifying the flow field simulation is insufficient, as we also need to reduce the fiber orientation model.

Fiber orientation angle is not merely a scalar depending on fluid velocity. To be precise, in any location (or small volume) there are large numbers of fibers oriented more or less randomly in different directions, and hence forming a distribution of fiber orientation angles. Modeling of the fiber orientation probability distribution (FOPD) leads to a four-dimensional equation for a 3D geometry, and to a three-dimensional equation for a 2D geometry. The detailed modeling of FOPD in the headbox and its free jet is considered in [17]. However, when seeking for the optimal fiber orientation profile over the width of a real paper machine, there are significant model reductions to be carried out. Consequently, the fiber orientation is considered simply as a misalignment angle with respect to the machine direction (MD), having a profile in the cross-machine direction (CD). In other words, in the HOCS Fiber simulator, only the expected value of the FOPD model is considered in each CD position. It is also assumed that the orientation angle is fully determined by the velocity components in MD and CD.

9.3.2 HOCS Fiber – A Trouble Shooting Tool

The last component in the headbox is the slice channel (see Figs. 9.2 and 9.3). The channel outflow is controlled by the height of the outlet boundary, called the slice opening.

The HOCS Fiber [16] makes basically two CFD-model based optimizations. Step 1 solves a diagnostics problem in order to identify what defects in the headbox cause the fiber orientation misalignment. Hence, the measured orientation angle profile determines the cost function, and the inlet flow speed is used as the control variable. After that, Step 2 searches for an

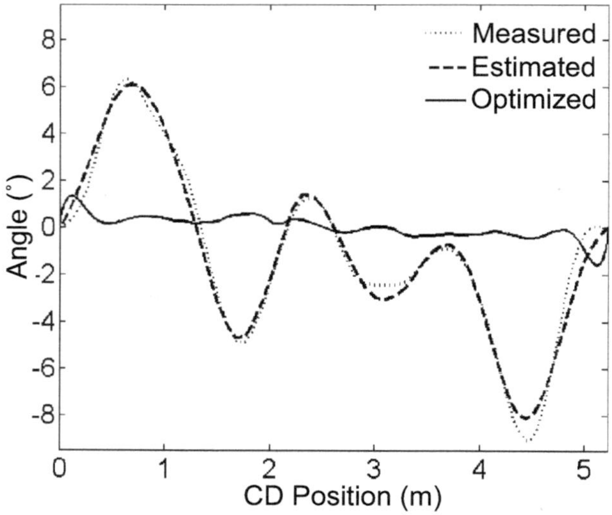

Fig. 9.6 Measured, estimated (Step 1) and optimized (Step 2) fiber orientation profiles

optimal control for the slice opening profile in order to obtain an even fiber orientation angle, that is, with angles equal to zero. When the optimal slice opening profile is found, it can be fed into actual slice opening control system of the paper machine.

Nowadays, the HOCS Fiber simulator is installed on the papermaking engineer's laptop, and hence the necessary control corrections can be calculated and implemented at the paper mill. Despite the model reductions, the accuracy of the HOCS Fiber has been confirmed: typically, for the first paper machine start-up, only one slice opening tuning proposed by the software is needed. HOCS Fiber has been successfully used at dozens of paper mills and one example of a real trouble shooting case is given in Fig. 9.6. As seen in the figure, the fiber orientation angle is much better after optimization than the original one measured at the mill. The original fiber orientation profile may cause problems in printing machines, but the optimized profile fulfills all the market requirements.

9.3.3 Depth-averaged Navier-Stokes Equations

In order to be able to optimize the fiber orientation for trouble shooting purposes, certain model simplifications have to be carried out. First of all, the FOPD model cannot be used. Instead, the expected value of the FOPD is assumed to be determined by the mean velocity field. Furthermore, instead of using the 3D geometry of the headbox slice channel, the depth-averaged Navier-Stokes equations are used. These equations are introduced next.

Let the velocity vector $\underline{U} = (u, v, w)$ and the static pressure p be a solution of the 3D Navier-Stokes equations. The depth-averaged pressure P and the velocity components U and V are defined as averaged values in the vertical direction over the depth of the slice channel as follows

$$\begin{aligned} U(x,z) &= \frac{1}{D(x,z)} \int_0^{D(x,z)} u(x,y,z)\,dy \\ V(x,z) &= \frac{1}{D(x,z)} \int_0^{D(x,z)} v(x,y,z)\,dy \\ P(x,z) &= \frac{1}{D(x,z)} \int_0^{D(x,z)} p(x,y,z)\,dy \end{aligned} \tag{9.1}$$

where $D = (x, z)$ is the depth of the slice channel depending on a position (x, z). Then, by integrating the 3D continuity equation over the depth and by using the definitions (9.1), the following depth-averaged continuity equation is obtained

$$\nabla \cdot (D\underline{V}) = 0 \tag{9.2}$$

where $\underline{V} = (U, V)$ is the reduced 2D velocity vector. The depth-averaged momentum equation is derived similarly, and it is

$$\begin{aligned} -\frac{1}{D}\nabla \cdot \left(2\mu D \underline{\underline{\varepsilon}}(\underline{V})\right) + \rho C(m)\,\underline{V} \cdot \nabla (D\underline{V}) \\ + \left[\frac{4(m+1)\mu}{D^2}\right]\underline{V} + \nabla P = 0 \end{aligned} \tag{9.3}$$

where

$$C(m) = \frac{2m+2}{2m+1}, \quad m \geq 2 \tag{9.4}$$

is a coefficient associated with the velocity profile in the depth direction. For example, $m = 2$ corresponds to a parabolic velocity profile, and $m = 7$ to a turbulent plug flow profile. See also [33] for depth-averaging of the k–ε turbulence model.

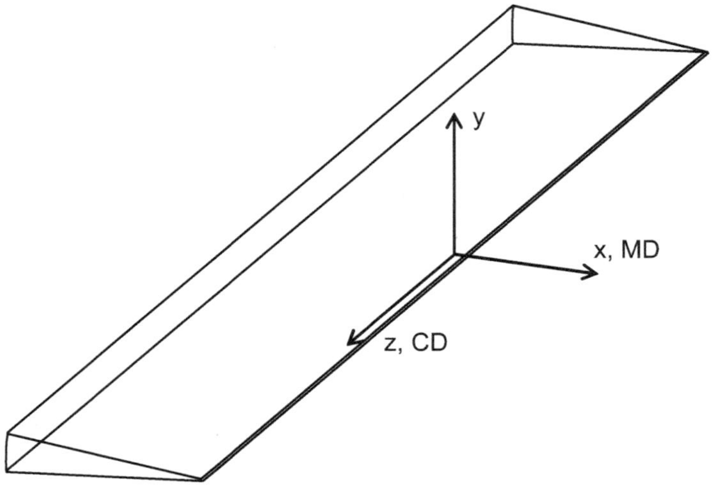

Fig. 9.7 The coordinate frame of the full 3D model

9.3.4 Validation of the Depth-averaged Navier-Stokes Equations

The reduced model, that is, the depth-averaged Navier-Stokes equations, has been validated by comparing the results to a complete CFD model which includes the headbox slice channel and its free jet in 3D. The coordinate frame of the model is shown in the sketch in Fig. 9.7. The x-axis follows the main flow direction, while the z-axis lies in the span-wise direction. The 3D CFD model utilized has been validated using wind tunnel experiments [31].

The fiber orientation angle profile is mostly determined by the CD velocity component, because variations in the MD velocity are only of the order of one per mille. Thus, the CD velocity is the most important component in the validation. The CD velocity profiles at the slice opening for Step 1 are presented in Fig. 9.8 (Step 1 is the diagnostics step determining flow rate profiles inside the headbox which creates the measured fiber orientation profile). The whole CD velocity field is presented in Fig. 9.9 to illustrate how the cross-directional flows develop in the slice channel. As can be seen in Fig. 9.8, the prediction given by HOCS Fiber agrees well with the result calculated using the whole 3D model, even though the reduced model does not take into account the effect of the jet.

After the diagnostics step, the fiber orientation is optimized. The CD profiles obtained from Step 2 are presented in Fig. 9.10. As can be seen, the difference between the reduced and the full model is now slightly bigger than in Step 1, but both models predict very small CD velocities resulting in optimal fiber orientation angle profile.

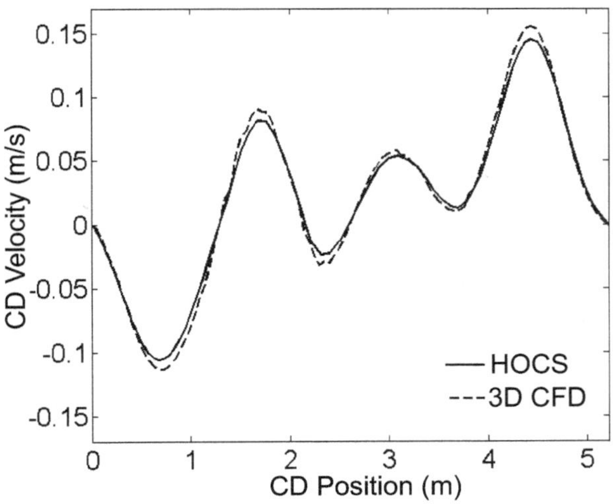

Fig. 9.8 Comparison of CD velocity for the diagnostics (Step 1) between HOCS Fiber and the three-dimensional CFD model

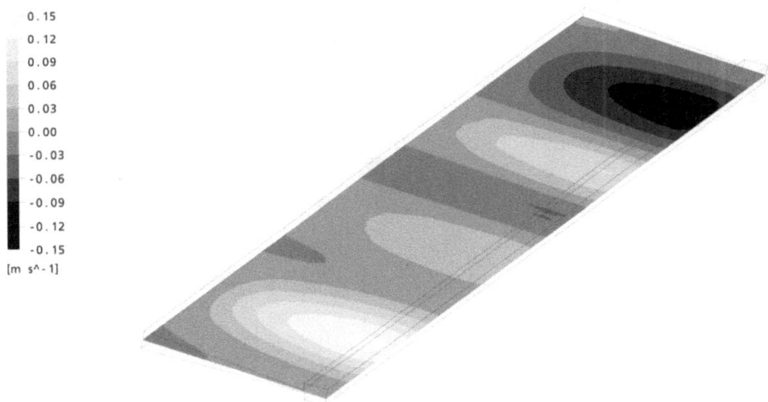

Fig. 9.9 CD velocity field (Step 1) in the slice channel predicted by the 3D model

The validation shows that the depth-averaged equations are sufficiently accurate for use in solving industrial fiber orientation trouble-shooting problems. The reduced 2D model is solved by using a stabilized, upwinding Petrov-Galerkin finite element method [3, 5]. One CFD solution takes only a few seconds on a laptop computer for a typical finite element mesh consisting of 5,000 elements. The CPU time is only 1/1000 compared to the full 3D model (2-3 million elements, turbulence model and free jet). HOCS Fiber can perform both optimization steps, that is, both the diagnostics and optimal control steps, in a couple of minutes. If the full 3D model had been

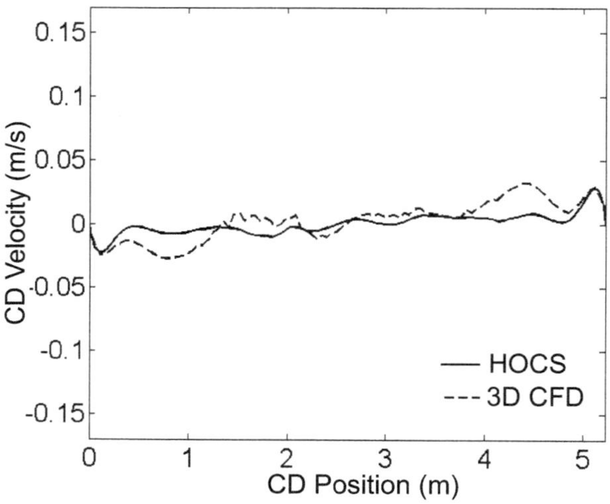

Fig. 9.10 Comparison of CD velocity for the optimal control (Step 2) between HOCS Fiber and the 3D CFD model

coupled with optimization by using finite difference approximation for the cost function gradient, this would lead to computing times of days or weeks, which is unreasonable. As illustrated in Fig. 9.1, compromises need to be made and CFD models need to be reduced when developing tools for control purposes.

9.4 Multi-objective Optimization of Papermaking

As presented in the previous sections, the unit-processes of the paper machine have been individually modeled and optimized for more than ten years. Nowadays, focus has extended to handle larger ensembles including the whole papermaking process. This poses a real challenge since a paper machine consists of a number of consecutive sub-ensembles as shown in Fig. 9.11. There are four main parts essential in the papermaking process: the headbox, wire section, press section and the drying section. Thus, the model of the whole process represents the combination of different unit-process models as a chain where the output of one unit-process is an input for the following unit-processes. In order to decrease the CPU time used, the elaborateness and the accuracy of the CFD models often have to be significantly reduced. Sometimes, the CFD model might even need to be replaced with a statistical or stochastic model. Thus, when handling multiphysical or multidisciplinary problems, more efficient simulation and optimization procedures are needed.

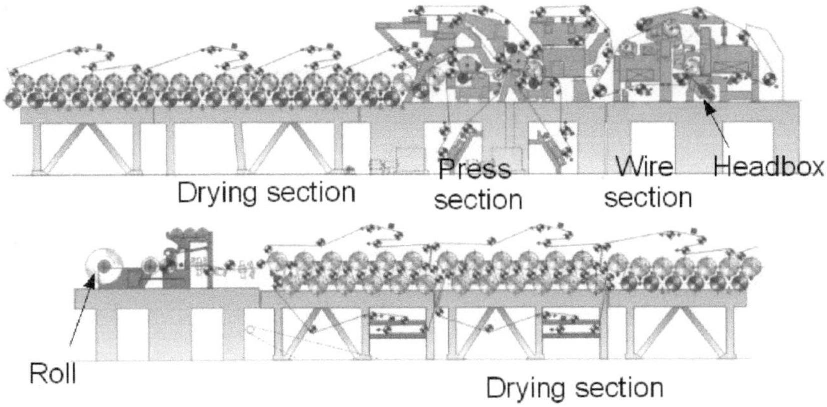

Fig. 9.11 Layout of a paper machine. The process starts from the headbox, then continues from right to left and ends-up as a finished paper on the roll. Courtesy of Metso Paper, Inc.

In the papermaking process there are always several requirements for the end product that should be fulfilled simultaneously. These targets are often conflicting. For example, by accelerating the machine speed in order to increase the amount of production, the probability of web breaks may be increased resulting in more downtime, which in turn, reduces the amount of production. Moreover, better runnability could be obtained by producing stronger paper from more expensive raw material, but this affects economy. Hence, numerous criteria need to be taken into account at the same time. However, when combining multiple objectives it is difficult to steer the optimization to the best solution. That is why we utilize multi-objective optimization.

9.4.1 Multi-objective Optimization

In general, a multi-objective optimization problem can be defined as follows [26]:

$$\begin{aligned} &\min\ \{f_1(\boldsymbol{x}), f_2(\boldsymbol{x}), \ldots, f_{nf}(\boldsymbol{x})\} \\ &\text{subject to } \boldsymbol{x} \in S \end{aligned} \tag{9.5}$$

where $\boldsymbol{x}$ is a vector of decision variables from the feasible set $S \subset \mathbb{R}^n$ defined by box, linear and nonlinear constraints. We denote a vector of objective functions, a so-called objective vector, by $\boldsymbol{f}(\boldsymbol{x}) = (f_1(\boldsymbol{x}), f_2(\boldsymbol{x}), \ldots, f_{nf}(\boldsymbol{x}))^T$. Furthermore, we define the image of the feasible set by $\boldsymbol{f}(S) = Z$ and call it a feasible objective set. The optimization problem (9.5) is formulated as a minimization problem, but if some objective function f_i is to be maximized, it is equivalent to considering minimization of $-f_i$.

In multi-objective optimization, optimality is understood in the sense of Pareto optimality [4]. A decision vector $\boldsymbol{x}' \in S$ is Pareto optimal if there does not exist another decision vector $\boldsymbol{x} \in S$ such that $f_i(\boldsymbol{x}) \leq f_i(\boldsymbol{x}')$ for all $i = 1, \ldots, nf$ and $f_j(\boldsymbol{x}) < f_j(\boldsymbol{x}')$ for at least one index j. The Pareto optimal solutions constitute a Pareto optimal set. From a mathematical point of view, all of them are equally good and they can be regarded as equally valid compromise solutions of the problem considered. Because vectors cannot be ordered completely, there exists no trivial mathematical tool in order to find the best solution in the Pareto optimal set and thus additional information is necessary. In our approach an expert of the problem known as a decision maker, participates in solution process by giving preferences regarding the best or most satisfying solution to be called the final one.

The methods developed for multi-objective optimization can be divided into four classes according to the role of the decision maker [26]. First, there are methods where no decision maker is available and where the final solution is some neutral compromise solution. The three other classes are a priori, a posteriori and interactive methods, where the decision maker participates in the solution process before, after, or iteratively during the process, respectively.

As mentioned the decision maker can participate in the solution process in one way or the other, and determine which one of the Pareto optimal solutions is the most desirable to be the final solution. It is often useful for the decision maker to know the ranges of objective function values in the Pareto optimal set. An ideal objective vector $\boldsymbol{z}^* \in \mathbb{R}^{nf}$ gives lower bounds for the objective functions in the Pareto optimal set and it is obtained by minimizing each objective function individually subject to the constraints. A vector strictly better than $\boldsymbol{z}^*$ can be called a utopian objective vector denoted by $\boldsymbol{z}^{**}$, that is, we set $z_i^{**} = z_i^* - \epsilon$ for $i = 1, \ldots, nf$, where ϵ is a small positive scalar. A nadir objective vector $\boldsymbol{z}^{nad}$ giving upper bounds of objective functions in the Pareto optimal set can be difficult to calculate, and, thus, its values are usually only approximated by using pay-off tables, for example. For more, see [26] and references therein.

Often multi-objective optimization problems are solved using scalarizing functions that form subproblems, as we call them. In the scalarization procedure, the original multi-objective optimization problem is formed as a single objective subproblem that can be solved with appropriate single objective optimizers. In literature, many different scalarizing functions have been presented, see e.g., [27, 28].

9.4.2 Modeling and Optimizing the Complete Papermaking Process

Although accurate simulation of the entire paper machine is still far ahead, a virtual papermaking line, as we call it, has been developed [20, 22]. It can be used for papermaking simulation as well as for tailored multi-objective optimization. The virtual papermaking line combines dissimilar unit-process models from different disciplines that include mathematical formulas ranging from simple algebraic equations to CFD models. It also includes models for moisture and heat transfer, and for paper quality properties. Simplifications of computationally expensive models are used, for instance, in water removal which means that an accurate multi-phase flow model of the dewatering phenomenon is replaced with a statistical model based on data produced by the accurate model. The latter models are especially useful in optimization, when tens, hundreds or even thousands of simulation model evaluations are needed.

We define the virtual papermaking line model as follows

$$\begin{cases} A_1(\boldsymbol{p}_1, \boldsymbol{q}_1) = 0 \\ A_2(\boldsymbol{p}_2, \boldsymbol{q}_1, \boldsymbol{q}_2) = 0 \\ \vdots \\ A_{nm}(\boldsymbol{p}_{nm}, \boldsymbol{q}_1, \dots, \boldsymbol{q}_{nm}) = 0 \end{cases} \tag{9.6}$$

where A_i for all $i = 1, \dots, nm$ stand for the unit-process models, $\boldsymbol{p}_i \in \mathbb{R}^{l_i}$ denotes a vector of the inputs, and $\boldsymbol{q}_i \in \mathbb{R}^{k_i}$ is a vector of the outputs (model states) for the i-th unit-process model A_i. We assume here that all the unit-process models and their outputs are continuously differentiable or if there is nonsmoothness involved, we assume at least H-differentiability [7].

Based on (9.6) multi-objective optimization problems related to papermaking process can be determined. Instead of the problem formulation (9.5) we use the following model-based optimization problem formulation

$$\begin{aligned} &\underset{\boldsymbol{x}}{\text{Optimize}} \{f_1(\boldsymbol{x}, \boldsymbol{q}_1, \dots, \boldsymbol{q}_{nm}), \dots, f_{nf}(\boldsymbol{x}, \boldsymbol{q}_1, \dots, \boldsymbol{q}_{nm})\} \\ &\text{subject to} \begin{cases} (9.6) \\ \boldsymbol{x} \in S \end{cases} \end{aligned} \tag{9.7}$$

where $\boldsymbol{x} \in S$ is a vector of the continuous decision variables (also called control or design variables) which is a selected set of the unit-process model inputs $\boldsymbol{p} = (\boldsymbol{p}_1, \dots, \boldsymbol{p}_{nm})^T$. The feasible set S is defined by the box constraints of the decision variables and the linear and non-linear constraint functions similarly to (9.5). In (9.7), we ignore those input parameters that are not chosen as decision variables, because they are constants during the optimization process. By $f_1(\boldsymbol{x}, \boldsymbol{q}_1, \dots, \boldsymbol{q}_{nm}), \dots, f_{nf}(\boldsymbol{x}, \boldsymbol{q}_1, \dots, \boldsymbol{q}_{nm})$ we denote the objectives which are to be optimized, i.e., minimized or maximized. Similarly to the unit-process models and their outputs, we make an assumption that the objectives are continuously differentiable or H-differentiable [7]. One should note that in (9.7) the objectives f_i, $i = 1, \ldots, nf$ depend on the decision variables and also, the unit-process model outputs. That is, the virtual papermaking line model (9.6) has to be solved every time the objectives are evaluated.

Next we discuss two features related to model-based optimization problems such as (9.7). These features are gradient evaluations and reliability of the used modeling approaches.

Optimization methods utilizing gradient information of the objectives (gradient-based optimizers) have often been found computationally efficient in engineering applications. However, gradient information needs to be calculated when using these methods. The finite difference approaches are widely used techniques for gradient evaluations, but a large number of decision variables increases the number of the function evaluations making the approach computationally time-consuming. Instead, more sophisticated techniques can be used. We present briefly a technique based on chain rule approach and implicit differentiation schema. For that we assume that objectives and unit-process models as well as model outputs are continuously differentiable or at least H-differentiable [7] in nonsmooth cases. Then gradient of f_i with respect to the vector $\boldsymbol{x}$ is

$$\begin{aligned} &\partial f_i(\boldsymbol{x}, \boldsymbol{q}_1, \ldots, \boldsymbol{q}_{nm}) = \partial_{\boldsymbol{x}} f_i(\boldsymbol{x}, \boldsymbol{q}_1, \ldots, \boldsymbol{q}_{nm}) + \\ &\sum_{j=1}^{nm} \partial_{\boldsymbol{q}_j} f_i(\boldsymbol{x}, \boldsymbol{q}_1, \ldots, \boldsymbol{q}_{nm}) \partial_{\boldsymbol{x}} \boldsymbol{q}_j(\boldsymbol{x}, \boldsymbol{q}_1, \ldots, \boldsymbol{q}_{j-1}) \end{aligned} \tag{9.8}$$

where differentials $\partial_{\boldsymbol{x}} f_i(\boldsymbol{x}, \boldsymbol{q}_1, \ldots, \boldsymbol{q}_{nm})$ and $\partial_{\boldsymbol{q}_j} f_i(\boldsymbol{x}, \boldsymbol{q}_1, \ldots, \boldsymbol{q}_{nm})$, for all $i = 1, \ldots, nf$ and $j = 1, \ldots, nm$ are assumed to be known and

$$\begin{aligned} &\partial_{\boldsymbol{x}} \boldsymbol{q}_j(\boldsymbol{x}, \boldsymbol{q}_1, \ldots, \boldsymbol{q}_{j-1}) = -\partial_{\boldsymbol{q}_j} A_j(\boldsymbol{x}, \boldsymbol{q}_1, \ldots, \boldsymbol{q}_j)^{-1} \Big(\partial_{\boldsymbol{x}} A_j(\boldsymbol{x}, \boldsymbol{q}_1, \ldots, \boldsymbol{q}_j) + \\ &\sum_{k=1}^{j-1} \partial_{\boldsymbol{q}_k} A_j(\boldsymbol{x}, \boldsymbol{q}_1, \ldots, \boldsymbol{q}_j) \partial_{\boldsymbol{x}} \boldsymbol{q}_k(\boldsymbol{x}, \boldsymbol{q}_1, \ldots, \boldsymbol{q}_{k-1}) \Big), \quad j = 1, \ldots, nm \, . \end{aligned} \tag{9.9}$$

In this way, gradient calculations can be done model-wise and gradients of objective functions can be put together after all the unit-processes have been solved. Thus, different modeling approaches can be easily combined together into the virtual papermaking line model. For more details on this technique, we refer to [21].

The virtual papermaking line (9.6) consists of unit-process models from different disciplines. When statistical or stochastic techniques are used, the unit-process models are based on experimental data and on conditions prevailing during data collection. That is, the models can give reliable results only if there has been enough varying modeling data. Furthermore, ranges of the modeling data should not be exceeded, which can cause problems

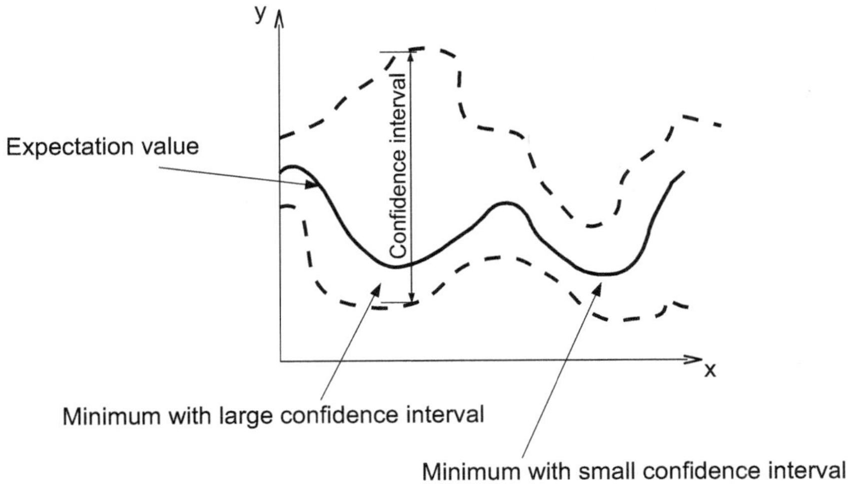

Fig. 9.12 Example of output and its confidence interval

when unit-process models are coupled together. In order to use the statistical unit-process models in simulation or optimization, some kind of reliability or uncertainty information related to the models is needed in addition to the primary model output. Moreover, the uncertainties regarding the models need to be brought into the optimization model. In this way, the optimization method used can take them into account and reliability of the results can be guaranteed. There are numerous optimization approaches that are able to utilize uncertainty information such as optimization under uncertainty [6, 19, 25], stochastic programming [2] and robust optimization [8, 30] among others.

Figure 9.12 illustrates on a simple example how model uncertainty may affect reliability of the optimization results. In this example for simplicity, we assume that the primary output is a function of only one input parameter. As seen, the output has two minima, which have very different confidence intervals denoted by dotted lines in the figure. The confidence interval gives limits in which the output value has a 99% probability, for example. The left minimum has a wider confidence interval than the right one, that is, the model gives more reliable solution at the right minimum. Hence, from the output value point of view, the solution candidates do not differ, but the latter minimum can be considered as more reliable. Therefore, systematic and efficient control of reliability is a crucial point in industrial optimizations.

Nowadays, we can define the multi-objective optimization problem related to papermaking such that it is able to make use of related unit-process uncertainties [23]. Then, we formulate objectives also for uncertainties and minimize them while optimizing the papermaking targets. Thus, the idea is to formulate the originally stochastic optimization problem in such a way that it can be solved efficiently as a deterministic problem using a gradient-based optimization algorithm. This kind of approach has been used in the following kind of paper quality optimization examples where the statistical modeling techniques have been involved [23].

9.4.3 Numerical Examples

Next we present two numerical examples, where the virtual papermaking line was utilized. Both examples were solved using appropriate gradient-based optimizer and gradients were evaluated with the help of the technique described above.

9.4.3.1 Example 1

This example illustrates the advantages of multi-objective optimization compared with trial-and-error simulation. We studied here a conflict between two paper quality properties, formation and tensile strength ratio. Formation is a small scale weight variation of a paper sheet and tensile strength ratio is the machine-directional tensile strength divided by the cross-directional one. Both of these properties were to be minimized, but because they were in conflict they could not reach their optimum at the same time. These conflicting targets were simulated using the above presented virtual papermaking line combining dissimilar unit-process models from different disciplines. The virtual papermaking line used included the CFD models, and there were also models for moisture and heat transfer, and naturally statistical models for paper quality properties were involved.

By doing trial-and-error simulations, a number of solutions were obtained as can be seen on the left-hand side of Table 9.1. These solution were obtained by using typical machine control values, which were determined using so-called engineering knowledge. That is, a person with expertise in papermaking provided the control variable values used in simulation. Based on his/her knowledge, she/he tried to find a solution which would fulfill the preferences. As can be seen in Table 9.1, the person made some simulations and obtained different kinds of solutions, but she/he could not be sure if any of these solutions was optimal. Trial-and-error simulations could have continued for as long as the expert had time or a satisfactory solution was obtained.

Instead we used a multi-objective optimization method to search the optimal control variable values. In this way, much better solutions were found because the method obtained only Pareto optimal solutions. Figure 9.13 shows all the solutions obtained (simulated and optimized). A few Pareto optimal solutions are presented in Table 9.1 on the right-hand side. As one can see, all the solutions were better than the solutions obtained by simulation.

Table 9.1 Simulated vs. optimized solutions

Simulated		Optimized	
Formation (g/m^2)	Tensile strength ratio	Formation (g/m^2)	Tensile strength ratio
0.388	2.731	0.449	2.000
0.439	2.496	0.300	2.233
0.497	2.414	0.366	2.002
0.563	2.414	0.318	2.037
0.563	2.630		
0.540	2.532		
0.518	2.234		
0.586	2.235		
0.479	2.233		
0.442	2.233		

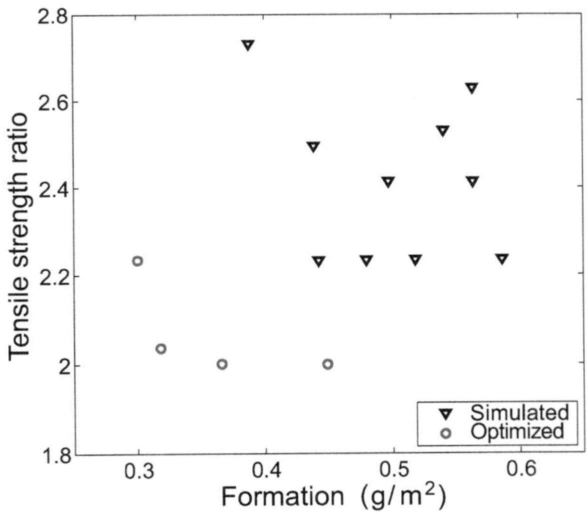

Fig. 9.13 Optimized and simulated conflicting paper quality properties

Thus, there was no need for the time-consuming trial-and-error simulations. Also the number of solutions needed to be generated in optimization before finding the most satisfying compromise was small compared with the simulations. Since all the solutions found in optimization were mathematically equally good Pareto optimal solutions, the papermaking expert's knowledge could be used in a case-specific way when selecting the most satisfying final solution.

Table 9.2 Six optimal compromise solutions

	Tensile strength ratio	Formation (g/m^2)	Basis weight (g/m^2)	Dry solids content (%)
Desired values	3.00	0.30 . . . 0.35	54.00	92.00
Compr.1	2.80	0.43	54.03	92.48
Compr.2	**3.00**	0.40	**54.00**	92.14
Compr.3	2.34	0.39	54.33	92.30
Compr.4	4.91	**0.35**	54.35	92.27
Compr.5	3.90	0.38	53.74	92.27
Compr.6	3.38	0.41	53.92	**92.12**

9.4.3.2 Example 2

A more complicated optimization example of papermaking targets was studied next. In this example, there were four conflicting process and quality targets: tensile strength ratio, formation, basis weight and dry solids content. Basis weight describes the mass of the paper per square meter, and dry solids content is measured from finished paper. All four papermaking targets were given the desired values and the deviation from these desired values were to be minimized. Table 9.2 presents the desired values of the optimization targets.

We used a multi-objective optimization method combined with virtual papermaking line to search the optimal solution, i.e., those control variable values that correspond to the optimal process and quality target values. Six Pareto-optimal compromise solutions were calculated and they are shown in Table 9.2. When comparing the solutions, it was apparent that the chosen targets were really conflicting, and hence all the targets could not reach the optimum simultaneously. For example, in Compromise 2, we obtained the best values for tensile strength ratio and basis weight (in bold face in Table 9.2), but, at the same time, formation and dry solids content were not so good. In addition, we can see that formation attained its best value in Compromise 4 and dry solids content was the best in Compromise 6, but, in these solutions, tensile strength ratio and basis weight were unfavorable. Nevertheless, all the solutions found were mathematically equally good Pareto optimal solutions, and the most satisfying final solution could be selected from these solutions using the papermaking expert's knowledge.

9.5 Towards Decision Support Systems

The virtual papermaking line integrates mathematical modeling with multi-objective optimization, and thus provides a basis for the development of an intelligent decision support system [20]. As we can see from presented examples, handling even a few conflicting targets at the same time is too complex for the human mind and therefore multi-objective optimization methods are needed. Furthermore, when the papermaking process can be controlled as a whole, relationships between the targets can be seen clearer. By means of such a system, a papermaking expert can take into account not only the individual requirements of the paper to be produced, but, at the same time, all the related aspects beginning from choice of raw materials and control of wood fiber pre-processing to as far as market situation and customer demands – all the facts affecting decision making in real life.

This kind of virtual papermaking line and decision support system demands a lot of convenience from the models used. When the main focus has extended to handle larger systems, even the whole papermaking process, the number of models as a chain and sub-ensembles gets bigger. This poses a real challenge since the CPU times cannot be extended too much. The CFD-models are usually time-consuming and that is why it is important to develop new computationally effective ways to model phenomena occurring in the systems such as virtual papermaking line. The accuracy of the CFD-models has to be reduced in order to shorten the computing time or the CFD-model might even sometimes need to be replaced with a statistical or stochastic model, for instance.

A flexible decision support tool is required for different types of usage and different users: from research and development engineers to managers of a paper mill. Moreover, the system must include interactive interfaces with automation and control softwares as well as direct connections to on-line measurements. This trend sets new challenges for modeling and optimization. At its best, the decision support system is able to provide new knowledge on the physical and economical mechanisms affecting papermaking, and ultimately leads to considerable financial benefits.

9.6 Conclusions

Depending on the requirements of the software tool to be developed, the choice of appropriate modeling approach is essential. This means making compromises between accuracy and extent of CFD, especially when coupled with optimization. In phenomenological research the road leads towards more detailed and accurate modeling, such as DNS, but when the aim is to control larger ensembles, there is a need for decision support systems.

In this chapter some examples of coupling optimization with CFD in the field of papermaking have been presented. They are chosen to represent different categories of modeling extent and accuracy. Decision support systems are coming in the near future, but optimal shape design and optimal control are possible today, and in fact, are in everyday use in the industry. All the examples presented here are from the paper industry but, nevertheless, independent of the industrial sector (motor, or aircraft industries, for example), CFD-related optimization problems set the same challenges everywhere. As a matter of fact, the industrial application in question is described by a state equation, that is, the CFD model. All the other aspects are common including optimization methods, needs for model reductions, and handling multiple criteria.

Acknowledgements Many colleagues of the authors have contributed results discussed in this paper, especially Pasi Tarvainen, Ph.D., and Eeva-Kaisa Rouhiainen, M.Sc., from Numerola Oy (Ltd). We are also thankful to Finnish Agency for Technology and Innovation (Tekes) and Metso Paper, Inc. for financial support of this work.

References

1. Baines, W., Nicholl, C., Cook, R., Mardon, J.: The taper-flow headboxes (a new design concept). Pulp and Paper Magazine of Canada pp. 139–148 (1956)
2. Birke, J.R., Louveaux, F.: Introduction to Stochastic Programming. Springer (1997)
3. Brooks, A., Hughes, T.: Streamline upwind/Petrov-Galerkin formulations for convection dominated flows with particular emphasis on the incompressible Navier-Stokes equations. Computer Methods Applied in Mechanics and Engineering **32**, 199–259 (1982)
4. Deb, K.: Multi-Objective Optimization using Evolutionary Algorithms. John Wiley & Sons, Inc. (2001)
5. Franca, L., Frey, S.: Stabilized finite element methods: II. the incompressible Navier-Stokes equations. Computer Methods Applied in Mechanics and Engineering **99**, 209–233 (1992)
6. Giunta, A., Eldred, M.S., Swiler, L.P., Trucano, T.G., Wotjkiewicz, S.F.J.: Perspective on optimization under uncertainty: algorithms and applications. Tech. Rep. 4451, AIAA, American Institute of Aeronautics and Astronautics (2004)
7. Gowda, M.S.: Inverse and implicit function theorems for H-differentiable and semismooth functions. Optimization Methods and Software **19**(5), 443–461 (2004)
8. Gunawan, S., Azarm, S.: Multi-objective robust optimization using a sensitivity region concept. Structural and Multidisciplinary Optimization **29**, 50–60 (2005)
9. Hadamard, J.: Leçons sur le calcul des variations. Gauthier-Villars (1910)
10. Hämäläinen, J.: The optimal shape design of the header. In: Proc. Finite Element Method in Simulation, no. R07/90 in CSC Research Reports, pp. 18–23. Jyväskylä, Finland (1990)
11. Hämäläinen, J.: Numerical modelling and simulation of fluid flow in headboxes of paper machines. Licentiate Thesis, University of Jyväskylä (1991). Reports on Applied Mathematics and Computing
12. Hämäläinen, J.: Mathematical modelling and simulation of fluid flows in the headbox of a paper machine. Ph.D. thesis, University of Jyväskylä, Department of Mathematics (1993)
13. Hämäläinen, J., Mäkinen, R., Tarvainen, P.: Optimal design of paper machine headboxes. International Journal for Numerical Methods in Fluids **34**, 685–700 (2000)
14. Hämäläinen, J., Malkamäki, T., Toivanen, J.: Genetic algorithms in shape optimization of a paper machine headbox. In: Proc. Evolutionary Algorithms in Engineering and Computer Science, EUROGEN 99, pp. 435–443. Jyväskylä, Finland (1999)

15. Hämäläinen, J., Tarvainen, P.: CFD-optimized headbox flows. Pulp & Paper Canada **103**(1), 39–41 (2002)
16. Hämäläinen, J., Tarvainen, P., Aspholm, P.: HOCS FIBRE – new tool for optimized fibre orientation angles. In: Proc. 91st annual meeting of Pulp and Paper Technical Association of Canada, PAPTAC 2005. Montreal, Canada (2005)
17. Hämäläinen, T., Hämäläinen, J.: Modelling of fibre orientation in the headbox jet. Journal of Pulp and Paper Science **33**(1), 49–53 (2007)
18. Haslinger, J., Neittaanmäki, P.: Finite Element Approximation for Optimal Shape, Material and Topology Design. John Wiley & Sons, Inc. (1996)
19. Jin, R., Du, X., Chen, W.: The use of metamodeling techniques for optimization under uncertainty. Structural and Multidisciplinary Optimization **25**, 99–116 (2003)
20. Karlsson, M., Hämäläinen, J.: A model-based decision-aid system to add value to papermaking. In: P. Neittaanmäki, T. Rossi, K. Majava, O. Pironneau (eds.) Proc. of the 6th European Congress on Computational Methods in Applied Sciences and Engineering (ECCOMAS 2006). Jyväskylä, Finland (2006)
21. Madetoja, E., Mäkelä, M.M.: On sensitivity analysis of nonsmooth multidisciplinary optimization problems in engineering process line applications. Structural and Multidisciplinary Optimization **31**(5), 355–362 (2006)
22. Madetoja, E., Rouhiainen, E.K., Tarvainen, P.: A decision support system for paper making based on simulation and optimization. Engineering with Computers (accepted for publication) (2007)
23. Madetoja, E., Tarvainen, P.: Multiobjective process line optimization under uncertainty applied to papermaking. Structural and Multidisciplinary Optimization (accepted for publication)
24. Mardon, J., Manson, D., Wilder, J., Monahan, R., Trufitt, A., Brown, E.: The design of manifold systems for paper machine headboxes, part II – taper flow manifolds. TAPPI **46** (1963)
25. Mattson, C.A., Messac, A.: Pareto frontier based concept selection under uncertainty, with visualization. Optimization and Engineering **6**(1), 85–115 (2005)
26. Miettinen, K.: Nonlinear Multiobjective Optimization. Kluwer Academic Publisher (1999)
27. Miettinen, K., Mäkelä, M.M.: On scalarizing functions in multiobjective optimization. OR Spectrum **24**, 193–213 (2002)
28. Miettinen, K., Mäkelä, M.M.: Synchronous approach in interactive multiobjective optimization. European Journal of Operational Research **170**, 909–922 (2006)
29. Mohammadi, B., Pironneau, O.: Applied Optimal Shape Optimization for Fluids. Oxford University Press (2001)
30. Odaka, Y., Furuya, H.: Robust structural optimization of plate wing corresponding to bifurcation in higher mode flutter. Structural and Multidisciplinary Optimization **30**, 437–446 (2005)
31. Parsheh, M.: Flow in contractions with applications to headboxes. Ph.D. thesis, Royal Institute of Technology (KTH), Department of Mechanics (2001)
32. Pironneau, O.: Optimal Shape Design for Elliptic Systems. Springer-Verlag (1984)
33. Rodi, W.: Turbulence Models and Their Application in Hydraulics. International Association for Hydraulic Research (1993)
34. Steuer, R.: Multiple Criteria Optimization: Theory, Computations, and Applications. John Wiley & Sons, Inc. (1986)
35. Trufitt, A.: Design aspect of manifold type flowspreaders. Pulp and Paper Technology Series, Joint Textbook Committee of the Paper Industry, C.P.P.A. and TAPPI (1975)

Index